I0500110

THE ESSENTIAL GUIDE TO MAINTENANCE & FACILITIES MANAGEMENT

(Special Edition)

By

Brendan Shine (B.Eng.)

ISBN-13: 978-1548604431

ISBN-10: 1548604437

This book is dedicated to my Son & Daughter
Jamie & Sophie

CONTENTS

ACKNOWLEDGMENTS

My thanks and gratitude goes to the people and organisations who helped me put this book together, especially for their valued contributions, suggestions, personal insights, shared experiences, and support.

PRAISE FOR 'THE ESSENTIAL GUIDE TO MAINTENANCE & FACILITIES MANAGEMENT'

- In our book recommendation 'The Essential Guide to Maintenance & Facilities Management', Brendan Shine discusses the fundamental principles of Maintenance and Facilities Management with more than 35 years 'on-the-job experience' in a very successful pharmaceutical company, author of 'The Essential Guide to Maintenance & Facilities Management' Brendan Shine has the right credentials for an informative read. Great reading for anyone involved in engineering! **Matthias Körber - Endress & Hauser**

- The gap between college theory and work experience is finally bridged in this book, it is a 'must read' for all graduate engineering students starting out on their careers. **Pat Reidy - Dunreidy Eng.**

- I have read your book, very practical information made very clear, well done on an excellent reference for engineering students. **Thomas Martin – Limerick Institute of Technology (LIT)**

- This is one of the most 'hands on' all-encompassing books I've ever read on these topics. Brendan covers all the bases in non-overpowering detail. This book is a 'must read' and should be part of every Building Managers' library. **Denny Hydrick – Association for Facilities Engineering (AFE)**

- This book provides 'what you need to know' regarding the engineering background knowledge and advanced skills that people should try to attain regardless of a person's chosen engineering discipline before entering or whilst in the workforce as future or current employers will appreciate such versatility in a person and the initiative shown in trying to further expand their knowledge management base. **Chris Speirs – Chemical Engineer – UCB**

- I have read your book and it is fair to say it is an excellent read, I will be forwarding same to our supervisors as I feel it is a valuable tool to help us progress. **Tony Murray – Project Mgr. Hertel**

- The **essential principles** of Maintenance and Facilities Management are made very clear in this book. It specifically targets the importance of 'fit for purpose' training and being prepared if a breakdown of a critical system occurs or if disaster strikes and how to deal with it. It is excellent reading. **Michael McDonnell – Civil Engineer - LCC**

- The essential principles of Maintenance and Facilities Management are made very clear in this book. It is excellent reading. **Aidan McNabola BEng – Engineering Director - AVARA**

- Excellent work with the book, I would make reading chapter 18 mandatory for all employees within a company. It is so true and should be applied both in and outside of work. Working environments would become less hostile and more respect shared. **Kevin Kennedy – Facilities Engineer – AES Automation**

- Brilliant book to reference from. I have my copy in my workshop and have used it regularly to prove and justify decisions that I make. **John Hartigan – Maintenance Electrician – HSE Connolly Hospital**

- Great read, very logical approach, and great source of information, I will be keeping it close at hand. **Dave Brown - Warehouse & Facilities Business Owner - Sercom Solutions**

- I found the book very informative and an easy read, I would highly recommend it. The real-world problem of understanding the difference between Facilities Management and Maintenance Management is solved in this book. **Donal McDonagh - Mechtronic Technician - Microsemi**

- People's health and safety is critical. The latest building codes and fire safety regulations must be implemented, adhered to, and maintained to the highest standards, regardless of the cost. This book must be in every 'Building Manager/Responsible Person's' office. **Bernadette Hayes – Contract Safety Consultant**

- Having neither an electrical or mechanical background, I now have a firm understanding of how a 'building' operates without ever picking up a tool, this book has it all, great source of information. **Jennifer McCarthy – Building Asset Manager – ASM**

- I just downloaded the book and it is jam packed with information. Congratulations. **Barry Higgins – General Manager – Aquachem**

"An investment in knowledge pays the best interest."
- Benjamin Franklin

Setac
Training

FOREWORD

This book is the right choice for you if you are looking for an all-encompassing, comprehensive guide to Maintenance and Facilities Management. It is also the right book if you are looking for practical tips and tools that you can apply to improve every facet of your engineering role. I have written this book because I work with people like you every day, where I share, support, and help others with their technical and non-technical knowledge. I train groups in how to manage Maintenance and Facilities disciplines effectively and coach them in their specific roles. I provide this support internationally, working extensively in Europe, the Middle East, Asia, and the United States.

Practical Engineering involves the collection and understanding of knowledge about your industry, the line of business in general, and how all the different functions smoothly interact with each other (Physics, Electrical, Mechanical, Chemical, Automation, Programming, Embedded systems and Electro-Technical) in delivering a quality service or quality product or both.

This book covers the tribal institutional knowledge of passing down information and experience (both tacit and explicit) to others by applying proven coaching methods such as Auditory, Visual and Kinaesthetic learning. Managing and retaining such experiential knowledge is becoming one of the major challenges facing companies. Experienced engineering personnel, who are resident in any workplace over many years and maybe, are about to leave, or retire from the business can sometimes intuitively predict or diagnose a design flaw or deficiency that even the vendor of the equipment or the designers of the entire installation did not envisage. **N.B.** They will also have past experiences with process safety incidents. Great intuition comes from crystallised intelligence. A person, who possibly has spent their whole career with a company (from hire to retire), may have accumulated a lifetime of institutional knowledge, practice and experience about complex systems and processes; keep in mind, that this **know how** will just disappear when they retire or leave, along with their client relationships and contacts.

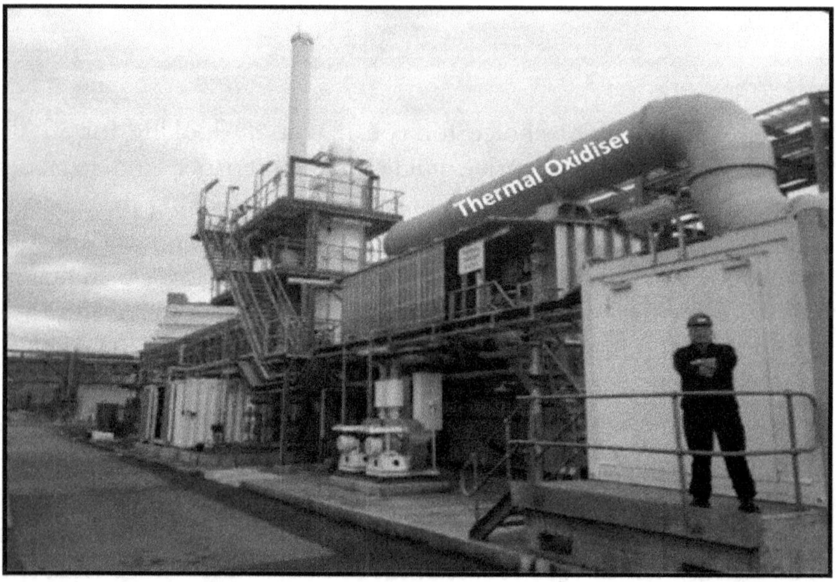

Knowledge Transfer is key. There must be an integrated approach to identifying, capturing, and retrieving this un-captured experience and expertise in individual workers via a **Knowledge Management** system; companies must address this challenge early on and ensure it is not lost. The sharing and managing of this knowledge will be paramount to the future success of any company.

Losing talented employees through retirement or attrition can be extremely expensive. To fight this growing problem, a company must focus on the personnel they currently have (ensure employee satisfaction) and utilise technology to make up where there is a gap.

A knowledge vacuum must be filled. Management must ensure that existing dedicated onsite personnel learn or come up to speed with the experienced person's level of knowledge; otherwise, it will present itself in costly process interruptions.

What are the differences between FM/AM/MM?

- **Facilities Management** is focused on providing the optimal working environment to the End User/Occupier's workplace needs and demands whilst at the same time ensuring maximum 'uptime, reliability, efficiency and redundancy' of the key utility services at a predetermined operating cost (OPEX) to the site's 'core and non-core' physical assets.

- **Asset Management's** prime interest is in enhancing the effectiveness and efficiency of its five 'core' components: (1. Critical assets. 2. Current state of the assets. 3. Required level of service. 4. Minimum life cycle cost. 5. Long term funding strategy.) and its associated production processes. The Asset Manager is not primarily concerned with the operating cost (OPEX) incurred by the asset.

- **Maintenance Management** is committed to maintaining the site's 'core' and 'non-core' assets in a safe, and cost-efficient manner which sustain a company's business needs. To operate and maintain a plant economically, it is essential to have a comprehensive Plant Asset Management System. Engineering Asset Management of ageing equipment requires effective maintenance investment management (CAPEX).

N.B. Maintenance involves keeping the workplace, its structures, equipment, machinery, and facilities in good repair and operating efficiently and safely. It includes many tasks including repairing, replacing, servicing, inspecting, and testing. When carrying out this essential maintenance work, you must ensure that all tasks are done in accordance with current legislation, regulations, and directives issued by your country's authorities and that an appropriate risk assessment is carried out, and that appropriate procedures are stipulated and followed.

Every effort has been made to ensure the content of this book is accurate at the time of publishing. The shared information in the following chapters is an accumulation of many thousands of hours of 'on the job' engineering and operational experiential knowledge combined.

The book is written to be read in its entirety to get an overall understanding and comprehension of Maintenance & Facilities Management, but the subject matter is split into sections that enable you to use each chapter separately to get whatever you need for reference and support on specific topics.

This is a very practical 'how to' book based on the author's work, research, and interaction with thousands of people throughout his career.

1. INTRODUCTION

Reducing costs, increasing efficiency, saving time, solving problems, improving machinery uptime, complying with regulatory targets, and freeing up capital is paramount to a business's survival. Companies are taking stock of their ageing equipment and how they can properly maintain and modernise versus replacing, and still make the equipment as efficient as possible, thus reducing overhead costs. Replacing a piece of expensive equipment because of its age is no longer an option.

A company's mission must be to engage the best people and technology. Technology changes frequently, as do the tools and knowledge needed to succeed, it is fundamentally changing the way we live and work and ultimately determines how businesses grow and innovate. If a company is not 'Digitally Ready' and is not technologically literate, they can be rest assured that a competitor will be.

Equipment designers continue to remove the **black art of fault finding** by using technology for self-diagnostics and to flag potential problems. Automation is constantly evolving to assist non-technical personnel in diagnosing faults. Technology offers a great opportunity for improving cycle times and efficiency, companies are now becoming increasingly engaged in a digital strategy e.g. Asset Maintenance health diagnostics is becoming much easier when an unexpected event occurs on a key component of your production line or plant, enabling you to act very quickly i.e. the failed component will provide a diagnosis code giving clear instructions on how to remedy the problem via your **smart phone app.** No more looking for, and retrieving hard copy documents.

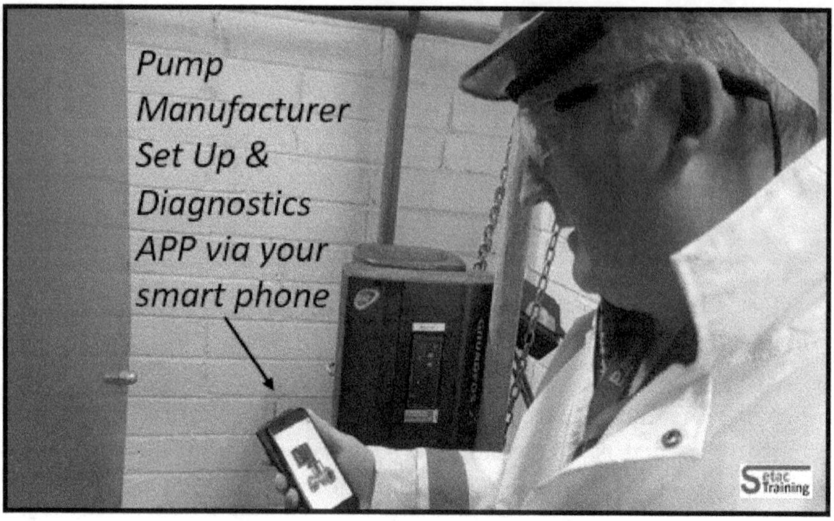

Reliable manufacturing and production processes via the **Industrial Internet of Things** (IIoT) is opening new possibilities in control, sensor, and actuator technology. IIoT digital services use the most modern and secure internet technologies and combine them with industrial production technologies to open new value-adding possibilities.

Strategic Digitalisation investments will play a major role in a company's future. Technology is transforming the way that buildings and infrastructure are designed, constructed, and operated e.g. BIM (Building Information Modelling). And it's helping to improve decision making and performance across the building and infrastructure lifecycle. Buildings and equipment assets communicate all the time, digitalisation enables you to listen and react.

Companies must embrace **Artificial Intelligence (AI)** techniques such as 'machine learning' to extract more value from data from self-calibrating instruments, self-diagnostics in predicting machine failure, and recognising product defects. Machine learning technologies, diagnostics & intelligent analysis, and smart camera devices will play a more active role in the factories of tomorrow. Machine learning systems don't just extract insights from the data they are fed as traditional analytics do, they change the underlying algorithm based on what they learn from the data. So, the "garbage in, garbage out" truism that applies to all analytic pursuits is truer than ever.

Big DATA will become King. AI will be fundamental to the future success of a company. A company must decide on its data management strategy, technology and best practices that will be essential for its continued success. The key to efficiently lowering your expenses without compromising the quality of your products is to derive accurate process data from reliable instruments. These assets often deliver much more data information than just a single process value i.e. other process values, self- diagnosis about the asset's health, or even the prediction, based on internally diagnosed device parameters of potential problems that might occur in the near future.

Automation technology provides the relevant technical information needed to operate, monitor, and perform diagnostics on equipment in 'real time'. 'Digital Command Centres' have operators and engineers using insights from data analytics (Sensor data is analysed by software e.g. sensors collect information like temperature, pressure, speed, vibration, and wear on equipment) to predict possible failures and schedule maintenance. Digital technologies provide us with an additional set of capabilities to help us improve the way we create value for, and with users. It exists not to replace people, but to assist them, it cannot entirely consider the **'human dynamics'** factor (e.g. our ability to focus, think logically, create structure, maintain objectivity).

Sometimes complex and seemingly fail proof technical systems can malfunction because of overlooked problems that interact with each other in unexpected ways. Any piece of equipment can be subject to a technological (or physical) failure. There will always be 'technical unknowns', it is impossible to foresee everything that could go wrong. When a unique set of circumstances exposes an unforeseen design flaw, personnel must use their technical skills and experience to rectify these issues as they arise.

'Diagnostic and Maintenance Assistants' are provided by machine manufacturers via Intelligent Controllers which use data analytics and machine learning to complement human decision making. State Based Control and self-learning diagnostic routines have raised a machine's controller ability to detect, annunciate, and describe problems with machinery. The control system can only do what it is programmed to do: alarm, put the equipment into 'hold, fail safe, or shutdown' mode. If any action occurs outside what it's designed to control, monitor, diagnose, and manage; the **human factor** must play its part. It will send warnings (an audible alarm, a visible flashing light, written text on a PC screen or all together) of impending danger but cannot make a person perform an action to counteract this danger. Operators must be cognisant of not letting distraction and boredom lead to error **due to inattention.** They must intervene and take swift action, which is the very reason their relevant knowledge, experience, skills, and abilities must come to the fore in operating and troubleshooting any bespoke control system issue.

There is no getting away from the proliferation of process automation. The 'control room operators' and 'pilots of the engineering process' will be the focal point as new technologies come to the fore for monitoring machine health and predicting failures.

Another key aspect in helping to reduce failures, reduce costs and system downtime is to train in-house personnel to do **specialist** work. The initial cost of the training may be high, but the benefits to the company will far out way the initial investment made overall. When you are responsible for maintaining a piece of equipment, you must take over **due care**, **custody,** and **control** of all its systems and take ownership, be accountable and responsible for all activities associated with it during its lifecycle.

Trouble Shooting and Fault-Finding methods must be learned, honed, maintained, practiced and adhered to. If adopted and implemented properly, will save many lives, prevent environmental & industrial accidents, save industry massive monetary losses per year in personnel injury claims, machinery down time and lost productivity. New trouble shooting and fault-finding techniques will need to be adopted as equipment and its associated technology is constantly advancing. The fundamentals though, will never change.

Knowledge, experience and good judgment are key elements in any engineering discipline. **Learn from others,** their knowledge and experience will prove invaluable. Learn from their successes and implement them. Just as important though, learn from their mistakes and try not to repeat them.

It is inevitable you will make mistakes, but you must be aware, admit, and learn from them and try to ensure they are not repeated.

N.B. *Think before you do, to question is to understand. Minimise distractions as much as possible, give yourself a chance to focus on what is important.*

The following chapters are guidelines and may contain terms or references you may not have heard of but can be accessed via the internet or a local library. There is an abundance of sites where, if needed, further information and research is readily available. Where terminology or abbreviations are used, such terminology is listed in the **Terms and Definitions** section at the back of this book and an explanation provided as to its meaning.

2. TRAINING AND COMPETENCY

The term **training** refers to the acquisition of knowledge, skills, and competencies because of teaching practical skills and knowledge that relate to specific useful competencies. You must recognise the need to continue training beyond initial qualifications and continue to maintain, upgrade, and update skills throughout your working life. Remember, personal development is about enhancing skills and knowledge.

Every employer should encourage the development of its employees. Employee training remains one of the most cost-effective ways to cut costs, reduce errors, save time, and improve the efficiency of a company's operations.

Training is never complete - Investing in training to stay abreast of industry's latest techniques and requirements is key. A rotational training program should be in place; this is where personnel are exposed to different engineering functions within the company, this prevents knowledge silos being created.

'Learnability' means developing the ability to learn new skills quickly. It's about having an agile mind which is always ready for new challenges. Being endlessly flexible is necessary to remaining employable. Constantly reskilling is part of ongoing learning.

From the most basic to the most advanced, anything you do is an improvable skill. Improvement comes from learning. The quality of how you approach what you do, the quality of how you do it, and the continual improvement and innovation of both.

Improvement issues and the changes required need to be identified. We are taught new information every day in colleges, the workplace or at home, but, it is important people are taught **how to think, and not what to think**. In any job, creative and imaginative thinking is highly regarded. The function of education is to teach a person to think intensively, and to think critically.

Employees come and go, skills become rusty, and poor operation leads to breakdowns. Staying focussed, selecting the correct information that must be learned and communicated to personnel as to its relevance will be important. If all work activities throughout your company are approached with the same mindset, information will be processed more deeply and be retained longer. Constantly strive to do things better and take responsibility for any action you undertake.

3 Key themes for the information age in the 21st century are:

1. **Knowledge –** is no load to bear, learning never exhausts the mind.
2. **Lifelong learning** – be self-motivated and take personal responsibility.
3. **Education** - be proactive in continuous learning.

Tip: Dedicate an average of 15-30 minutes (1-2%) of your busy day to learning something new or to learn something you don't understand. You'll be amazed in 12 months' time the amount of knowledge you've accumulated, and how much it will help you, in both your personal and professional life.

The most valuable asset any company has is its people. They are the driving force behind its success, but, if they lack essential knowledge and competencies, their value and the company strength will be seriously diminished.

For a company to remain competitive, it will need to keep abreast of developments in **new technology** and to exploit the benefits it offers. With new technology comes new challenges. Investment in new technology on the factory floor will be seriously undermined without the corresponding investment in the technical awareness of personnel. **Such awareness comes from training.**

Investing in the correct training now will give a business the leading edge it needs to compete in today's market place. By raising the level of technical understanding, people will be empowered to play a more important and proactive role. A company's success can be assured through increased efficiency, and reduced downtime.

Technical training should encompass theoretical knowledge, building design, system design, familiarity with the installation, evaluation of system performance, troubleshooting, maintenance, documentation, and safety of as many engineering disciplines as possible e.g. electrical, mechanical, hydraulic, and embedded systems.

Employers are looking for versatility in employees who are multi - disciplined and will cover all aspects of machine maintenance. For example, just because you have no experience working with **pneumatic systems** doesn't mean you can't learn about it and become very proficient in the discipline. Choose a career that you have great interest in, as you may spend 40 years doing it. **Attitude and Motivation** are the key words here, don't wait around for someone else to progress your career, you must make it happen yourself.

Use the internet, search for free **online** training courses and technical information being provided by some the world market leaders in their respective fields such as Siemens, ABB, Festo, Endress & Hauser, Spirax Sarco websites e.g. electrical control, hydraulic systems, pneumatic systems, robotics, instrumentation, steam systems. Stay up to date and ensure any relevant training received pertinent to your job doesn't lapse. Review old notes and training modules to refresh your mind on a subject.

Your training should be based on:

- **Ability and Skills** – Have you the ability, and attained the proper skills required to carry out maintenance on bespoke equipment?
- **Competency** – Have you the abilities to satisfactorily perform the required functions of your job?

- **Knowledge** – Have you a clear understanding of the systems operations and capabilities and all the knowledge needed to run the plant safely, efficiently, and confidently.

Training should ensure that you have good technical knowledge of the equipment documentation provided by the vendor and are capable of operating and maintaining the equipment without danger to you or the equipment without the vendor's supervision. This training normally takes place before and during the installation and commissioning phases. Ensure the vendor provides a good training plan to encompass all relevant personnel. The more onsite people who are trained, the better chance of resolving a major fault if one develops through the accumulated knowledge of the collective.

People learn in different ways: (Seeing and hearing are two ways memory is created.)

1. **Visual Learning -** Various areas of the brain work together in a multitude of ways to produce the images that we see with our eyes and that are encoded by our brains e.g.

Creating a training document is very easy to do. Use your camera phone and take pictures of the equipment you are responsible for. Create a PowerPoint Presentation (PPT) and insert the pictures, then add the most relevant operational text. By creating such a document, it helps you to understand the system better.

Tip: Pictures are normally positioned on the LHS, and text on the RHS.

PPT presentations assist trouble shooting. Both technical and non-technical personnel can see all the normal operational running parameters at a glance. If there is an equipment 'break down' issue, they can compare the **Normal Readings** on the PPT slide with the abnormal readings on the HMI and they may be able to diagnose the issue very quickly.

N.B. Ensure all newly created training ppt. presentations are uploaded onto a shared 'Knowledge Management' hard drive for all personnel to access.

PPT slide example of a CDI unit on a Purified Water skid:

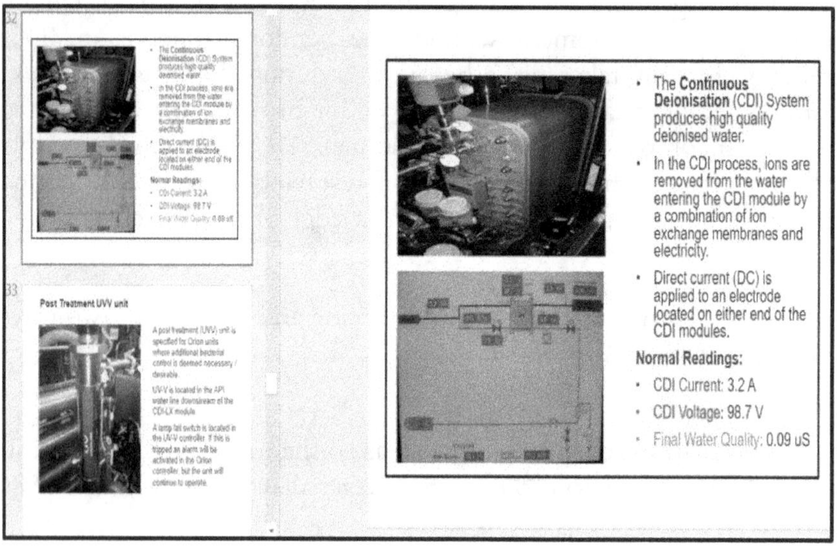

2. Kinaesthetic Learning: or tactile learning is a learning style in which learning takes place by you carrying out physical activities, rather than listening to a lecture, or watching demonstrations. There is no substitute for **training by doing**. Aristotle wrote "the things we have to learn before we do them, we learn by doing them." Although written circa 350 BC, still holds true to this day.

3. Auditory learning: is a learning style which helps you to learn through listening. An auditory learner depends on hearing and speaking as a main way of learning. Auditory learners are those who find it easiest to remember what they hear. They concentrate best by receiving new or difficult information by listening to themselves or someone else talking, and they replay the information in their heads. They remember the key words and phrases.

4. Reading/Writing: You learn well by reading the material you want to learn. The act of writing helps you to clarify your thoughts and remember things better. Writing stimulates your brain to 'Concentrate! Focus! Listen! Don't miss this detail!'.

5. Mnemonics: is any learning technique that aids information retention or retrieval (remembering) in the human memory e.g.

Order of Math operations:

Please **E**xcuse **M**y **D**ear **A**unt **S**ally

(**P**arentheses, **E**xponents, **M**ultiply, **D**ivide, **A**dd, and **S**ubtract)

6. Chunking: is a term referring to the process of taking individual pieces of information (chunks) and grouping them into larger units. By grouping each piece into a large whole, you can improve the amount of information you can remember e.g.

If you need to remember a list of things—such as buying figs, lettuce, oranges, apples, and tomatoes—you can create a word out of the first letters (e.g. "FLOAT"), which is easier to remember than the individual items.

Frequency of Training

Cognitive tasks normally decay faster than physical tasks, to prevent this happening:

- More frequent refresher training is needed for those types of tasks to avoid loss of knowledge.
- Use bursts of **knowledge bites** training to maintain that knowledge.

Tip: When and where allowable, **record** the person giving the training using your smart phone so you can review more easily what they were teaching in the following hours, days, and months. Studies have shown that immediately after the average person has listened to someone talk, they remember only about 50% of what they've heard—no matter how carefully they thought they were listening. Within 48 hours, whatever information they've retained decreases to 25%.

Training Equipment & Human Performance Analysis

Ensure the training equipment is as similar as possible to the real-life equipment. Providing training on 3D simulation systems and models as well as standard operating procedures facilitates greater personnel performance and retention. **We learn, understand, and retain 80% of what we practice and experience.**

A company must ensure that all engineering knowledge does not reside with one person. This type of management is narrow minded and can leave the company badly exposed if something happens to this person. A company's training vision must include as many people as possible to never allow such an issue to arise. Every eventuality cannot be catered for; this fact must be recognised.

Written and verbal assessments must be carried out to prove competencies. These assessments must be given and corrected only by the **Subject Matter Experts (SMEs).** Whether it is given by the vendor of the equipment or an onsite specialist, the trainee can only be passed competent by the system experts. All this training must be documented, auditable, and flagged when refresher training is required. If a piece of equipment has been upgraded, personnel must be **up skilled** accordingly.

If you feel the training you received on a piece of equipment is not adequate, **you must say so**. The onus is both on the company and you to ensure proper training is given and received. If the system vendors' training is not good enough or deficient, the vendor must be written to, or told verbally. The depth of knowledge of experienced trainers is key. Trainers must be in touch with day to day issues and their solutions, and also the practical aspects of the person being trained (e.g. roles and responsibilities). Any course given should be updated in-line with the latest regulations, technical developments, and customer feedback. Accurate training will increase your capability to respond rapidly and efficiently in any given situation.

Training starts with you and your personal development.

- What must I do to improve?
- How do I go about improving my skills and abilities?
- Do I have a plan and deadlines in place in trying to achieve the goals I set for myself?

These fundamental questions must be asked and answered honestly. It must start with you and then ask others for their professional advice and opinions to help you gauge how other people perceive you regarding how you go about doing your daily work activities.

Don't allow your education to lapse, keep up to date with the latest skills and knowledge required for your current role. What you know is crucial, it shows your current employer that you are a lifelong learner by gaining advanced skills and expanding your general business knowledge. Potential future employers will also appreciate such initiatives.

Training is a wise investment, the benefits both financially and operationally to a company will far outweigh any costs involved. A company must maximise the benefits of training. It is imperative to determine not only those technologies in which staff have a knowledge deficit, but also to identify what level of training is required for each person to perform successfully and efficiently his or her role within the plant. Efficient learning can put newly acquired knowledge and skills into practice, ensuring that skills acquired in the class room transfer readily into the work place. It's so important to ensure your knowledge is underpinned with formal training. Recognise your strengths, and work on your weaknesses. Never be afraid to ask questions or take additional training. **Don't vegetate**. Go on as many engineering courses and seminars as possible to enhance your opportunities. Meet other people and find out what their companies are doing. It is also a very good way of networking, setting up contacts, and being able to ask others for their expertise, experience, and advice on universal issues e.g. vendor selection, equipment selection, value for money.

Each employee in the company should have a designated folder assigned to them with all their relevant personal details along with their original copies of their initial qualifications. All subsequent **in house** or **off-site** training with all certificates of courses attended should be tracked via a **Training Matrix**: Standard Operating Procedures (SOP's), Training Methods, Internal training, and new skills acquired. The HR department will safely file and store this folder for future reference (e.g. Audit purposes) and update as an employee's new qualifications are attained and future personal circumstances change.

Improvements from learning means doing things better, faster, and cheaper. Setting goals and having the vision, confidence, persistence, and perseverance to pursue and achieve them will form part of any successful career.

N.B. "Data Integrity is key." Good literacy and numeracy skills are essential to any engineering discipline. One of the single most important actions any person carries out as part of their daily work routine is signing their name and date (physically or electronically) to any work completed. This is where a considerable amount of errors emanates from and subsequently becomes a **knock-on** effect for other parameters being recorded: writing is not legible, incorrect data, wrong date and time is written in. The key is to stop, think, and perform a quick scan of the document before data is added, signed, and dated. If that type of thinking can be propagated to all day to day work activities, deviations and mistakes will be immediately reduced across any business.

Competency

Goal: To increase the level of competency, to measure and improve the knowledge and skills of all engineering practitioners.

Competent employees are essential to the success of any company both now and in the future, in other words, **doing things smarter and better**.

Personnel must be adequately or well qualified, having skills and abilities suited for the tasks they are undertaking, and they must always be within their ***envelope of competence.*** Competencies must be addressed to optimise performance and help meet the company's perceived challenges. Formal measures of **certified competence** are increasingly demanded.

Competencies are:

- A state, or quality of being adequately qualified, or having the ability to perform a specific role
- The ability to undertake responsibilities, and perform activities to a relevant standard
- The combination of knowledge, experience, and skills that lead to high performance.

Certified Competence Example:

CompEx

"*Competency Validation of personnel* ***whose duties include an understanding of the risks associated with working in*** *explosive atmospheres*"

CERTIFICATE
of
CORE COMPETENCE

This certificate has been awarded to:

Brendan Shine

Certificate No: 00085RP

On successful completion of the Competence Validation Assessments for the
Explosive Atmosphere – Responsible Persons Module
in relation to the following units:

EX14	Requirements of the Responsible Persons duties in explosive atmospheres	Rev (1)

EEMUA

This unit is based on the IEC Standards 60079 : Parts 10, 14 & 17

(PLEASE NOTE THAT FULL DETAILS OF ALL THE COMPEX UNITS ARE GIVEN OVERLEAF)

Martin D. Jones
JTLimited | CompEx Operations Manager

JT LIMITED
Certification Body

Date Awarded:	**16 March 2017**
Expiry Date:	**16 March 2022**

This certificate is valid up to five years from the date of the award

To verify this certificate call:
+44 (0)800 085 23 08
or email **info@compex.org.uk**

This certificate remains the property of JTLimited who reserve the right to withdraw this certificate if evidence shows failure to comply with the requirements of the CompEx Scheme.

© JTL935RP Rev1 09/16

Are you developing your competency or just training? Proficiency embedded in an organisation will improve performance and ensure that internal competence is developed, demonstrable, and sustainable.

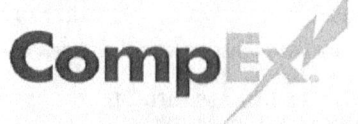

The progress of technology remains an ongoing challenge as any sizeable piece of equipment is made up of electrical, electronic, mechanical, hydraulic, and pneumatic modules operated via a PLC with either a HMI or a SCADA PC attached. There is a constant flow of new technologies in the engineering field requiring new knowledge and skills. New technologies arrive, requiring competency in an ever-increasing range of technologies. It is estimated that it takes **five to seven years of on the job** experience to develop a contributing engineering practitioner. What must be identified are the **competency gaps**: skill or knowledge areas in which the ability of the individual or organisation might not be sufficient to meet the business needs.

There are very reputable companies that work with a business as a training partner. They will review the current competency levels of staff against those required. They will then help identify deficits in the skills and competencies required of personnel and can develop customised training solutions to bridge this shortfall ensuring a company achieves its strategic training targets.

Other key competencies required by Employer's:

- Teamwork
- Responsibility
- Commitment to career
- Commercial awareness
- Career motivation
- Decision making
- Communication
- Leadership
- Trustworthiness & Ethics
- Results orientation
- Problem solving
- Organisation

3. SAFETY & ENVIRONMENT

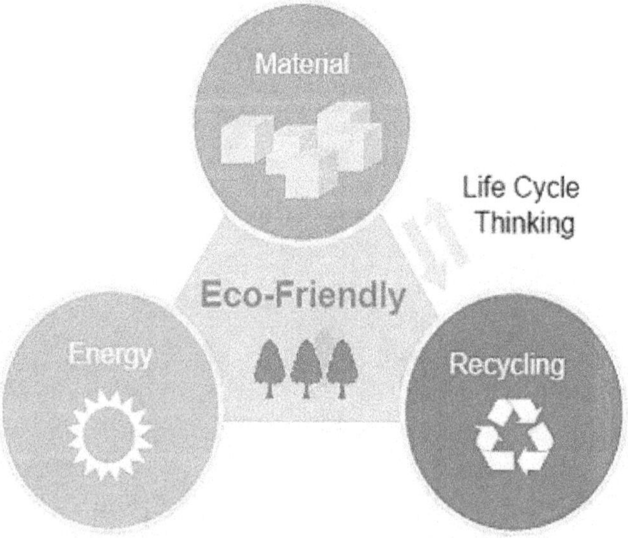

Safety First: All accidents, whether in industrial, commercial, marine, domestic, or agricultural areas are preventable; most accidents are caused by failure to observe basic safety rules and precautions and also, a lack of safety oversight. This chapter is not meant to replace regulations or their requirements issued by your country's authorities, it can instead be used as a tool to complement regulatory requirements and support workplace safety.

Safety Statement: A company has a legal duty to provide a safety statement. It is a protective and preventative measure to ensure the safety, health and welfare of all employees while they work, the safety and health of other people who might be at the workplace, including customers, visitors, and members of the public.

Safety Risk Management: is a key component of a successful safety management system required to assess the risks associated with identified hazards, and to develop and implement effective mitigation. Initial causes of accidents rarely originate from purely technical aspects. Insufficient risk analysis, organisational failure, change management, insufficient or poorly adapted training, absence or non-compliance with guidelines, poor maintenance, non-replacement of ageing equipment and utility services, inspection, or vigilance are often the real reasons. A site containing older equipment on which ageing is not properly managed is more prone to accidents. **Corrosion** (especially under insulation) is the leading cause of ageing-related accidents in pipelines and piping whether they are buried underground or fitted on overhead pipe racks.

Accidentology has illustrated numerous examples of accidents related to an overall organisational failure and poor or unsuitable operating conditions. Maintaining a sense of vulnerability is an essential characteristic of a good process safety culture. Currently, **Safety Management** in industrial facilities is governed by a set of well-established principles:

- Organisation of duties and responsibilities of staff including subcontractors

- Regular and adapted training of staff
- Identification and assessment of accident risks
- Total control of processes by using guidelines and procedures enabling operations under the best possible safety conditions in both steady and transient state
- Work management, analysis of risk beforehand upon receipt of project including consultation with all involved players, accreditation of the players and organisation and supervision of project
- Use of organisational measures to manage change in facilities
- Feedback management within the same group
- Inspection of discrepancies observed between the overall organisation of the working of the plant and practices followed
- Involvement of the management in safety management.

Studies and analysis have shown that maintenance is a large contributing factor in accidents. It is estimated in Europe, 15-20% of all workplace accidents relate to maintenance and in several sectors over half of all accidents are maintenance related. 10-15% of fatal accidents at work can be attributed to maintenance operations. Therefore, maintenance must be carried out properly, taking into consideration workers' health and safety. In the construction industry worldwide, statistics show that a worker is killed on average every 10 minutes. That is >52,560 people per year. The **"Fatal Four"** causes were:

1. Falls.

2. Struck by an object.

3. Electrocution.

4. Caught-in/between.

N.B. No job or deadline is so important that it may risk injury to people or damage to property or equipment due to work done in haste or improper installation. If a serious accident or incident occurs on your site which resulted in injury to person(s), damage to property or serious damage to the environment, the HSA must be notified. e.g. Accidents and dangerous occurrences are required to be reported to the Authority in line with the Safety, Health and Welfare at Work (Reporting of Accidents and Dangerous Occurrences) Regulations 2016 (S.I. No. 370 of 2016). Accident reporting, and investigation is a legal requirement of the General Applications Regulations 2007 and the Safety, Health & Welfare at Work Act 2005. When the HSA inspector arrives onsite, he/she will ask the following questions (as a minimum):

1. About the health of the injured person(s).
2. Was there a proper documented risk assessment completed before carrying out the work task.
3. Did the person(s) follow the correct work procedure for the task being carried out & were they suitably qualified and trained to do.
4. About the site emergency plan.
5. What environmental damage has occurred.
6. What damage has been done to property.

Work Place Errors - While "system failures" play a part in some incidents, human error plays a big part in many accidents/incidents. 96% of work place errors involve human factors. **Errors:**

- Are preventable
- Cause serious and life changing injuries
- Cost lives
- Result in fines and penalties
- Can cause damage to the environment
- Cause reduction in morale
- Destroy personal reputations and self-confidence

- Destroy business reputations and trust
- Cost money

"Zero incidents" must be the goal. Special focus must be centred around the area of **Behaviour and Culture**. Systems and Equipment can only do so much. Only by changing our behaviour can we move to the higher levels of safety performance required.

Don't take risks, take responsibility. In recognising that **humans make mistakes** we must build in **system safeguards** to:

- Capture
- Control
- Alarm

System Safeguards: are measures taken to protect someone or something, or to prevent something undesirable. We must build in barriers and layers of protection to prevent an issue or mistake from going undetected.

Safety Protection relies on multiple layers of protection. Each layer of protection is never perfect, so it is essential that we ensure that all the protective layers remain in place in order to deal with the latent hazards that exist. If we allow any of the layers to be removed, we increase the likelihood of a hazard resulting in a loss i.e. Accident or Incident.

Swiss Cheese Model of accident causation:

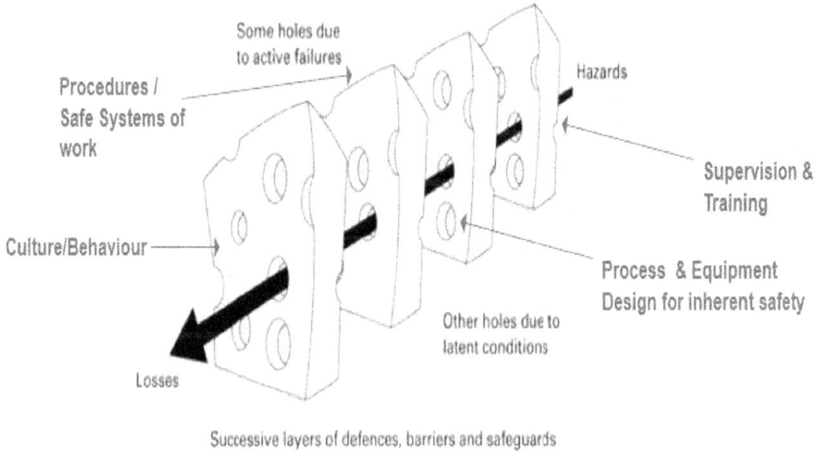

There are two types of defences – **the 'Hard and the Soft'**

Hard defences are best:

- Technical controlled devices
- Automation & Electrical Interlocks
- Mechanical barriers
- Mechanical Interlocks
- Audible & Visible Alarms

Soft defences: Supervision and Training; Culture and Behaviour; a combination of paper e.g. daily operational logbooks, and people driven 'safe systems of work'.

N.B. We must remember that no barrier is ever fool proof. They will all have flaws or weaknesses. After all, humans have designed them!

During maintenance activities, direct contact between the worker and machine is inevitable; it is an activity where workers will be in close contact with machinery and associated processes. Enforcement of safety and best engineering practices must be adhered to. This can be best achieved if there is a good workplace induction training programme given

to all personnel whether it is delivered 'face to face' by a site EHS (Environmental, Health & Safety) contact, by PowerPoint presentation or video. This induction should include pertinent site environmental, health and safety practices pertaining to the site's activities and will have at the end of the training session, a written assessment to prove competency.

These signed training documents will be reviewed and filed away by the site EHS contact, where they will be kept on record for future reference and to flag when **site induction training** for personnel must reoccur (normally every 12 months). The purpose of site induction is to provide the worker with information on the site prior to commencing work e.g. welfare facilities, contractor rules, emergency preparedness measures, site specific hazards, principal supervisor for construction stage (PSCS) rules, names and details of supervisors, safety representatives, first aiders or other responsible persons in emergencies.

A company must be committed to carrying out all activities in a conscientious manner which protects the health and safety of its workers, the facility itself and the environment. The ***first rule of safety*** *is to work in a safe environment. 'Being safe' is in your hands.*

W.O.R.K.I.N.G. with Awareness

When working, always maintain a 'Heightened Situational Awareness'.

Organise and verbalise; Plan the Work and Work the Plan.

Recognise the dangers; A 'See Something, Say Something' approach must be employed. Risk assess, eliminate, or mitigate accordingly.

Keep your discipline and remember your training; read and understand all associated O&M manuals.

If in Doubt, Ask; talk to a peer, call a vendor/external specialist for full factory technical support and advice.

Never take a chance, base any proposed action on engineering fact.

Getting to go home safely is key.

Never lose Focus … Your life depends on it !!!

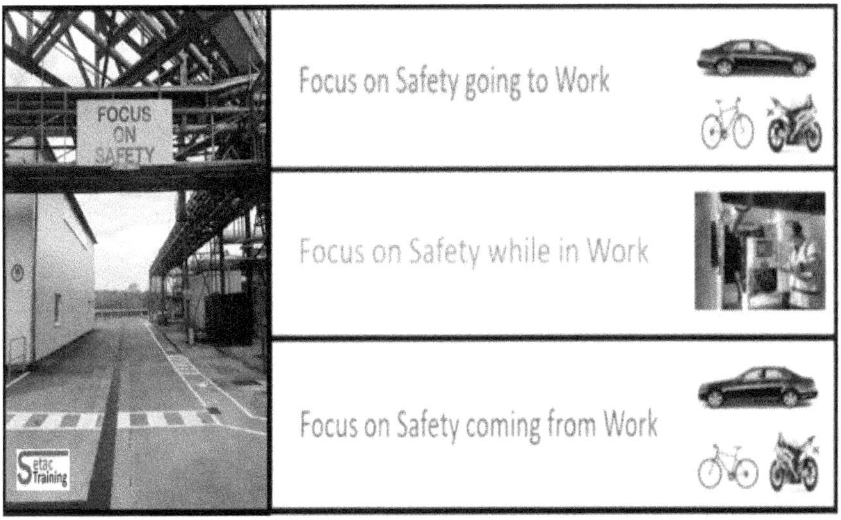

Toolbox talks provide a convenient and effective method of communicating and reinforcing the safety message throughout the workforce and when used properly, can significantly enhance the development of a **safe working culture**. The cost of implementing a regular toolbox talk system is minimal, 10 -15 minutes a week. The benefits will include greater awareness, with the potential to reduce accident rates and possibly even save a life.

Good communication is essential prior to carrying out any maintenance or installation work, the area manager or production supervisor, must be informed in a timely manner by the person proposing to do the work of the impending task.

A **Permit-to-Work** (PTW) form must be issued by the system owner prior to any non-routine work being carried out in their area. This may include a:

- Cold Work Permit
- Hot Work permit
- Line Break Permit
- Confined Space Entry Permit
- Excavation Permit

or any other work activity that the system owner may deem a Permit-to-Work necessary on safety grounds.

Permit-to-Work definition:

A Permit-to-Work system will identify hazards, specify control measures, and describe work procedures. It is designed to plan and control all types of work which are potentially hazardous in nature and which could adversely affect the safety of personnel, the environment, or plant and equipment. It requires communication and coordination between site management, supervisory personnel, project/engineering personnel, operations personnel, and contractors, to ensure that all work is completed in a safe, and environmentally responsible manner.

Permit-to-Work Example (Risk Assessment Prompts, Roles and Responsibilities): Hot Work

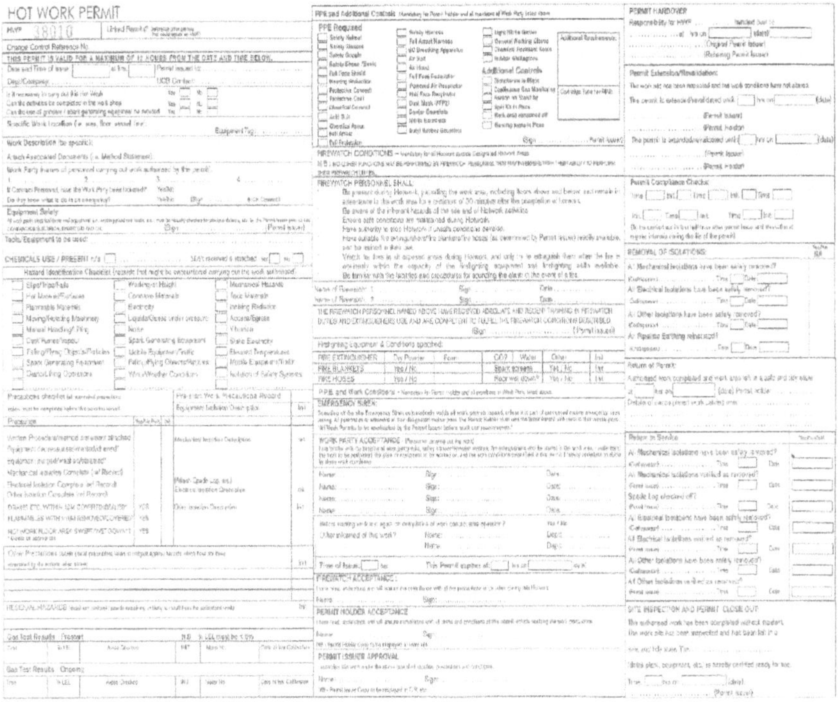

Permit-to-Work (Detailed) Example: Hot Work

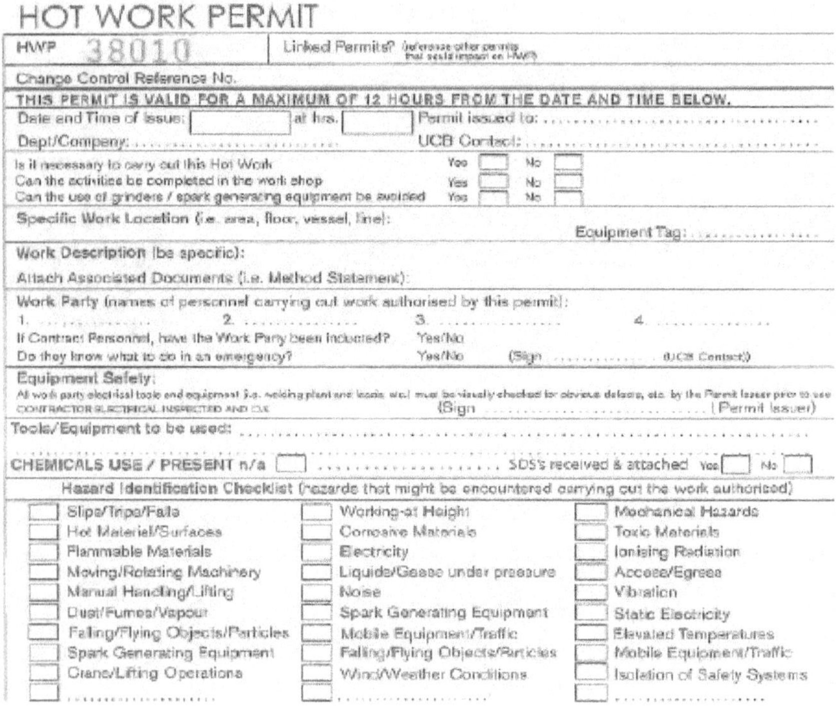
HOT WORK PERMIT

HWP 38010 | Linked Permits? (reference other permits that could impact on HWP)

Change Control Reference No.

THIS PERMIT IS VALID FOR A MAXIMUM OF 12 HOURS FROM THE DATE AND TIME BELOW.

Date and Time of Issue: [ ] at hrs. [ ] Permit issued to: ....

Dept/Company: .... UCB Contact: ....

Is it necessary to carry out this Hot Work Yes [ ] No [ ]
Can the activities be completed in the work shop Yes [ ] No [ ]
Can the use of grinders / spark generating equipment be avoided Yes [ ] No [ ]

Specific Work Location (i.e. area, floor, vessel, line):

Equipment Tag: ....

Work Description (be specific):

Attach Associated Documents (i.e. Method Statement):

Work Party (names of personnel carrying out work authorised by this permit):
1. .... 2. .... 3. .... 4. ....
If Contract Personnel, have the Work Party been inducted? Yes/No
Do they know what to do in an emergency? Yes/No (Sign .... (UCB Contact))

Equipment Safety:
All work party electrical tools and equipment (i.e. welding plant and leads etc.) must be visually checked for obvious defects, etc. by the Permit Issuer prior to use
CONTRACTOR ELECTRICAL INSPECTED AND O.K. (Sign .... (Permit Issuer)

Tools/Equipment to be used: ....

CHEMICALS USE / PRESENT n/a [ ] .... SDS's received & attached Yes [ ] No [ ]

Hazard Identification Checklist (hazards that might be encountered carrying out the work authorised)

[ ] Slips/Trips/Falls	[ ] Working-at Height	[ ] Mechanical Hazards
[ ] Hot Material/Surfaces	[ ] Corrosive Materials	[ ] Toxic Materials
[ ] Flammable Materials	[ ] Electricity	[ ] Ionising Radiation
[ ] Moving/Rotating Machinery	[ ] Liquids/Gases under pressure	[ ] Access/Egress
[ ] Manual Handling/Lifting	[ ] Noise	[ ] Vibration
[ ] Dust/Fumes/Vapour	[ ] Spark Generating Equipment	[ ] Static Electricity
[ ] Falling/Flying Objects/Particles	[ ] Mobile Equipment/Traffic	[ ] Elevated Temperatures
[ ] Spark Generating Equipment	[ ] Falling/Flying Objects/Particles	[ ] Mobile Equipment/Traffic
[ ] Crane/Lifting Operations	[ ] Wind/Weather Conditions	[ ] Isolation of Safety Systems
[ ]	[ ]	[ ] ....

Safety data provides knowledge, knowledge gives people the ability to make sound decisions, and sound decisions lead to a much higher probability of success.

The system owner may also need to be furnished with a well written, detailed **Method Statement** by the person proposing to do the installation or maintenance of exactly what work is to be done and how it has been risk assessed. This method statement must leave no room for guesswork on the permit issuer's behalf.

Method Statement:

This formally describes the necessary **Safe System of Work** prior to a task being undertaken.

Method Statement procedure must:

- Assess the hazards through risk assessment associated with the proposed system of work prior to that work commencing.
- Describe the controls necessary to eliminate identified hazards, or to reduce the associated risk to an acceptable level.

N.B. 'Design out Hazards, Design in Controls'

- Describe the emergency measures required to mitigate the consequences of an associated emergency, should it arise.
- Describe the training, roles, responsibilities, and general welfare arrangements which must be in place prior to the work being undertaken.

When the method statement is completed, reviewed, and approved, it must be communicated to all stakeholders and personnel carrying out the task. Everyone involved must fully understand the proposed system of work, the associated hazards, the identified controls, and the actions to be taken in the event of an emergency as documented on the method statement.

If a permit to work has been issued and work has commenced, and an extra person needs to be added to complete the task and was not part of the original assigned team, their name must be added by the person who issued the permit. The permit issuer must make sure that the person is fully aware of all associated work and dangers pertaining to the ongoing task. *At all times,* personnel must be aware of the dangers around them. The work area must be 'cordoned off' with red and white barrier tape. Erect safety barriers, maintenance work in progress signs and permit to work forms visible for all to see, read, and thus understand the nature of work being carried out.

Make sure that an appropriate **fire extinguisher** is selected, readily available, and suitable for purpose (DO NOT use fire extinguishers as coat hooks for jackets, sweaters, or any other piece of clothing, they must always be clearly visible). Be familiar with its operation, inspect the fire extinguisher and ensure to obey the recommendations on the instruction plate.

N.B. These procedures are put in place to protect other personnel in the immediate area and the person doing the work and never underestimate how essential this preparatory work is.

Slips, Trips, and Falls:

Consider the following contributing factors when managing the risk from slips, trips and falls:

- **Floor surface** - it must be suitable for the work activity and kept in good condition
- **Floor contamination and obstacles** - prevent contamination (e.g. wet floors, greasy floors, paper wrappings) and improve housekeeping and you will reduce if not eliminate the risk
- **Floor cleaning** - stop pedestrian access to wet floors with a retractable belt barrier system & signage, spot clean where possible
- **Environmental aspects** - ensure adequate lighting for walkways and level changes, good entrance design (e.g. canopies) can help
- **People** - their behaviour and physical attributes can be influencing factors. A positive attitude to health and safety by all in the workplace will create a safer working environment. Taking account of physical disabilities such as poor vision, limited mobility etc. in planning and design will help reduce the risk.
- **Footwear** - wearing suitable footwear for the environment and work activities will help reduce the risk.

Key Preventive Measures include:

Employers:

- Have a system in place to manage health and safety
- Risk assess and identify high risk areas
- Consult with and involve staff
- Get the flooring right from the start, it must be suitable for the work activities
- Have a system in place to deal with spills without delay
- Encourage good housekeeping
- Encourage a 'see it, sort it' mentality
- Clearly mark slopes and changes in level
- Review your cleaning procedures to ensure they do not contribute to slips and trips
- Provide slip resistant footwear where other controls are not adequate, trial footwear in the workplace to ensure it is effective.

Always keep a work area clean and tidy; ensure items are not left thrown around to become tripping hazards. You are never too busy or don't have the time to keep a work area clean, tidy, and hazard free. If you are in control of the task you are carrying out, you will always make time available for such activities. At the end of each shift, ensure the work area is cleaned up and everything including tools, power tools etc. are restored to their proper designated place.

A **shadow board** for example, gives a clear indicator if a hand or power tool is missing. This makes it easy to know what goes where and have confidence that everything is where it should be.

The key point is that maintaining cleanliness should be part of the daily work routine, not an occasional activity initiated when things get too messy. You must treat a work area, equipment, and material with respect. Keeping areas & equipment clean and tidy creates a better environment to work in; it also leaves a good impression on others, both inside and outside a facility. **N.B.** All oils, sludge, slurries or drained off liquids are to be disposed of per the pertinent environmental regulations.

Every person is ultimately responsible for their own safety.

You must take the extra 5 minutes before you carry out a task to satisfy yourself that you have all the properly maintained safety equipment as determined by the permit to work (PPE – Personal Protective Equipment) in place both on yourself (e.g. hard hat, safety glasses, Hi-Viz vest, safety boots, ear plugs, gloves, dust mask) and in the immediate work area where the work is to be carried out. Ensure you are wearing suitable work wear for the hot or harsh freezing weather where the task is being undertaken. **Body temperature and clear thinking go 'Hand in Hand'** e.g. People who are cold make more mistakes. A reduction in body temperature also results in impaired body function. This is seen most easily in low temperatures when you try to do something like tie a shoe-lace or do up an awkward button with fingers that are clumsy with cold. The nerve cells that transmit impulses work more slowly as do the muscles controlling your fingers.

Example of Template:

JOB TITLE	NAME OF EMPLOYEE	TRAINING REQUIREMENTS	PPE REQUIREMENTS
Supervisor	Joe Bloggs	A, B, C, F, H, I, J.	A, B, C, D, E, F, G
Fitter	John Mills	A, C, F, G, H, I.	A, B, C, D, E, F, G, J.
Fitter	Mike Dann	A, C, F, G, H, I.	A, B, C, D, E, F, G, J.
Fitter	Jim Keane	A, C, F, G, H, I.	A, B, C, D, E, F, G, J.

TRAINING REQUIREMENTS:

A=Safe Pass, **B**=First Aid, **C**=Manual Handling, **D**=MEWP, **E**=Confined Space, F=Abrasive Wheels, **G**=Working at Height,

H = Site Induction others please specify, **I**= Chemical Handling,

J= Rigger.

PPE REQUIREMENTS:

A=Hard Hat, **B**=Safety Boots, **C**=Safety Glasses, **D**=High Visibility Vest, **E**=Ear Defenders, F=Gloves, **G**=Full Face Screen or safety goggles **H**=Chemical resistant gloves, **I**= Chemical Splash Suit,

J= Full Face filtered mask for solvent vapour.

PPE Example:

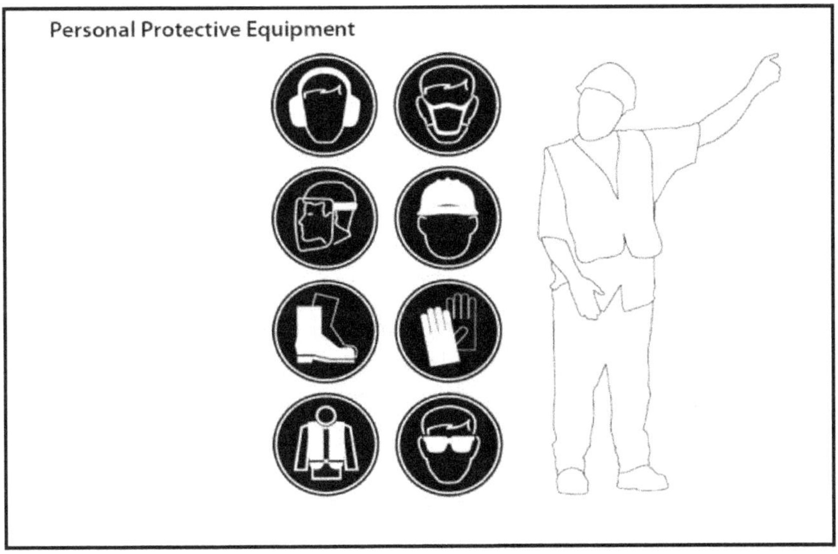

Site Safety Sign Example:

Keep safety equipment certified and in good condition. If damaged, worn out, or has passed the manufacturers date of safety guarantee (e.g. check the safety guarantee date stamped on a safety helmet) it must be replaced, not just because it is site regulations, but, because it is good safety practice and it may someday save you from severe injury or death.

Always remember, **Self-Preservation** is key, just because the permit issuer may not have indicated on the PTW the appropriate PPE, it should be highlighted and worn accordingly regardless.

Where an operation is likely to expose any employee on site to an average noise level of 80 dB(A) and above, an assessment shall be carried out by the EHS department, and records maintained for company inspection. In such circumstances, the company must keep stocks of adequate ear defenders. Ear defenders also drown out excessive noise, especially when trying to resolve a problem. **'Clear Mind, Clear Thinking'**.

N.B. Personnel must observe all valid health & safety provisions and regulations pertaining to their country.

Engineering Work Requests

A written 'Work Request' must be generated by the department who are experiencing equipment failure (may be written in paper hard copy or electronically via Computerised Maintenance

Management Software (CMMS) in soft copy). This **work request** should have all the associated information clearly written on it: the name of the person requesting the repair work; the unique identification number of the equipment; a full and clear description of what the failure is; and any other relevant details. The **work request** will then be processed by the engineering department and subsequently assigned to a person(s) who will investigate the problem and carry out the repair work. The detail of all works carried out, spare parts used, photos, date, time, and hours spent doing the work must also be recorded.

When the work is successfully completed, all work carried out on the piece of equipment must be signed off (either hand written or electronically) by both the person who completed the work and handed back the piece of equipment, **fit for purpose** and the person/designate who generated the **work request,** inspected the work carried out and accepted that the equipment has been tested and is **fit for purpose**.

Machine Status Stack Lighting System Example:

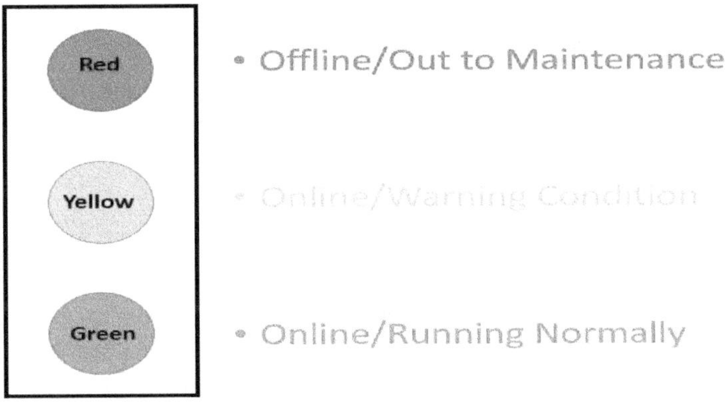

The completed **work request** must then be documented and available for future reference. Having access to this documented information and being able to reference it easily in the event of an occurrence of a similar type issue, can save hours of down time and not having to go through what others have already experienced.

Lock Out/Tag Out System:

A company must ensure a **Lock Out/Tag Out System** is in place on site and **Live by It**, if a LO/TO system does not exist, insist that one is implemented. All affected employees must be trained in proper LO/TO procedures and be retrained regularly. It may someday save their life, or the lives of their colleagues. A person must never remove a lock that is not theirs from a piece of equipment, unless the other person who originally fitted the lock is completely unavailable and then, only remove the lock when the area supervisor and possibly the EHS officer, have verified that it is safe to do so, and both have authorised the removal.

The **'FATAL FIVE'** Main Causes of LOCKOUT / TAGOUT **Incidents:**

(1) Failure to stop equipment.

(2) Failure to disconnect from power source.

(3) Failure to dissipate residual energy.

(4) Accidental restarting of equipment.

(5) Failure to clear work areas before restarting.

Main Energy Hazards and Controls

Electrical energy: Only authorised qualified electricians can isolate mains power and auxiliary circuits and discharge stored energy.

All power source isolation points must be:

- Accessible.

- Be within line of sight.
- On the same level as the equipment.
- In the same room (i.e. cannot be in tunnels, hallways, or alternate floors).

Static Electricity: Earth bonding removed must be noted and replaced on completion of work

Mechanical Energy Hazards

- **Kinetic** (in motion): Run down high & low speed elements
- **Potential** (stored): Release stored energy (Springs, Loads…)
- **Hydraulic / Pneumatic**: Relieve (drain/ vent) pressure

If after isolation, there is still a risk of the equipment moving; fit a proper device (e.g. Locking pin; chock or scotch block).

Thermal Energy (Heat and cold)

- Bring material to ambient temperature and drain
- Control it by using electrical LOTO and line isolation

Tip:

Chemicals: General Requirement - Ref: Material Safety Data Sheet (MSDS). **N.B.** - "Identify Hazards, Assess the Risks, Control Exposure."

LO/TO/TO (Lock Out/Tag Out/Try Out) PURPOSE

To safeguard employees, contractors, and visitors from:

- unexpected energisation, or start-up of machinery and equipment
- release of hazardous energy during project, service, or maintenance activities
- uncontrolled release of materials into the working environment
- The purpose of the lock is the physical control of the hazard
- The purpose of the tag is to provide a visual identification that the energy/hazard source has been isolated and to warn others not to operate the equipment. It also provides details of who, why, and when the device was isolated
- Lockout is a first means of protection; warning tags only supplement the use of locks
- The purpose of Try Out or Verify is to verify the effectiveness of the isolation.

One Golden Rule of Lock Out/Tag Out/Try Out (LO/TO/TO): If you are at risk, then you must have control of the isolation.

Lock Out/Tag Out Procedure Definition:

LO/TO is a safety procedure which is used in industry to ensure that equipment is properly shut off and not started up again prior to the completion of maintenance or servicing work. This procedure is put in place to outline practices, precautions, safety measures, and hazards involved in the Isolation and Lock Out/Tag Out of any equipment or system that contains Hazardous Energy. The procedure of Lock Out/Tag Out is to be used to control Hazardous Energy. All isolations and Lock Out/Tag Outs of equipment and systems on-site shall be controlled with the Permit to Work system.

If a piece of equipment has been locked out/tagged out via an isolock, the LO/TO must never be removed unless all concerned parties remove their own personal locks and 'sign off' on the lock out/tag out. The permit issuer will then complete the final sign off on the PTW once they have checked the equipment and are satisfied everything is in order both from a safety and operational perspective. The piece of equipment will then be returned to service.

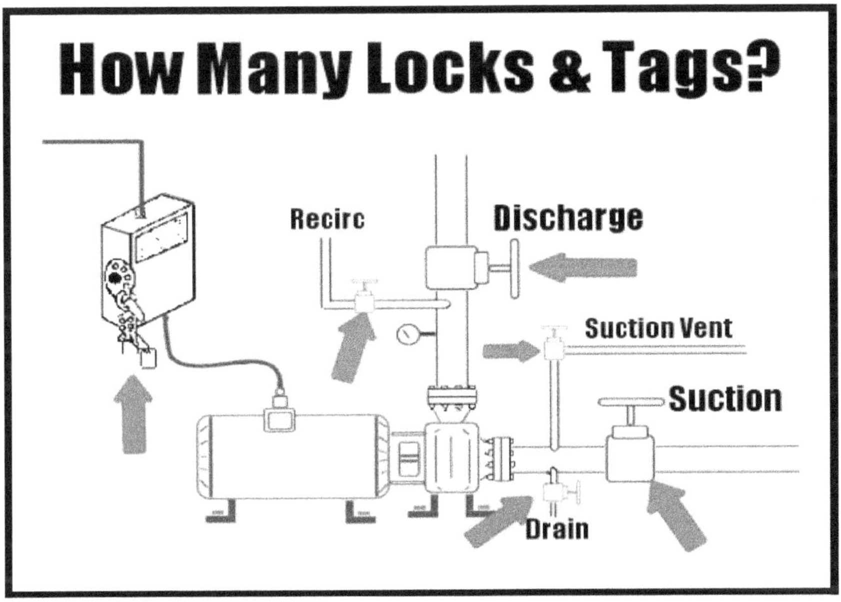

Remember to 'lock out' and 'tag out' all hazardous energy sources.Tag System

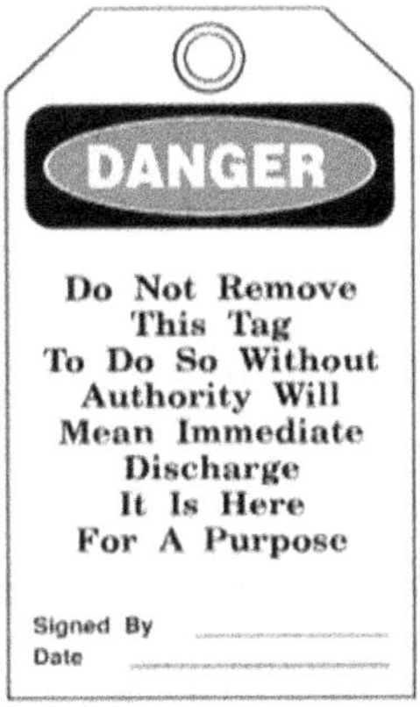

Good communication must take place during shift changes: a proper verbal and written handover between outgoing and incoming shift personnel pointing out any breakdowns; planned maintenance work done; any anomalies on equipment (no root cause found) during the previous shift; and any work outstanding to be completed which is still **locked out/tagged out** (e.g. the personal locks and tags of the leaving shift must be replaced with the locks and tags of the arriving shift). Any relieving Permit Issuer must countersign when he/she takes over, to ensure that responsibility is clearly maintained and that the relevant incoming personnel are correctly briefed about 'work carried out to date' and then their names added to the 'live' PTW. The outgoing shift must make sure that the incoming shift understands the maintenance process and the associated hazards.

If a piece of equipment has been taken out of service and was **not** properly isolated electrically or mechanically (no lock out/tag out in place) or properly communicated through the proper channels, **ask questions**.... Even if it means calling people who are off site, find out why it has been taken 'out of service' without the correct procedure being adhered to.

In a suspicious situation, exhaust all avenues of communication/ checks. Is it a shared utility? Has a person from another area or building turned the piece of equipment off without going through the proper procedures and channels? Try to find out why it was taken out of service in the first place and get agreement from peers/manager (if available) before deciding to put the piece of equipment **back to service**.

The person, who did **not** isolate the equipment properly, must be part of the investigation and it should be explained to them that this type of work practice is unacceptable. They may need more training on such safety practices, pointing out also that their behaviour could endanger their work colleagues' lives and cause unnecessary downtime trying to establish why it was switched off in the first place.

N.B. When working on pressurised compressed air/water/steam systems, ensure that any pressure is isolated and safely vented to atmospheric pressure. Consider **double isolation** and the locking and labelling of closed valves. Do not assume the system has depressurised even when the pressure gauge reads zero.

N.B. Never assume anything, always check.

Human error can prove very costly to both personnel and plant. 42% of 'all' industrial accidents are the result of human error. Process and equipment failures account for the other 58%. Normally a critical chain of events occur, that if overlooked, may lead to disaster.

Sometimes you may have to make judgements and decisions based only on anecdotal information in a dangerous environment. You may have to interact with people when seeking more specific information. This leads to more critical time being lost, whilst at the same time trying to ensure proper procedures are being followed. Your focus must be to stop an emerging abnormal situation from leading to a catastrophic accident.

On average, 30% of the assets in a plant are obsolete. If a piece of machinery is taken out of service for a period and remains in situ, ensure it is still maintained even though it is not in use. Electrically and mechanically isolate. Drain any liquids that have the potential to leak e.g. if a machine is left exposed to freezing air, water may freeze and split pipes.

If a piece of machinery is no longer in use and has been removed from site but all the various utility lines are still in place: 'live' electrical cables; compressed air; high pressure steam etc; remove them where possible, back to source.

If this is not possible, clearly LO/TO all these utility lines and label each line. Never underestimate the importance of these activities. An improperly isolated utility line where electricity, compressed air, high pressure steam etc. may exist, can pose serious risk to both personnel and plant should it be turned on or worked on accidentally. Identify the risk and separate yourself and others from danger.

Never take any person's word that they have isolated a piece of equipment and it is safe to work on. Check it. You must satisfy yourself, that everything is in order and that all proper safety and operational procedures have been adhered to.

NEVER TAKE A CHANCE – You're never too busy to be careful. Take a minute. Safety first!!!

EU Pressure Equipment Directive

The Pressure Equipment Directive (PED) came into force in Europe in 1999. Since 2002, compliance with the Directive has been mandatory. The PED applies to pressure systems used in industry, with some exceptions (e.g. military and nuclear installations). The Directive aims to remove technical barriers to the trade and supply of pressure systems within Europe and to ensure that pressure systems are safe.

Pressure systems coming under the scope of the PED include:

- Refrigeration and air-conditioning systems
- Compressed Air Systems
- Steam Plant
- Gas systems including carbon-dioxide and nitrogen (e.g. for packaging or cooling)

Pressure systems are indispensable to all kinds of industries. The responsibility for compliance with the PED falls mainly to the manufacturer, supplier, or installer of the pressure system. However, employers have a responsibility to ensure that the procurement of new equipment or the modification of existing equipment complies with the PED.

Working at Height:

"Working at Height is a well-documented, well-regulated risk and there is no excuse for getting it wrong." i.e. the correct equipment for work at height is being used and good supervision of the work is maintained on site at all times. You 'personally' must always take extra precautions, wear a proper well fitted body safety harness and ensure you are **clamped on/tied off** to a solid strong permanent fixture that can take your weight in the event of a fall. Use properly erected **tagged out** scaffolds by certified scaffolders, or MEWP's (Mobile Elevated Working Platforms) where possible.

When Working at Height:

1. MEWP operators should have attended a recognised operator training course and received a certificate, card or 'licence', listing the categories of MEWP the bearer is trained to operate. The expiry date of the training licence or card should be checked.
2. Always be aware of other personnel and their activities in the general work area. Have proper signage erected and cordon off the work area with red and white barrier tape, ensure cordons do not hinder emergency egress. MEWP's are excellent vehicles to use when working at heights, remember also, they are big and powerful and can cause serious injury to personnel or damage to property if mishandled. Do not attempt to operate one unless you are certified to do so.
3. Be aware and ever vigilant of overhead power lines when working with or moving MEWP's, scaffolds or ladders, striking a power line can cause serious injury and could be fatal.
4. Before ascending to the working height, ensure that the ground the MEWP's, scaffolds, ladders are positioned on is solid, flat, clear of debris and can take the weight without fear of it collapsing and causing the MEWP, scaffold, or ladder to topple over. Remember, even rolling over a stone on a roadway can be amplified many times when in an aerial platform at an elevated height, resulting in severe shaking. Whilst working within the basket of a MEWP, personnel must wear a full body safety harness with **fall arrest retractable lanyard** securely attached to an anchor point within the basket. (Exception: check regulations when personnel are working in an aerial lift over water).

Forklifts:

Operating a forklift is a specialised skill that requires expert training and on-going refresher training in order to prevent site accidents that can otherwise be prevented with the correct knowledge and expertise. Only designated certified licenced employees can operate a forklift in their particular application.

When **pedestrians and fork lifts** operate together in a confined area, there are considerable risks involved. If an accident occurs, the damage can be devastating.

Ensure there are safe-guards in place that alerts pedestrians and drivers of each other early on (even through walls). Wherever it is feasible to segregate your operating areas, it is best practice to do so. In areas where physical segregation often isn't possible, there needs to be robust **Safe Systems of Work** in place to reduce the risk.

These must include the following measures:

- PPE – Hi-Viz workwear - you can be seen during the day, night and in foggy conditions.
- Traffic management
- Clear communication systems
- Clear signage
- Safe working procedures

N.B: You are ultimately responsible for your own safety. ***You must never take a chance with your life or that of your colleagues.*** All it takes is 10 seconds of thoughtlessness and lack of awareness e.g.

- Going up the outside of a scaffold instead of the safe designated way to retrieve a tool, falling back onto solid ground, leaving you paralysed for the rest of your life or possibly dead.
- Close to 20% of all forklift accidents involve a pedestrian being struck by a forklift.
- When driving, using your mobile phone makes you four times more likely to crash.

- Temporary works (TW) should be given the same degree of care and attention in construction projects as permanent works.

Tip: When using an extension ladder, keep it the right distance from the wall, **use the 4-to-1 rule**: for each 4 feet of distance between the ground and the upper point of contact (such as the wall or roof), move the base of the ladder out 1 foot. Place it on a level, firm surface and foot or tie at the bottom. Extend the ladder a minimum of 3.44 feet above the upper landing and tie off or anchor at the top, on the stiles not on the rungs.

Confined Space:

A **Confined space** is "any place, including any chamber, tank, vat, silo, pit, trench, pipe, sewer, flue, well, or other similar space in which, by virtue of its **enclosed space**, there arises a reasonably foreseeable risk of a **hazardous atmosphere** or a **dangerous condition.**" Confined spaces are significantly more hazardous than normal workplaces. Extra precautions must be taken when working in a confined space. **Entry into confined spaces can be unforgiving as any errors can be fatal.** Permit required confined spaces are still the leading cause of multiple fatalities in the workplace. A seemingly insignificant error or oversight while working in a confined space can result in a tragic accident. There are more than ten 'Confined Space' fatalities per year in the UK alone.

Safety must be based on:

1. A sound Management System
2. Full understanding of all the hazards
3. Thorough training
4. Rigorous adherence to procedures
5. Tested Emergency Procedures

Is there another way of doing the task rather than the confined space option? The following factors must be considered and physically checked before vessel entry is to be carried out:

1. Previous contents
2. Residues
3. Contamination
4. Oxygen deficiency and oxygen enrichment
5. Machinery
6. Physical dimensions
7. Work to be carried out:
    a. cleaning chemicals
    b. sources of ignition
8. Dangers from outside

Could hazardous vapours or gases be formed by the work being carried out?

When working in confined spaces the following points must be implemented/addressed:

- Issue of a permit to work
- Work scope and method
- Appropriate PPE as dictated by the PTW
- Nominated Supervisor
- Tally man/Attendant
- Method of communication
- Rescue procedures and equipment
- Training (mandatory)

- Tools and equipment to be used, including low voltage or pneumatic
- Lighting requirements, including standby/emergency
- Explosion proof fittings
- Ventilation
- Access
- Bonding to prevent both electrical shock and static discharge
- Work cycles, to reduce risk of personnel heat exhaustion
- Fire safety and extinguisher requirements.

Personnel shall not enter or commence work in any excavation, tank, vessel, pipe or chamber or other enclosed space, until a valid permit to work (confined space permit) has been issued by the Issuing Authority. Where operations result in a **dangerous atmosphere** arising during the monitoring of the work activity, the PTW issuing authority must be informed and all personnel removed from the area.

No new activity shall be introduced into a confined space without the permission and signed approval of the PTW issuing authority. Whilst work is ongoing within a confined space, the company will be required to provide a properly trained standby/attendant person. All personnel who have to enter confined spaces must have undertaken the training appropriate to this task.

No site personnel can enter a confined space to carry out a work activity in that confined space unless there is, in respect of that confined space, suitable and sufficient arrangements for the rescue of persons in the event of an emergency. A confined space register which includes an assessment of known and potential hazards for all **Confined Spaces** on site must be provided. Detailed risk assessments, which define the manner in which the risks identified for the confined space and the work to be carried out within the confined space are to be controlled must also be provided. A standard test process for vessel entry would be, e.g.:

Atmospheric Test (pre-entry):

A test performed by a competent person to determine the suitability of the atmosphere for breathing by persons and a safe environment with respect to the flammable, combustible, and explosive potential prior to entry into the confined space. Results of this test will be recorded on the Confined Space Entry Permit/Work Permit prior to the work being carried out and further tests checked, and written recordings made whilst the work is ongoing.

Before entering any vessel e.g., a tank, silo, or underground drain; a **Confined Space entry permit** must be completed and fully signed off by the system owner and all personnel involved in the work. All involved must make sure any associated gaseous/electrical/ mechanical apparatus attached to the vessel is totally identified, isolated, removed/spaded/blanked off, locked off with each individual having their own respective key for any **isolock** installed.

Both the person entering the confined space and the second person acting as a **lifeline** must ensure there is proper access/egress maintained constantly. The person entering the confined space must at least be watched at all times from outside and all appropriate precautions must be taken to ensure that they can be assisted effectively and immediately. Ensure also, a constant means of communication exists between the two parties while the confined space entry permit is active.

When working alone in remote areas, personnel must **always** inform a **site designate** of their location and the anticipated duration of their task before commencing work. If a person does not and something unforeseen happens to them, they may not be missed for some time.

The air a person breathes in the Earth's atmosphere is comprised of approximately 78% Nitrogen, 20.9 % Oxygen and the other 1% or so made up of special gases. If the oxygen % level drops, it can lead to a person almost instantly collapsing or be fatal.

N.B. Never make even a partial entry into a confined space unless it has been tested and proven safe. **Never** even consider the possibility of holding your breath and entering the confined space.

If you find yourself in the unfortunate circumstance where a work colleague has collapsed in a silo or tank and you are the attendant person. **You must never go in after them,** as you may suffer the same fate, instead call for emergency assistance, and try to winch them out safely.

Remember, by just putting your head into an open pipe or tank, you can be overcome and collapse in seconds if the oxygen content is in any way diminished.

Human beings must breathe oxygen to survive and begin to suffer adverse health effects when the oxygen level of their breathing air drops below 19.5% oxygen. Below 19.5% oxygen, air is considered oxygen-deficient. At concentrations of 16% to 19.5%, workers engaged in any form of exertion can rapidly develop problems as their tissues fail to obtain the oxygen necessary to function properly. Increased breathing rates, accelerated heartbeat, and impaired thinking or coordination occur more quickly in an oxygen-deficient environment. Even a momentary loss of coordination may be devastating to a worker if it occurs while the worker is performing a potentially dangerous activity, such as climbing a ladder. Concentrations of 12% to 16% oxygen cause tachypnoea (increased breathing rates), tachycardia (accelerated heartbeat), and impaired attention, thinking, and coordination, even in people who are resting.

At oxygen levels of 10% to 14%, faulty judgment, intermittent respiration, and exhaustion can be expected even with minimal exertion. Breathing air containing 6% to 10% oxygen, results in nausea, vomiting, lethargic movements, and perhaps unconsciousness. Breathing air containing less than 6% oxygen produces convulsions, then apnoea (cessation of breathing), followed by cardiac standstill. These symptoms occur immediately. Even if a worker survives the hypoxic conditions, organs may show evidence of hypoxic damage, which may be irreversible.

The person entering the confined space must wear a full body safety harness with **fall arrest retractable lanyard** securely attached to an anchor point e.g. Davit arm, mobile tripod with properly certified winching gear attached to the fixture so a person can be removed from the confined space easily in the event of them being injured or collapsing. They must also wear all other appropriate PPE as dictated by the PTW.

Think and be aware before entering a confined space. You must never take unnecessary risks and put your life or the life of your colleagues in danger.

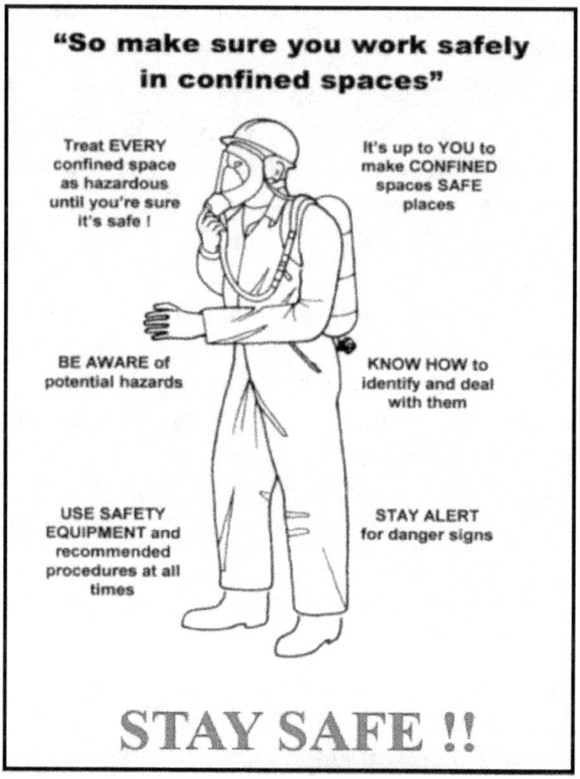

Excavation and Openings Works:

- **N.B.** Prior to the start of any excavation on site, **engineering must be consulted** and the presence of all potentially hazardous overhead and underground utility services shall be checked (e.g. a HUM detector: a cable-locating device set on power mode). Where **'Live'** utility services are present, hand excavation must be carried out until the location of the service has been identified, recorded, and made safe. Excavation work must be carried out carefully and follow all recognised safe digging practices.

- Personnel must erect suitable solid edge protection (double handrails) around excavations or openings. During the hours of darkness, any excavations, openings, or obstructions near or on roadways and walkways must be indicated by a sufficient number of warning lamps.
- The sides of all excavations should be properly shored, battered or stepped to prevent collapse. No excavation work shall commence unless there are adequate resources present to ensure the stability of the excavation. Excavations shall be inspected prior to commencement, or re-commencement of the work to ensure the excavation is still in a safe condition.

Mobile Crane Operations

Key points to observe with the safe setting up and operation of mobile cranes include:

- Personnel Training and Competency
- Risk Assessment & Management
- Pre-Job Planning and Inspection
- Correct Crane and Equipment for Application
- Onsite Arrival and Review
- Onsite Planning
- **Correct Setup of Crane**
  - Outrigger Operation & Packing
  - Trial Operation
- Safe Continual Operation.
- Safe completion of operation.
- Safety Equipment

Machinery:

Fundamental criteria that must be addressed and implemented when designing and working with machinery to ensure safe, environmentally friendly, reliable and cost-efficient operation should be:

1. To provide properly designed, fully documented safe automation technology.
2. To ensure **fit for purpose** reliable electrical, mechanical, control and instrumentation equipment is installed and properly maintained.
3. To ensure relevant specific real time information with proper context is being generated and is user friendly.
4. To ensure specific knowledge, information and skills are communicated clearly and associated abilities attained in a way that personnel can understand and apply to their specific work.
5. To ensure personnel can disseminate and understand real time information easily and act accordingly and also have the ability to make multiple decisions at any point in time.

When considering safety in relation to a piece of equipment:

- Ensure that proper emergency safety equipment is installed to mitigate the consequences of an emergency event.
- The piece of equipment must be capable of operating to its designed specifications and conditions with an appropriate safety factor.
- The piece of equipment must be protected from conditions that might occur outside of its operating conditions or capabilities and within safe limits of its operation e.g. an air compressor has to supply compressed air to a production process. The compressor itself has been **incorrectly** installed in an area where sub-zero temperature conditions may exist from time to time. The compressors' minimum protection start up temperature of > 2**°C** above zero was not taken into consideration. Compressor will not start below 2**°C**, thus not allowing production processes to start up.

- The piece of equipment must have appropriate safety features that protect personnel while installing, operating or maintaining it. Ensure hands and clothing are protected. Do not wear loose clothing (e.g. ties or carves) or jewellery and tie back long hair. **All of these could become entangled in a pulley or be dragged into a rotating mechanism.** The equipment must be installed in a manner that allows for these safety requirements. Stay clear of all rotating parts and of all moving parts, ensure safety guards are always in place and properly secured.

- Chips or debris may fly off objects when they are being struck (safety glasses must be worn). Before objects are struck, try to ensure no other personnel will be struck and possibly injured by flying debris.

- Ensure you **support equipment properly** when working on or beneath it: any type of machinery: car; truck; tractor; trailer. **You must not, under any circumstances**, depend solely on a car/truck jack to support the vehicles weight if you have to physically go underneath it to investigate or fix something. A secondary **fail-safe support** must be in place e.g. stacked concrete blocks or an RSJ properly positioned under a **weight bearing** part of its frame which can take the vehicles' weight if the car/truck jack fails or falls to one side while you're still underneath it. Ensure also to brace the wheels of the vehicle with a block of wood/concrete to prevent it from rolling while working on it.

- You **must not** mount or dismount a piece of equipment other than at designated locations that have handholds or steps. You **must not** stand on components mounted on the equipment e.g. solenoid valves, probes, filters …. to elevate yourself to a higher position. Use an adequate ladder or a work platform. Secure the climbing equipment so that it does not move.

- Do not carry tools or supplies when mounting/dismounting a piece of equipment. Use a hand line to raise and lower hand tools or supplies.

- Death may result if safety mechanisms are removed or nonfunctional. Removed safety mechanisms (e.g. for cleaning, maintenance and repair work) may pose a health hazard or cause a fatal accident. Replace the removed safety mechanisms immediately on completion of the assigned work and test them if necessary. Check the safety mechanisms on a regular basis.

- Equipment that has been in service may contain trapped pressure and/or residual media even after washing. Opening and/or disassembling of machine components should **only be carried out by trained personnel** in strict observance of the manufacturers' instructions. The same applies to the safety precautions which need to be taken when handling the residual media itself.

N.B. **Do not** use **emergency fire alarm break glass, emergency stop buttons, electrical isolating switches** …. as coat hooks for a jacket, sweater or any other piece of clothing. Not alone is it obstructing a piece of emergency equipment, but it is also blinding other people to its actual location in the event of an actual emergency. A person could also inadvertently activate the device by simply leaning against the piece of clothing. If an emergency stop is activated or any other piece of emergency equipment in **error**, ensure it is reported to the area manager.

Set up an Environmental, Health & Safety (EHS) Systems Software Package

A dedicated systems software package is a computer-based tool which helps a company to easily manage and track their key functions on Environmental, Health & Safety issues:

- Accidents
- Incidents
- Near misses
- Environmental Impacts
- Non Conformance
- Safety risk assessments
- CAPA's

- Monitoring
- Objectives, targets and plans

The system allows full visibility and traceability on all company EHS activities.

Key Benefits

- Analysis of system effectiveness through KPI Reporting
- Creates a knowledge base to identify reoccurring problems
- Reduce EHS issues through effective analysis
- Improve information quality with key information captured during issue resolution
- Provides a unique historical record of all EHS issues for both company and regulatory audits.

Key Features:

- Record of all Incidents/Accidents
- Root Cause Analysis reporting along with the time taken to satisfactorily complete investigation and close out
- Creates CAPA's
- Fully Configurable System designed to meet a company's unique requirements.
- Management Reporting Suite allowing interrogation of data - (DATA mining).

Emergency Response Preparation:

"Knowledge is Everything, and Information is critical in an Emergency."

Emergency Phone Numbers:

Ensure the emergency medical numbers, especially for onsite

First Aid personnel, are available and easily accessible, as well as local hospitals and doctors.

Role play and preparatory training on site for accident and incident scenarios should be mandatory e.g. know where the Defibrillators are located and how to use them. Knowing who to call (e.g. security, shift manager or main site utility (Electricity, Natural Gas, Water) providers emergency **on call** phone numbers), what to do (e.g. know the location of fire extinguishers and selecting the correct one to use) and how to escalate an emergency (e.g. activate site fire alarm via Manual Call Point) must be known by all personnel in the workplace. These escalation procedures should also be part of any proposed works' method statement.

Personnel must implement all provisions valid for their country and pertaining to the opening and repairing of electrical/mechanical devices. All instructions regarding the repairing of electrical and mechanical devices must be adhered to without fail.

Personnel must absolutely and without fail, read, and understand the relevant piece of equipment's O&M manuals before carrying out its instructions. Only trained qualified personnel can carry out work on a piece of equipment. The detailed maintenance instructions and safety precautions, as per the Technical Operation and Maintenance Manual of any piece of equipment, must be followed at all times.

Never be tempted to cut corners on safety. Ensure proper maintenance intervals are adhered to as recommended by the manufacturers of the equipment supplied to the company. **Do not take shortcuts**; this may lead to failures later. It can also leave the equipment in an unsafe condition, putting other people's lives at risk.

Never become complacent about hazards in a workplace.

N.B. When working on hazardous liquids or gases in a **pipeline**, consider what is in the pipeline or what may have been in the pipeline at some previous time.

Consider:

1. Flammable materials.
2. Substances hazardous to health.
3. Extremes of temperature.

Incident/Accident:

The 5 key psychological factors in incidents and accidents:

1. Unsolicited advice.
2. Unqualified assumptions.
3. Complacency.
4. Over confidence.
5. Emotional Outbursts.

None of these factors have any place in a critical decision-making process, always use a FACT – based approach. Make sure you base decisions on good factual data.

If an incident/accident has taken place, the investigation must begin immediately ensuring to capture as much information as possible in the first 30 minutes.

1. Determine extent of injury/damage/disruption.
2. Determine if the incident/accident is under-control.
3. Determine if the incident/accident has had or may have any potential knock on impacts e.g. any chemical substance in use or present at the scene.
4. Cordon off the area.
5. Determine what resources are required to deal with any or all the issues.

6. If the incident/accident is of such a nature that it has, or may have, off site implications, it may be necessary to gather an incident management team on site.
7. Ensure only **authorised personnel** enter the cordoned off area to gather the evidence/facts.
8. If several people witnessed it, ask everyone involved to immediately write down and document exactly what they remember happened and what they were doing at the time of the incident/accident e.g. operating a mixer via a PC screen or manually opening a hand valve allowing liquid into a tank.
9. Take photographs, videos, and measurements. Record anything out of the ordinary as soon as possible. This type of recorded information can prove important during the subsequent investigation and may help find the **root cause** of what caused the incident/accident in the first place.

Right first time

Whatever work is carried out, **do it right first time**, regardless of the time pressures. Emphasis on 'Production over Protection' can prove to be a company's biggest weakness. Do not put a piece of equipment **back to service** unless 100% satisfied that it is safe to use, fit for purpose and fully operational.

N.B. To be a focal point of an **accident/incident/corporate manslaughter investigation** because of something you did can be a very lonely place, especially if people are injured or dead, environmental damage has occurred, and serious monetary loss has been incurred on the company. Making excuses like 'I didn't think', 'I was watching the clock', 'I was under pressure to get the job done' while someone lies dead in a morgue because of your actions will be of **cold comfort**.

4. MAINTENANCE MANAGEMENT

Developing a Maintenance Department Structure

An Engineering department must be committed to maintaining the equipment and site facilities which sustain the company's business needs. This must be done in a safe and regulatory compliant manner. New, more creative approaches are needed to make budgets go further and minimise costs and waste. Modern manufacturing processes rely heavily on innovative, integral, flexible equipment to achieve quality, efficiency, and cost effectiveness. Equipment reliability has never been as important as it is now. **Lean Manufacturing** means that plant downtime must be minimised and avoided wherever possible. Maintenance Engineers with the proper specialist skill sets must increasingly become Reliability Engineers and focus on Asset Lifecycle Management and Continuous Improvement.

Correct operation and regular routine maintenance is the best way to ensure plant machinery remains at peak efficiency thereby extending its working life whilst keeping running costs to a minimum. The best way to promote maximum efficiency is through the development of the skills and awareness of your maintenance team by attending **Original Equipment Manufacturers** (OEM) training courses. The engineering department must be committed to the objective of employing the best available practices to ensure all team members are competently trained on all aspects of their jobs. Looking at it from the **outside in**, some people believe everything in maintenance is waste, they see it as a non-core activity, since none of it directly contributes to the customer's needs and wants. While a perfect factory, with perfect people with perfect equipment would need zero maintenance effort, real factories and real equipment must run efficiently while real people need to function effectively.

The key is to ensure valuable maintenance resources are not spent on inappropriate maintenance strategies and poor execution of daily maintenance activities. Regardless, the facts of the matter are, companies are under increasing pressure to reduce maintenance costs and plant downtime.

Engineering **Asset Management** of ageing equipment requires effective maintenance investment management in Protection, Inspection, Detection, and Correction which will extend the service life of installations and, above all, limit the occurrence of accidents.

The 11 most commonly used Precision Maintenance Programmes synonymous with Asset Reliability are:

1. Preventive Maintenance - (PM)
2. Predictive Maintenance - (PdM)
3. Condition-based Maintenance - (CbM)
4. Condition Monitoring - (CM)
5. Corrective Maintenance - (A.K.A. - Reactive Maintenance)
6. Total Productive Maintenance (TPM)
7. Reliability Centered Maintenance - (RCM)
8. Maintenance, Repair & Overhaul (MRO)
9. Overall Equipment Effectiveness (OEE)
10. Lean Maintenance
11. Time-based Maintenance (TbM)

1. **Preventive Maintenance - (PM) -** Preventive Maintenance (or Preventative Maintenance) is maintenance that is regularly performed on a piece of equipment to lessen the likelihood of it failing, it can either be initiated on a 'usage or time trigger', based on the manufacturer's suggested maintenance schedule. Instead of waiting for the machine to malfunction or stop working completely, maintenance and inspections are scheduled at regular intervals. This enables you to discover things that may become issues before they become an issue. Another benefit to this approach includes setting up orders for replacement parts well in advance of when you will need them. A plan must be in place to do Preventive Maintenance! Preventive Maintenance can have a significant impact in many ways. Besides increased asset life, it can reduce the need for redundant equipment due to reductions in unplanned downtime. Reducing redundant equipment and the need for earlier replacement of equipment has a financial impact that goes all the way to company financial statements and operating efficiency.

2. **Predictive Maintenance - (PdM)** is where engineering personnel can accurately predict the health and status of assets which

enables informed decisions to be made regarding maintenance cycles and interventions. Predicting the status of equipment health via Condition Monitoring, leads to direct cost saving benefits, and improves your bottom line. It is important to gather the 'hidden data', manage the field data effectively, and decipher how this data can be brought together as 'Information' to make informed decisions regarding asset status. The 'Predictive' component of Predictive Maintenance stems from the goal of predicting the future trend of the equipment's condition. This approach uses principles of statistical process control using for example a Computerised Maintenance Management System (CMMS) to determine at what point in the future maintenance activities will be appropriate. Most Predictive inspections are performed while equipment is in service thereby minimising disruption of normal system operations. Adoption of PdM can result in substantial cost savings and higher system reliability. **N.B.** The ultimate goal of the Predictive Maintenance approach is to perform maintenance at a scheduled point in time when the maintenance activity is most cost-effective and before the equipment loses performance within a threshold.

3. **Condition-Based Maintenance** - **(CBM)** is a maintenance strategy that monitors the actual condition of the asset to decide what maintenance needs to be done. CBM dictates that maintenance should only be performed when certain indicators show signs of decreasing performance or upcoming failure. It is a maintenance programme that recommends maintenance decisions based on the information collected through **Condition Monitoring**: analysis of values reported by sensors (e.g. pressure sensors; vibration sensors; temperature sensors; voltage and ampere meters; flow meters) will result in a decision if a physical maintenance intervention is required or not. Its main goal is to increase reliability and availability of machinery, while minimising downtime, labour and repair costs. CbM can be performed by scheduling downtime, labour and materials based on machinery health. Ideally Condition-based Maintenance will allow the maintenance personnel to do only the right things, minimising spare parts cost, system downtime and time spent on maintenance.

Example: the differences between **PM, PdM & CbM** using a simple 20-Watt LED bulb with a 1,600 lumens light output (Fig.1).

N.B. This explanation can be applied to almost any piece of equipment:

Fig.1

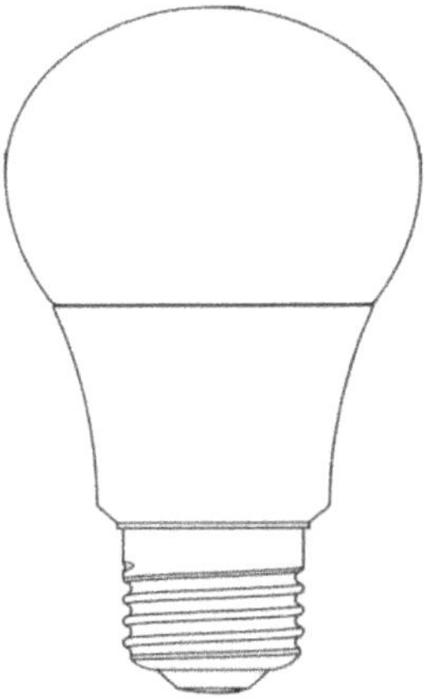

- The life span of the LED bulb is approximately 5.5 years or 50,000+ hours of continual use according to its manufacturer. So just before the 5.5 years' expiration date, under scheduled maintenance, you replace it with a new one. This is called **Preventive Maintenance (PM).**
- You routinely monitor the LED bulbs' operation, after 4 years, you notice it starts to flicker. So, you are predicting that it is going to fail and decide to change it with a new one and scheduled for "Just-In Time" maintenance. This is called **Predictive Maintenance (PdM).**
- After 6 years of continual use, the light output from the 20-Watt LED bulb is measured using a lux meter and its light output is still approximately 1,600 lumens which is the minimum required on the working plain, the bulb is not replaced and remains in operation. In the seventh year, the light output is measured again,

it is noted that the 1,600 lumens is not being maintained. It is now at 1,400 lumens and that the light output is deteriorating (e.g. bulb discolourisation). It is replaced with a new one to restore the required light output, this is called **Condition Based Maintenance (CBM).**

4. **Condition Monitoring - (CM)** is the process of monitoring a parameter of condition in machinery, such that a significant change is indicative of a developing failure **e.g.** overheating is monitored for early signs of impending failure much like that of a 'human being' when experiencing a high body temperature, medicine will be taken to counteract it. The same analogy applies when a machine is running constantly at an 'above normal' operating temperature, action must be taken, otherwise it will shut down. **N.B.** Electromechanical devices **will fail** if not maintained properly. Equipment can be monitored using sophisticated instrumentation such as vibration analysis and ultrasonic monitoring equipment or the human senses. Where instrumentation is used, actual limits can be imposed to trigger maintenance activity. It is a major component of Predictive Maintenance and allows maintenance to be scheduled or other actions to be taken to avoid the consequences of a failure before the failure occurs.

As part of an overall **Asset Management** strategy**,** a **Condition Monitoring** (CM) programme has many benefits including Overall Equipment Effectiveness (OEE), reduced Mean Time To Repair (MTTR); increase Return on Net Assets (RONA); reduced unplanned and planned downtime duration and lower inventory cost.

A company must strive to improve **early asset failure recognition** to maximise equipment reliability.

N.B. Condition Monitoring is more cost effective than a 'Run to Failure' maintenance policy as it minimises direct maintenance costs from the following:

- Unnecessary downtime
- Upstream and downstream equipment damage
- Allows for more cost effective planned maintenance

- Reduces the need for call-in or working unscheduled overtime on breakdowns
- Minimises the need to hold excessive amounts of spare parts.

Saves energy, waste, lost production e.g.

- Undetected misalignment condition of motors coupled to drives using excessive energy through Vibration Analysis
- IR Thermography Surveys can indicate loss of temperature e.g. through poor insulation/refractory on boilers
- IR Thermography can highlight energy or production losses through detection of line restriction
- Ultrasonic condition monitoring can detect even incipient faults in bearings as well as adequacy of lubricant
- Oil Sample Analysis can reduce the need for unnecessary oil changes and costly environmental disposal of waste oil.

Condition Monitoring reduces indirect costs from lost production due to unscheduled downtime from an in-service failure.

The 4 most commonly used technologies in Condition Monitoring Programmes are:

- Infrared Thermography/Thermal Imaging
- Vibration Analysis
- Oil Analysis
- Ultrasonic Monitoring

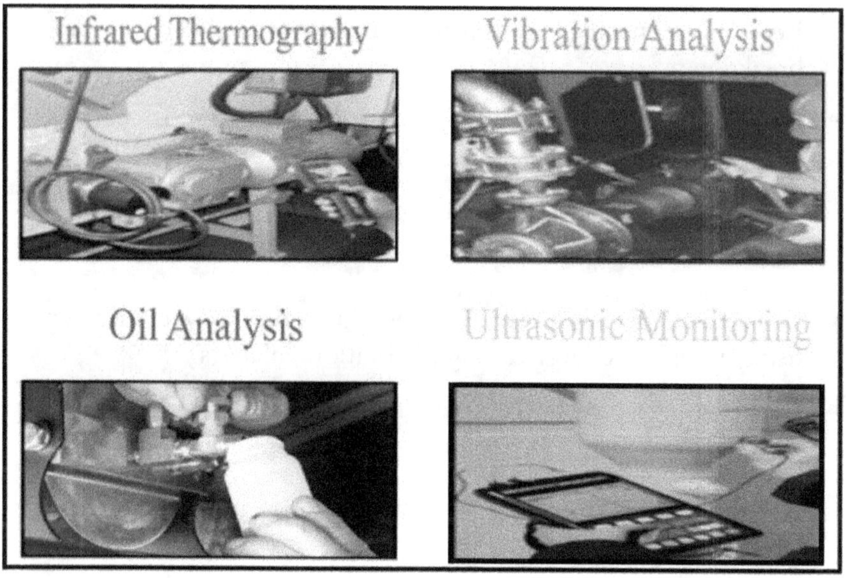

- Infrared Thermography: Every machine/object, process, system, or body constantly emits thermal energy to the environment in the form of invisible radiant energy. As an object heats up, it will radiate more and more energy from its surface. The technique for making infrared radiation visible is called Thermal Imaging or Infrared Thermography. It detects variations of temperatures in electrical, mechanical, infrastructure and process equipment. **Thermal Imaging** is an important part of any **Predictive Maintenance** programme.

How does it work?

Thermography is the science of obtaining images of the heat distribution of an object. It is a passive technique which examines the energy emitted by an object without any need to stimulate or touch it in any way. Since inspections can be made whilst the component is under full operational conditions, there is no loss of production or downtime.

Thermography Benefits

a) An effective and efficient way of establishing the condition of the equipment under survey during normal operation.

b) Development of faults can be pinpointed at an early stage, before breakdown occurs.

c) Allows planned corrective action to be taken, thus reducing downtime.

d) An effective Thermography programme reduces overall maintenance costs by identifying faults at an early stage and thus reducing the material and labour repair costs.

e) Infra-red thermal imaging inspections will benefit a company by providing real time data e.g. temperature readings, to determine defects that can lead to equipment failure **e.g.** motor running too hot. Applications include infra-red testing and inspection of electrical distribution systems, motors, heat exchangers .... in helping to diagnose abnormal **hot spots** in pieces of equipment e.g. M.C.B. failure (poor electrical connections), damaged motor bearing, or **cold spots** in a partially blocked heat exchanger leading to inefficient operation. An infrared camera can be used for a broad range of **Predictive Maintenance (PdM)** applications enabling problem areas to be clearly seen on a crisp clear image.

• Vibration Analysis is an essential **Predictive Maintenance** technique: it measures change in vibration intensity on mechanical equipment when machine conditions begin to degrade. It detects equipment vibration levels affected by factors such as misalignments, unbalance, displacement, velocity, looseness,

eccentricity, defective bearings, resonance, electrical problems, and aerodynamic/hydraulic forces. Equipment misalignment alone results in wasted energy and causes premature equipment breakdown. The best way to reduce maintenance and operating costs of rotating machinery is to align the machinery correctly in the first place. The goal is to identify changes in the condition of a machine that will indicate a potential problem. An effective condition-monitoring programme will help to detect faults at an early stage such as when rolling element bearing defects are still minor and do not affect the performance of the machine. This allows time to analyse the problem while the machine stays online and production processes continue normally. This advance warning allows planning of the repair at a convenient time.

- **Oil Analysis:** detects contamination (e.g. metal shavings) or degradation of oil which indicates machine wear. **Tip:** AN OIL SIGHT GLASS & LEVEL MONITOR - provides continuous and instantaneous monitoring of oil clarity, water contamination and oil level. This all-in-one tool allows for a quick and easy inspection to easily assess what is going on in your equipment. Correct oil lubrication is critical for the efficient operation of equipment and must be maintained according to the manufacturers' instructions. One of the keys to keeping a factory running is **Lubricants** – correct lubrication is essential for keeping machinery working, from the smallest motor to a multimillion euro machine. To ensure sufficient lubrication, Oil Analysis is an essential **Predictive Maintenance** technique which provides advanced warning of abnormal conditions as oil degradation will lead to accidental failures and unanticipated downtime. If a lubricant becomes contaminated or becomes deficient in quality, the machine will suffer the consequences. Lubricating with two different types of lubricant can lead to a degradation of the lubricant. If these problems go undetected, it will affect the machine's functionality, often resulting in devastating damage to the machine. Therefore, an effective oil analysis testing programme will increase the reliability of your machinery, and reduce maintenance costs associated with oil changes, repairs, and machine downtime. **Tip:** Review the type and grade of lubrication used on motors and machinery. Using the wrong lubricant can add 5% to energy costs. Additionally, some high performance lubricants can reduce energy costs by 5%.

- Ultrasonic Monitoring

This technology can be considered an integrated technology since it can be used with infrared and vibration inspections as well as standalone to perform a multiplicity of inspection activities. Instruments based on this technology can monitor a wide range of plant operations. Many failures and repairs that commonly occur in the industrial setting can be prevented with ultrasonic sensor technology, a highly effective, non-destructive, **Predictive Maintenance** method.

Airborne ultrasound instruments are becoming an important part of **Condition Monitoring**, fugitive emissions, and energy conservation programmes. Their versatility, ease of use and portability enable managers to effectively plan and implement inspection procedures. Instruments based on airborne ultrasound, sense high frequency sounds produced by leaks, electrical emissions, and mechanical operations. By locating leaks, detecting high-voltage electrical emissions, and sensing early warning of mechanical failure (e.g. metal fatigue - Ultrasound uses very high pitched sound waves to image materials and can be used to find cracks inside materials) these instruments contribute to cost reductions, improved system efficiencies, reduced production downtime, improved quality control and safety, and decreased work hours by improving troubleshooting capabilities. A lack of training and understanding, irregular monitoring and a lack of commitment to the **Predictive Maintenance** programme will lead to poor results. However, proper implementation of ultrasonic technology will increase reliability, decrease troubleshooting time, and decrease time spent 'firefighting' by your operation and maintenance staff. For optimum effectiveness, it is recommended that all three major technologies – **Vibration, Infrared, and Ultrasound** be used as part of a comprehensive inspection programme in monitoring critical components at your plant.

Laser Alignment & Predictive Maintenance - A misaligned machine causes stress to both bearings and shafts. As an effect of this, the seals open up, allowing lubrication leakage and contamination to enter. All together, the bearings lifetime is dramatically shortened. Shaft alignment is recommended for almost every industry, as it enables machine uptime to be significantly improved and maintenance costs to be reduced.

Example: Laser alignment of a 'motor coupled to a pump' arrangement.

Fact: Even with flexible couplings, shaft misalignment causes bearings, seals & couplings to wear faster.

Shaft alignment is the process to align two or more shafts with each other to within a tolerated margin. It is an absolute requirement for machinery before the machinery is put in service.

Shaft Alignment & Pipe Stress: Often during pump installations, pipe and pump flanges are not in alignment. When mated together, the pipe stress pulls on the pump flange, this forces the frame of the pump to distort vertically and/or horizontally. The resulting deflection of the pump casing results in internal misalignment of the bearings, inducing vibration, and premature failures. Sometimes it is thermal growth that causes pipe stress and the resulting strain changes the alignment at the coupling. It is important to check for pipe stress during pipe installation. A good laser alignment system can help you do this. Eliminating pipe stress will greatly improve the reliability of your machines. Poor shaft alignment also increases energy consumption. Laser Alignment is by far the most accurate & efficient method of eliminating these problems.

The benefits of **Predictive Maintenance** are tremendous savings, spares can be ordered in advance of equipment failure and problems can be identified in the early stages so that damage will not occur to costlier spares such as shafts, impellers, fans, rotors etc.

5. **Corrective Maintenance -** (A.K.A. - Reactive Maintenance - action is initiated when the unscheduled event of an equipment failure occurs) can be defined as a maintenance task performed to identify, isolate, and rectify a fault so that the failed equipment, machine, or system can be restored to an operational condition within the tolerances or limits established for in-service operations. Corrective maintenance is carried out on all items where the consequences of failure or wearing out are not significant and the cost of this maintenance is not greater than preventive maintenance.

6. **Total Productive Maintenance (TPM)** is a programme for planning and achieving minimal machine downtime. TPM is often the first place to begin the journey towards Lean Transformation because it creates a manufacturing environment of stability, reliability, and capability. A stable infrastructure is a critical foundation that must be built before more tightly coupled Lean practices can be successfully introduced and more importantly sustained. Equipment and tools are literally put on 'proactive' maintenance schedules to keep them running efficiently and with greatly reduced downtime. Machine operators take far greater responsibility for their machines upkeep. Maintenance technicians are liberated from mundane, routine maintenance, enabling them to focus on urgent repairs and proactive maintenance activities. A solid TPM programme allows you to plan your downtime and keep breakdowns to a minimum.

7. **Reliability Centered Maintenance - (RCM)** is a process to ensure that assets continue to do what their users require in their

present operating context. It is generally used to achieve improvements in fields such as the establishment of safe minimum levels of maintenance, changes to operating procedures & strategies and the establishment of capital maintenance regimes and plans. Successful implementation of RCM will lead to increase in cost effectiveness, machine uptime and a greater understanding of the level of risk that the organisation is presently managing.

8. **Maintenance, Repair & Overhaul (MRO)** involves fixing any sort of mechanical, plumbing or electrical device should it become out of order or broken (known as repair, unscheduled or casualty maintenance). It also includes performing routine actions which keep the device in working order (known as scheduled maintenance) or prevents trouble from arising (Preventive Maintenance).

9. **Overall Equipment Effectiveness (OEE)** is a critical methodology to drive improved efficiency, equipment availability, performance, higher quality, and reduced costs for companies who are looking to maximise productivity while minimising operational costs. It is a powerful KPI in providing a metric that can be used by operations, production, engineering, maintenance, quality, and continuous improvement teams. Companies are now seeing the benefits of OEE from the energy consumption viewpoint as well. OEE highlights the impact of issues such as unplanned stoppages, slow running equipment or high reject rates. The first step in the process is to establish the baseline OEE metric, this invariable reveals significant scope for improvement initiatives. Utilising a proven problem-solving methodology, improvement initiatives are implemented to improve the OEE metric and consequently the energy efficiency of the plant.

10. **Lean Maintenance** - Maintenance engagement with a lean manufacturing programme accelerates the development of an organisation's lean capability. Maintenance plays a role in supporting **Lean Manufacturing** and this requires a change in emphasis from a traditional approach to maintenance. **5S** is the name given to the

Lean Manufacturing method for the clearing out of all unnecessary things to allow room for the acquisition of tools and parts in the fastest and easiest manner.

A comparison of **5S methodology** with an evaluation and optimisation of a Preventive Maintenance programme at a plant quickly shows how similar these processes are. 5S is the name of a workplace organisation method that uses a list of five Japanese words: **Seiri, Seiton, Seiso, Seiketsu, and Shitsuke** Translated into Roman script **(Sort, Straighten, Sweep, Standardise, Sustain)**. The list describes how to organise a work space for efficiency and effectiveness by identifying and storing the items used, maintaining the area and items, and sustaining the new order. The decision-making process usually comes from a dialogue about standardisation, which builds understanding among employees of how they should do the work. Lean manufacturing requires a greater level of reliability and dependability from equipment. Maintenance plays an important role in securing higher levels of customer added value.

11.**Time-based Maintenance (TbM)** is maintenance performed on equipment based on a calendar schedule. This means that time is the maintenance trigger for this type of maintenance. Time-based maintenance is planned maintenance, as it must be scheduled in advance.

Tip: every hour spent planning the work saves three to five hours in execution.

Organisations must use the expertise of their maintenance & engineering functions to release new value from their operations by:

- Targeting weak technology to address the causes of management firefighting
- Leading the drive to deliver higher levels of customer added value

- Rapid delivery of new product and service offerings
- Provide a bridge between today's transformation processes and tomorrow's customer
- Winning operational capability
- Organisations must understand the value adding role which maintenance can play and transform the Maintenance role in 'How to stay in control of costs and raise the competence of their maintenance team to meet the challenge.'

Reliability

Improving reliability of equipment can improve profitability by preventing costly shutdowns and eliminating costly business interruptions. An essential part of any business plan must be to provide new solutions in reducing energy consumption, minimise downtime and to react intelligently to real problems as they occur.

A Business must fully understand how and why a plant and its bespoke assets function:

1. Plant Reliability and Compliance.
2. Training and Educating Personnel.
3. The employing of detailed, proven, disciplined principles and techniques in tandem with strategic system intelligence in:
    - Troubleshooting
    - Fault Finding
    - Problem Solving
    - Decision Making

1. Safety and Risk must be managed and costs minimised over extended asset lifecycles.
2. Money invested in maintenance must generate maximum value for the organisation.

A Maintenance Practitioners' key objectives are to:

Ensure proper equipment maintenance is carried out as inadequate or inferior maintenance practices often cause slow, irreversible equipment degradation and ultimately lost revenue. When a new asset is acquired, it must be properly serviced, maintained and future proofed for **'Deterioration prevention'** and that costs money.

The major factors affecting the deterioration of equipment include exposure to chemicals or solvents, likely accumulation of dust or dirt, likelihood of water ingress, exposure to excessive ambient temperature, risk of mechanical damage, lack of lubrication, exposure to undue vibration and susceptibility to corrosion **e.g.** the corrosion of structural steel is an electrochemical process that requires the simultaneous presence of moisture and oxygen. In the absence of either, corrosion does not occur. Essentially, the iron in the steel is oxidised to produce rust, which occupies approximately 6 times the volume of the original material consumed in the process.

Tip: When removing badly rusted bolts, nuts, bleed fittings; BE PATIENT, use plenty of good quality penetrating spray and leave to soak in (possibly hours) before loosening, reapply as necessary. Stripping threads or shearing off a nut, bolt or bleed fitting while trying to remove it, can prove to be very expensive and difficult to repair or replace. Remember, "Rust never sleeps, treat and protect now, or Pay later."

Corrosion Under Insulation (CUI)

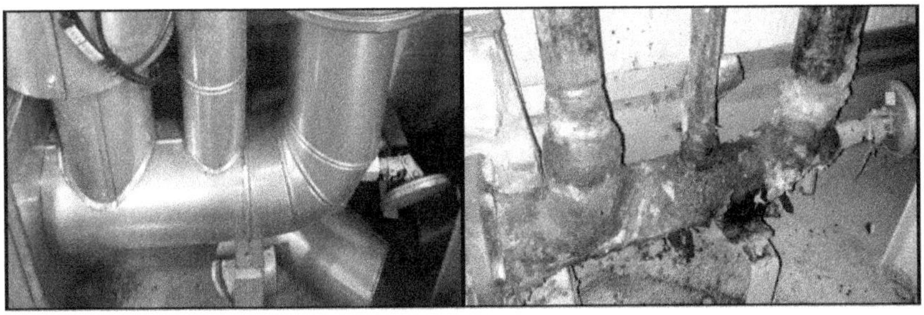

Over time, most metals tend to deteriorate due to corrosion, Exposure to certain liquids, for example, water, water vapour, acids, bases, ammonia, salts, and heavy metal ions - can induce corrosion. Corrosion Under Insulation is one of the biggest problems associated with high temperature insulated steel. The highly aggressive environment created beneath insulation materials, in conjunction with the lack of visible evidence of corrosion, means that if not correctly managed, CUI can lead to catastrophic and expensive failures.

Why insulate?

Insulation is used to minimise heat loss, reduce costs, and improve efficiency. It may also be employed to minimise heat gain or to protect personnel from the risk of injury from hot or cold surfaces. Traditional insulation systems typically consist of an insulating material such as mineral wool or calcium silicate, which is then protected by an outer layer of cladding. Thin metal sheet or composite wraps are the most common cladding materials.

Risk of Water Ingress

To help combat the risk of CUI, insulation systems are designed and installed with great care given to sealing joints, terminations, and protrusions. Despite these efforts, water ingress over time is inevitable. Mechanical damage, degradation of sealants, rainwater, deluge systems and atmospheric moisture will all contribute to water ingress through the insulation system, resulting in a warm, damp corrosive environment against the steelwork. It is therefore essential that an appropriate protective coating is applied to the steelwork first to offer long term protection before the insulation is fitted.

Increased Corrosion Rates

Once water penetrates an insulation material, a highly corrosive environment can be created at the interface between insulation and steel substrate. Moisture is often unable to escape, leading to prolonged periods of moisture contact and further build-up of corrosive contaminants. This raises the boiling point of the water, extending the risk of corrosion to higher temperatures and vastly increasing the corrosion rates of inadequately protected steelwork.

N.B. An inspection program must be in place e.g. regular thickness measurements at sensitive points and Hydraulic tests (apart from piping, a significant number of corrosion-related accidents involve vats and tanks).

Deterioration is what happens naturally to anything that is not taken care of. A popular approach is to reduce expenditures on equipment maintenance. However, **this is a very short-sighted view**, as deferred investments often resurface and can cost the business five to ten times more than the original maintenance scope of work. Inspection can be seen as a cost to business, and with pressure applied to save money by doing less or using lower skilled operatives. This can result in short term person-hour savings but degrade the asset and expose the business to greater risks of failure and poor reliability. Worse still, by focussing on individual items of equipment and inspection person-hour costs only, rather than the total cost of inspection, the business cost may be higher along with a greater risk of an incident. Oversight and maintenance standards must never be allowed to deteriorate regardless of how old a piece of equipment is.

N.B. It is acknowledged industry-wide that cutting costs on maintenance is a false economy.

Senior Management must:

- Ensure the culture of the company changes from the unpredictability of reactive work to a **best in class** model of using more planning, scheduling and preventive work.
- Carry out a skills analysis programme with the onsite engineering team to establish strengths and weaknesses.
- Ensure **Cross-skilling, up-skilling, and knowledge sharing** among the onsite engineering team to enable them to carry out diagnostics and repairs to keep machinery up-and-running and reduce the costs associated with downtime. This will ensure no inappropriate maintenance works are carried out and no unauthorised modifications or adjustments are made. All maintenance works must be completed in accordance with the manufacturer's recommendations.
- Maximise the uptime of the business's processes by ensuring quality reliable equipment is installed with redundancy (where possible) thus ensuring optimum lifetime of all machinery in a predictable, safe, and low-cost manner.

Good Management (Soft Skills)

There are many facets to a good management structure and the future success of a business. Along with the necessary qualifications for the role, **Engineering Management entails**:

- ➢ Finance, planning, and budget control.
- ➢ Having good ICT skills.
- ➢ Knowing how the business operates and an understanding of what other departments do. This will put you in a better position to style your work to be of maximum benefit to the business itself.
- ➢ Show integrity, always do the right thing.
- ➢ Having good personal people skills and being a good communicator.
- ➢ Having good organisational skills and can handle the inevitable pressure and stress that can accompany such a role.

- Having good conflict management skills.
- Instilling a good, honest work ethic and good decision making, "you need to say what you mean and do what you say", to do that; you must practice what you preach.
- Being calm and responsive, addressing others with respect. ***Talk to people, not at them.***
- Having the ability to motivate your staff and, in turn, get them to maximise their (and the business's) productivity in tandem with using good 'Time Management' techniques.
- Asking team members what obstacles (if any) block their ability to commit or participate fully in their **day to day activities** and work to remove the barriers and address the issues.
- Making unpopular decisions and taking the necessary steps to implement them to achieve the company's KPI targets.
- Respecting everyone's opinions, this will greatly improve the odds of having a successful, and positive working environment.
- Taking action to evaluate and minimise risks to the employee's and the company.
- Seeing to it that your own and others' work is accurate and complete. Working with diligence and care and ensuring all commitments are fulfilled.
- Having excellent writing and grammar skills, and having the ability to write concisely, clearly, and logically.
- Having the ability to adapt to changes while keeping focus on goals and applying knowledge to new circumstances.
- Can enlist the support and cooperation of others and encourage them to be proactive.
- Can make decisions and take responsibility for them.

E.g. Nobody wants an autocratic boss hovering over them, micromanaging, and watching their every move. It is up to the manager to praise employees' good work and offer suggestions in a professional, civilised manner.

Employees want to emulate the **higher-ups**, so set the standard of expectation. Even if you're the one in charge, others may not automatically see you as such. To get others to accept you as their leader, you'll need to prove that you're worthy of your status. People in general, admire and appreciate another person's achievements, not the **title** they have. 'Authority is something you are given; respect is something you earn!' As a manager, you may want things done your way, but you can only guide others in the direction you want them to go. When you treat staff like servants, you eat away at their pride and self-esteem, and they will ultimately despise working under you.

When you entrust a responsible and capable person with an important task, give them some breathing room to perform it. If you are new to a management role, you must resist the overriding urge to fix things, to tell people what to do, to continue to be the technical expert, to have all the answers; instead by asking team members to come up with a new project means empowering them and giving them the time and space to think creatively without constantly barraging them with questions. Step back and demonstrate that you have confidence in them, this will allow for a more efficient and less hostile working environment. For example, you may have multiple team leaders working on a range of projects at any one time. Some team leaders may like the title of **team leader** but maybe are unsure about their decision making and people skills. As their manager you must reassure them how certain you are of their capabilities and you are available to give direction and advice if necessary, at any time. Value your team's opinions, being a leader leaves much room for interpretation, sometimes, what ***you*** think is the best answer to a given dilemma might not necessarily be the best way to proceed according to those around you.

Thus, accepting a leadership position includes having to make difficult choices, and being able to accept that someone else may come up with a better plan of action. **A great leader makes choices based on the business's best interests, not their own.** It's only through a high level of trust and empathy, effective and active listening both verbal and non-verbal communication, that a real understanding of your team emerges.

Just because you're the one in charge, doesn't mean you know everything about your company's activities. While you may be more experienced and can likely provide more insight than others, you should always strive to learn new things.

Performance Management is important to any organisation where:

1. Work is planned, and expectations are set.
2. Performance of work is monitored.
3. Staff ability to perform is developed and enhanced.
4. Performance is rated or measured and the ratings summarised.
5. Top performance is rewarded.

Performance Management involving 'face to face' meetings must be held at least once every six months between the manager and all team members (monthly meetings with new employees for the first 6 months and beyond if necessary). Direct feedback can then be given on their past 6 month's performance using these five key discussion points above as reference. The time dedicated to this task cannot be underestimated and is real **value added.**

A dedicated folder or soft copy should be created, maintained, and updated by the supervisor/manager as required for each team member which will include the normal **day to day** department activities. The following points can also be used as additional discussion points which may need to be addressed:

- Safety practices
- Professional behaviour
- Absenteeism
- Attitude
- HR issues
- Interaction and communication with other team members
- Willingness to share 'work related' information
- Being helpful to others

- Time keeping
- Department Accidents/Incidents
- Department Compliance issues
- Good Documentation Practices
- Training attendance

A team of well-trained professionals is the key to running a good engineering department. The manager must make sure all their needs are catered for as much as possible both personally and professionally. Understanding the importance of maintaining good team morale, trying to ensure their needs are met and the implementation of a good team work environment cannot be stressed enough.

Whether it's by:

- Salaries and remuneration packages.
- Maintaining modern clean toilet, shower areas and restrooms. The state of a restroom can have a measurable effect on the health of a business.
- Maintaining modern canteen and workshop area.
- Providing proper PPE.
- Suitable 'fit for purpose' water proof clothing.
- Purchasing upgraded quality tools and equipment.
- Increased training.
- Applying new technology.

A manager must show that they are willing to fight for their team. They won't always be successful, but it's important that he/she acts as their advocate.

Good Engineering Management (Hard Skills):

The following fundamental structures should be in place:

A. Project Management
B. Original Equipment Manufacturers.
C. A properly indexed and 'User friendly' Engineering library.
D. Good Budget Strategy.
E. Good Service Level Agreements with vendors and contractors.
F. A well stocked and efficient Engineering stores area.
G. Vendor 'CMMS Maintenance Software' System Package.
H. Site Energy Management Strategy.
I. Emergency back up systems and contingency plans to ensure business continuity.
J. Vendor 'Calibration Management Software' System Package.
K. Basics of Finance.

A. Project Management Example

Project Management (PM): is the discipline of initiating, planning, executing, controlling, and closing the work of a team to achieve specific goals and meet specific success criteria.

An engineering **'Turn-Key Project'** is a project that is scoped, designed, built, installed, commissioned and handed over to a company in a fit for purpose, versatile, reliable, energy efficient and ready-to-use condition.

In the following example, an expertise is required which is not available from a single individual; a suitable turnkey project provider is now selected and employed by the company.

E.g. a new diesel powered 1,000 kVA electricity generator is to be installed in a plant to supply **back up** essential power to critical onsite equipment in the event of mains electricity power loss from the incoming national grid system. The various works will involve structural, mechanical, electrical and IT expertise to install.

A turnkey project such as this involves the following elements depending on scope and size:

- Project administration
- Design and engineering services
- Subcontracting of works
- Procurement and expediting of materials and equipment
- Inspection of equipment prior to delivery
- Management of actual equipment installation
- Control of schedule and quality
- Commissioning and completion
- Performance-guarantee testing
- Handover Documentation Package.

The turnkey provider provides the following warranties and guarantees and accepts liabilities which include:

- Warranties for the timeliness of deliveries of equipment, of erection and of completion times of structural, mechanical, and electrical works.
- Warranties for workmanship according to specifications and guarantees that correct standards will be used.
- Liability for property and equipment.
- Correct safety standards being implemented.
- Contractor undertakes to assure that Mechanical and Electrical performance will be maintained for a definite period.
- **Fitness for Purpose** - Warranties and guarantees that the complete installation is fit for its intended purpose.

B. Original Equipment Manufacturers (OEM's)

Original Equipment Manufacturers (OEM's) and **Project Management** (PM) companies welcome early involvement in the design process, allowing a company to get advice from their knowledgeable and experienced experts **who**:

- Provide a family of services delivered by a multi-disciplinary team of Design, Structural, Electrical, Mechanical, IT and control professionals.
- Understand the full capabilities of modern control systems and how to ensure a company achieves the optimum cost/ performance balance of installed equipment.
- Act as a focal point for the project, coordinating information and technical capabilities.
- Can assess, design, implement, audit and manage new and existing control and information networks and the security technology, policies and procedures for those networks and the personnel that use them.

Original Equipment Manufacturers must help maintenance personnel by ensuring their machines and associated systems are easy to troubleshoot, modify and repair during the machines lifecycle, in other words, **maintainable**.

C. Create an Engineering Library

A technical file should be compiled for all projects completed by the Engineering Projects Department. Technical files are created both in 'Hard' and 'Soft' copies to provide a historical archive for future reference. A copy is issued to the System Owner on completion of a project, with a master copy retained in the site library archive room.

Compiling a Project Technical File:

A typical contents page for a Project Technical File describes the various documentation requirements included under the relevant headings.

Approved Documentation:

Includes signed copies of the following:

- Project Initiation Form (PIF)
- Scope Documents (e.g. URS, FDS)
- Copy of approved Change Control (ref. Site Change Control Procedure)
- Copy of approved Capital Request Forms
- Any other miscellaneous approved documentation

Drawings:

- System 'As-Built P&IDs
- System 'As-Built' G.A.'s (General Arrangement Drawings)
- System 'As-Built' Electrical and Instrumentation Drawings / schematics
- System 'As-Built' Mechanical Drawings
- Other Miscellaneous Drawings relevant to the project

Commissioning Test Packs for:

Mechanical Systems:

- Commissioning Reports
- 'As-Built' Isometric drawings
- Pressure Test Certificates
- NDT (Non-Destructive Testing) Certificates
- Welder Competency Certificates
- Material certificates

- Miscellaneous

Electrical & Instrumentation Systems:

- As-Built Electrical and Instrumentation Schematics
- As-Built Instrument Loop Sheets
- Electrical Bonding and Electrical Test Sheets
- Equipment Data Sheets

Automation Systems:

- Control URS / FDS (As-Built)
- System Architecture Schematics
- (SDS) Software Design Specification
- Commissioning Test Protocols
- FAT / SAT Reports

Certificates of Conformity:

- Calibration Certificates
- Compliance Certificates
- PSSR (Pressure Systems Safety Regulations)
- PED (Pressure Equipment Directive) Certificates
- Material Certificates of Conformity
- Miscellaneous Certification (e.g. Noise)

Purchasing Data:

- Project Purchase Orders
- Purchasing / Vendor Data
- Equipment and Instrumentation Data Sheets

- Miscellaneous

Maintenance:

- Operation & Maintenance Manuals
- Calibration List / Updated Master Instrument List
- Equipment Index
- Spare Parts Lists
- RFID/Scanner technology
- Warranty Certificates
- Miscellaneous

EH&S Data:

- HAZOP / Risk Assessment Data
- Miscellaneous

Correspondence / Minutes:

- Project Handover Record
- Miscellaneous
- Labelling of Project Technical File
- Project Technical File Distribution

As part of the project handover an Engineering Project Handover Form must be completed and included in the Project Technical File.

Hardcopies of the Project Technical File should be made and circulated as required to:

1. The System Owner
2. Engineering / Maintenance
3. Library Archive (One Copy Retained)

A soft copy of the Project Technical File is maintained on the Project Engineering Hard Drive.

Compiling a Maintenance O&M technical file:

A typical contents page for a Maintenance Technical File describes the various documentation requirements included under the relevant headings. Normally this file is kept in the maintenance office/workshop area.

When creating a maintenance hard copy folder of relevant associated information pertaining to any piece of equipment on site for other personnel to reference or use, the following format can be used using typical chapter headings and then subsequently stored in the engineering library for easy retrieval.

- System overview.
- Control Philosophy - User Requirement Specification/ Functional Design Specification/ System Design Specification.
- Safety & Environment
- Piping & Instrumentation Diagrams.
- Process & Instrumentation Drawings.
- Process Flow Diagrams.
- Operation & Maintenance Manuals.
- 'As Built' Electrical Drawings.
- 'As Built' Mechanical drawings.
- Standard Operating Procedures/Work Instructions.
- Copies of Installation Qualifications/Operational Qualifications & Performance Qualifications.

- Routine/Preventative Maintenance Schedule.
- Spare Parts.
- Fault Finding/Operational issues
- Contact Information.

All information, if possible, should be electronically scanned and stored in a dedicated Engineering **hard drive** for easy access from any PC across site via the company's intranet. This may seem very time consuming but it can be carried out at minimal cost, the long term benefits to the company will far out way the initial time it takes to create it. As new information becomes available, the engineering library should be managed and populated by an 'Engineering Responsible Person' with write access, all others will have read access only.

Tip: If you don't have a PC, you'll probably have a 'Smart Phone', take photos of everything e.g. the equipment you're working on, the faulty part, it's associated O&M manual, and any hard copy hand written notes and drawings you make. When you gain access to a PC, upload the camera phone photos to a USB stick, you have now created your first digitally captured engineering folder in your engineering library.

Always ensure a copy of the 'As built' set of electrical drawings resides in each machine's bespoke electrical panels across site.

Once all the key engineering information is in place, the actual methods, and systems of how to monitor and maintain all site equipment must be set up:

Work Instructions (W/I's):

Each piece of equipment will have O&M manuals supplied which normally include installation, configuration, calibration, diagnostics, trouble shooting, service and maintenance instructions. Use these and the equipment manufacturers' expert advice to set up weekly/monthly/yearly work instructions to be **carried out** and

signed off by onsite trained maintenance personnel for reference and service history purposes. The maintenance and servicing instructions make it possible for site personnel to implement preventative maintenance measures e.g. information is provided relating to periodical replacement of wearing parts in order to avoid damage to equipment.

N.B. The O&M manuals must be read before working with the equipment. For personal and system safety and for optimum equipment performance, make sure personnel thoroughly understand the contents before installing, using, or maintaining the equipment.

Daily Logbooks:

Set up daily logbooks on key systems (hard copy or electronically), again, using the equipment manufacturers' expertise to advise on what key operating parameters should be recorded and the acceptable tolerances the piece of equipment should be running between e.g. (7 Bar +/- 1 Bar).

Standard Operating Procedures (SOP's):

Engineering SOP's can range from **Security and Access control** to **Management of Site drawings**. SOP's may also involve the use of a specialist vendor/contractor with the assistance of onsite trained personnel in carrying out the assigned tasks of the SOP itself **e.g.** Operability and maintainability of a Fire detection/sprinkler alarm system. The department manager must ensure that training needs are identified, and that training is performed before personnel carry out the task outlined within the scope of the SOP.

D. Good Budget Strategy

A good engineering budget is one of the most financially wise plans to have in organising and controlling financial resources as well as setting and realising goals. A budget allows the engineering department to figure out how much money it has to spend every month/year and where the department is spending it in a clear and documented fashion. As such, a budget is one of the most important steps to take toward maximising the power of the company's money.

Try to work from a **zero-based budget** platform. It is a technique of planning and decision-making which reverses the working process of traditional budgeting. In traditional incremental budgeting, departmental managers justify only increases over the previous year's budget and what has been already spent is automatically sanctioned. No reference is made to the previous level of expenditure.

By contrast, in zero-based budgeting, every department function is reviewed comprehensively, and all expenditures must be approved, rather than only increases. Zero-based budgeting requires the budget request to be justified in complete detail by each division manager starting from the zero-base. The zero-base is indifferent to whether the total budget is increasing or decreasing.

Advantages of zero-based budgeting:

a. Efficient allocation of resources, as it is based on needs and benefits.

b. Drives managers to find cost effective ways to improve operations.

c. Detects inflated budgets.

d. Useful for service departments where the output is difficult to identify.

e. Increases staff motivation by providing greater initiative and responsibility in decision-making.

f. Increases communication and coordination within the organisation.

g. Identifies and eliminates wasteful and obsolete operations.

h. Identifies opportunities for outsourcing.

i. Forces cost centre's to identify their mission and their relationship to overall goals.

E. Good Service Level Agreements (SLA's) with vendors and contractors

A Service Level Agreement is a part of a service contract where the level of service is formally defined. It is a negotiated agreement between the company and the vendor/contractor and records a common understanding about services, priorities, responsibilities, guarantees, and warranties. Each area of service scope should have the 'level of service' defined. The SLA may specify the levels of availability, serviceability, performance, operation or any other attributes of the service.

In advance of the commencement of any work on site, site management must firstly approve the proposed vendor/contractor; the following requirements must be fulfilled:

- Approval of all contractors/ service providers for use prior to coming on-site
- Forwarding the Contract Company's Safety Statement & Method Statements/ Safety Data Sheets for the proposed work, if applicable to the site responsible person, who will obtain the necessary approvals
- **Forwarding company details** of the Contract Company's relevant insurances, tax certificate, Quality documents and employees' training records including Confined Space, M.E.W.P., Manual Handling, Forklift and all applicable training requirements
- Forwarding any other relevant Safety Training applicable to the work being proposed on site to the site responsible person

- The Contract Company must also advise if any Subcontractors are to be used. If subcontractors are proposed, full details of insurance and environment, health and safety performance and competence must also be provided.

Only after the requested information is received and signed off by all the sites' responsible parties can the proposed Contractor/Service Provider commence works on site. Contract Companies must only use persons whom they have assessed to be competent and suitably qualified to carry out the proposed work on site.

F. Set up a well stocked and efficient Engineering stores area

Companies are well aware of the cost benefits that can be attributed to a successful maintenance strategy. Part of that strategy will be to determine and assess the risks associated with the purchasing and storage of industrial spares and consumables based on their expected service life and their quality in performance versus their cost price. It is important that personnel know and understand the products, services and materials that are suitable for purpose, reliable, work better, and last longer.

G. Set up a Maintenance Software System Package

A dedicated maintenance computer-based software system or **Computerised Maintenance Management System** (CMMS) empowers the Engineering Department to manage its key functions and derive value for the business. Keep detailed records of work done. This makes it much easier to see where a business needs improvement as well as identify areas where a business is doing well – something that is very important to show stakeholders & senior management. It may be tempting to use spread sheets or a home-grown database to keep track of work due and work done. However, the volume of data to be managed will soon make it difficult to review trends or keep accurate track of work done. Much useful information will not be recorded and there is a tendency to miss work that needs to be done. This is where **Computerised**

Maintenance Management Software (CMMS) can help.

A maintenance software system allows the protection of a company's investment in assets and equipment. The system will help to manage maintenance, reduce downtime, control maintenance costs and minimise the investment in engineering spare parts. This type of system often becomes the showcase for maintenance departments during client and regulatory audits; it allows full visibility and traceability on all engineering and maintenance activities.

Maintenance Software System Package example:

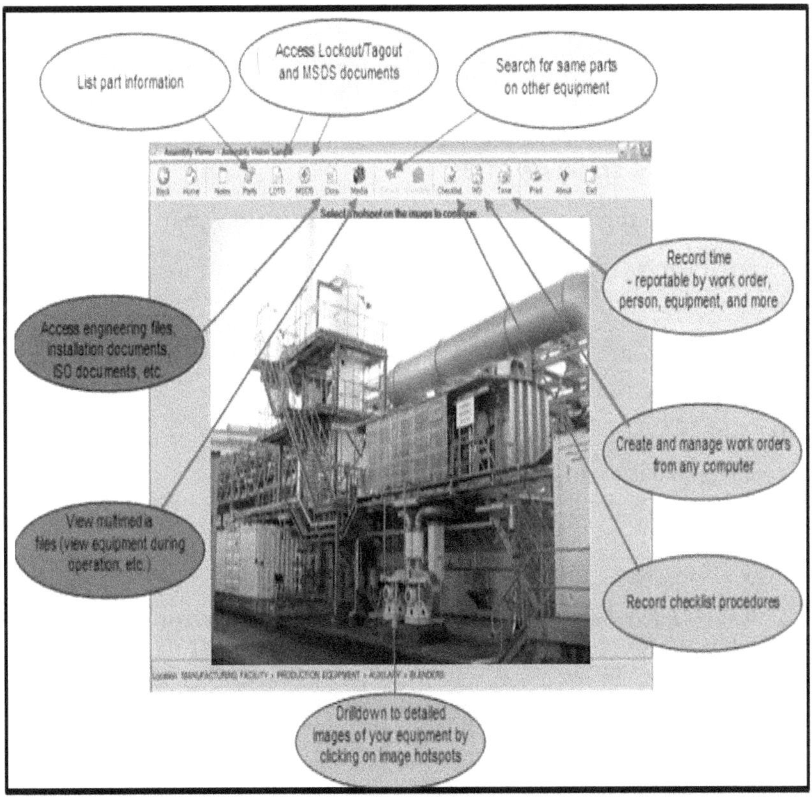

Key Benefits

- Manage both planned and unplanned maintenance work for both internal personnel and contractors
- Improve the understanding of equipment behaviour, monitor recurring problems and costs
- Gain a clear visibility on workload through effective resource planning
- Analyse maintenance effectiveness through KPI Reporting
- Prioritise maintenance activities through a simple graphic interface that highlights essential work
- Create a knowledge base to identify reoccurring problems with equipment and to reduce time to repair
- Reduce downtime through effective analysis of problems through a **root cause analysis tool**
- Manage a company's investment in engineering spares
- Improve information quality with key information captured during fault resolution.
- Extend the life of assets through the effective management of preventative maintenance
- Provide a unique historical record for all assets for both company and regulatory audits.

Key Features:

- Full Asset Register storing key equipment information
- Control Preventative Maintenance using both time and usage based intervals
- Create planned and unplanned work and distribute work through a simple interface
- Record completion of work along with time taken, downtime and defect causes
- Create follow-up work orders to ensure that no activity is forgotten

- Root Cause Analysis during Close Out
- Fully Configurable System designed to meet a company's unique requirements
- Downtime Analysis Reporting
- Fault Analysis Reporting
- Management Reporting Suite allowing interrogation of data

Key Factors for Success

- A system owner responsible for the Maintenance System
- Adoption of system by the Maintenance Department
- Management buy-in
- An Implementation approach
- Spares Parts Management
- Purchasing.

H. Set up a Site Energy Management Strategy

Energy is a plant's biggest controllable operating cost. Costs must be controlled by optimising the efficiency of the plant and its associated machinery to obtain the maximum production output and to minimise downtime.

I. Emergency backup systems and contingency plans to ensure business continuity

Preparation and time dedicated to reviewing and anticipating 'Worst case scenarios' by key knowledgeable users of all site equipment and activities cannot be underestimated. More importantly **the layers of protection** needed to ensure the worst case scenarios never materialise, must be put in place.

J. Set up a Calibration Management Software System Package

This involves scheduling, documenting, and controlling calibration activities. Maintaining an accurate instrument database and regular calibration and maintenance of field instrumentation is a must for many industries as they systematically prove the performance of their quality critical devices. The task of **in situ** calibration itself is often challenging. This involves scheduling the activities, making arrangements with production personnel to make plant available, planning the resources, mobilising the technician who then carries out the instrument calibration and documents the results; all activities are time consuming, costly, and challenging. Yet the key to ensuring the quality of the process whilst at the same time minimising costs for unnecessary maintenance are dependent upon the people who have control over this process.

Calibration Management Software is a high performance computer based software tool that meets all requirements regarding calibration management and implementation. Calibration Management Software allows a company to efficiently maintain and calibrate on-site instrumentation. It fulfils the high quality demanded of auditors and at the same time reduces complexity, time and costs associated with the management of the calibration activity. **CM Software example:**

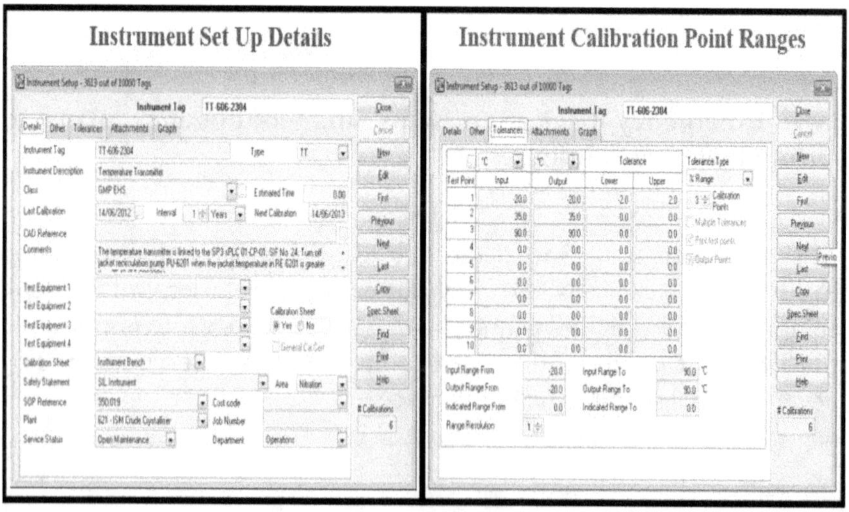

Advantages:

- Controls the maintenance and calibration schedules, issuing work orders and reports as defined by maintenance personnel
- Can attach important documents such as SOP's, Health and Safety sheets, loop diagrams to each tag
- Incorporates audit trail and high security features
- Upgradable to full compliance of relevant regulations (Electronic Records and signature) with full validation services without loss of any historical data
- Roaming laptop version is optionally available to allow technicians to work remotely on plant independent of the network
- Produces management reports for cost analysis, historical data analysis, work due listings and overdue listings
- Can print certificates related to discrete calibration or loop calibration
- Is able to consider unplanned requests for work
- Full records sheet for each tag allowing all important data to be logged and recorded
- Work areas can be arranged in plant locations and operating modules
- E-mail control system which can be set to automatically alert pre-determined personnel of reports actions and deviations when action is necessary
- Allows a company to control cost and review calibration schedules accordingly.

Lifecycle Management

Calibration Management Software will improve maintenance planning, improve the efficiency of a company's instrumentation department, reduce the cost of maintenance and at the same time satisfy the requirements of legislators and auditors.

K. Basics of Finance

Financial Management is a managerial activity concerned with planning and controlling of a company's financial resources to generate returns on its invested funds. The following is a basic understanding of Accounting and where budgets fit into the overall scheme.

- **Going Concern:** is the ability of a business to meet its financial obligations when they fall due.
- **Prudence Concept:** do not overestimate the amount of revenues recognised or underestimate the amount of expenses. You should also be conservative in recording the amount of assets, and not underestimate liabilities. The result should be conservatively-stated financial statements.
- **Accruals or Matching Concept:** The matching concept is used under accrual accounting. Accrual Accounting simply means recognising revenues when they are earned, and expenses when they are incurred. In other words, expenses incurred in a period must be matched against revenues accrued in that period.
- **Consistency Concept:** once you adopt an accounting principle or method, continue to follow it consistently in future accounting periods. This ensures that there is a consistent approach to the treatment of varied items e.g. Depreciation (an accounting method of allocating the cost of a tangible asset over its useful life and is used to account for declines in value. Businesses depreciate long-term assets for both tax and accounting purposes).

Capital Expenditure V Operating Expense

- **Capital Expenditure:** money spent by a business or organisation on acquiring or maintaining fixed assets, such as land, buildings, and equipment. Expenditure that yields benefits over multiple periods could be considered capital in nature. This is not written-off in the period it is incurred but over the life of the asset (Depreciation) e.g. New Equipment.

- **Operating Expense:** Expenditure that yields benefits only in the immediate period or periods could be considered operating expense. This is written-off in the period it is incurred as the benefit has been gained in that period e.g. Energy Costs.
- Sometimes distinguishing between expenditure of a Capital nature versus Operating is a tricky business but the Accounting Regulations do provide guidance e.g. Repairs & Maintenance, Labour Costs (work on Capital).

Effective Cost Control – Good Practice

- Cost control is the practice of identifying and reducing business expenses to increase profits
- **Ownership of Costs!** Costs in YOUR cost centre belong to YOU and are YOUR responsibility.
- **Good Detailed Planning** based on solid and realistic business assumptions. A good Budget can then be "flexed" throughout the year to give a more realistic forecast should business assumptions change.
- **Regular Review** of Actual Costs Performance versus Budget & Forecast. (Monthly Management Review)
- **Early detection** of potential over-runs or savings. These should be communicated to Finance as soon as possible and included in the next Quarterly Rolling Forecast.

RFC – Rolling Forecast

The idea is that instead of managing the business based on a static budget that was created in the prior year, rolling forecasts are used to revisit and update budgeting assumptions throughout the year. Businesses establish a set period, such as quarters or months, to update their forecast.

Discretionary V Non-Discretionary Expenditure

These are terms used when dividing up your Cost Centre budget into groups for controlling the costs. However, all costs are controllable!

- **Discretionary Expenditure** - costs that you as a manager have direct control over which are:
  a. All Overheads
  b. Personnel Expense

- **Non-Discretionary Expenditure** - costs that are imposed on you as a result of business demands:
  a. Depreciation
  b. Allocations of costs to your cost centre

Tips for Cost Control

- Re-visit your Budget. Know and understand what you budgeted for and the basis of your assumptions.
- Run the Finance report monthly. The Finance department will advise when you when the books are closed so that the cost centre information is complete.
- Look out for variances to your budget and see if it is genuine or being off-set elsewhere. If required, ask Finance to **re-class** it.
- Use the drill-down functionality in your Finance system to get the details behind the actual postings. If there is something you don't understand, ask Finance to explain what it is.

If a variance has occurred and you expect it to continue, see if savings can be made in other areas. If not then note it, advise finance and include it in the next forecast as an expected **over-run** or **saving**.

5. CALIBRATION AND VALIDATION

Calibration is a process of comparing the accuracy of an instrument reading to known standards. It is a demonstration that an instrument produces results within specified limits against measurement standards which are traceable to 'international or national standards' over an appropriate range of measurement. This is achieved by comparing the instrument against a 'standard certified instrument' of higher accuracy to detect, correlate, adjust, rectify, and document the accuracy of the instrument being compared. The results of the comparisons are recorded in calibration certificates.

WHY IS CALIBRATION REQUIRED?

- Instruments and their reliable measurement accuracy are necessary for the proper and safe operation of modern buildings and industrial plants, they ensure process operability, efficiency, reliability and safety. Fluctuating measurements can not only have an impact on process stability and operating costs but potentially have legal and regulatory consequences.
- Instrumentation is an indispensable part of the production line. In order to fulfil regulatory compliance, a company must perform periodically traceable calibrations on instruments to ensure product quality. Periodic calibrations to specified tolerances using approved procedures are an important element of any **quality system.**

It makes sense that calibration is required for a new instrument. We want to make sure the instrument is providing accurate indication or output signal when it is installed. But why can't we just leave it alone if the instrument is operating properly and continues to provide the indication we expect? Periodic calibrations are performed to detect and correct instrument error. Even if a periodic calibration reveals the instrument is perfect and no adjustment is required, it would not be known unless a calibration was performed. And even if adjustments are not required for several consecutive calibrations, the calibration check must still be performed at the next scheduled due date.

Instrument error can occur due to a variety of factors: e.g. drift, environment, electrical supply, addition of components to the output loop, process changes. Since a calibration is performed by comparing or applying a known signal to the instrument under test, errors are detected by performing a calibration. An error is the algebraic difference between the indication and the actual value of the measured variable.

In the industrial environment, sensors are used to measure an extremely wide range of environmental factors including temperature, motion, positioning, acceleration, weight, humidity, chemical composition, gases or pressure or liquid flow. **They allow you to "see" the invisible.** Typical examples include ultrasonic liquid-level transmitters for continuous level measurement and alarm switching at a water treatment plant; or pressure, temperature, and level sensors in the monitoring of hydraulic fluids and coolants in machine tool and piping applications; or inductive proximity switches in automotive assembly lines. In fact, today, there is virtually nothing that cannot be measured or sensed and used as an input to a system (data flow starts in the field at the instruments).

Selecting the correct instrument type and material of construction during the design phase is key e.g. sensor material is corrosion resistant.

Commissioning an instrument can be a daunting prospect if you're not familiar with the process. Ensuring that the instrument is correctly installed and commissioned as per the manufacturer's specifications and is operated according to their instructions should give years of trouble free operation. (Tip: seek manufacturer's 'technical support', they will have extensive knowledge, commissioning and service expertise).

E.g. Radar liquid-level detector (transmits and receives high frequency radio waves (microwaves) to measure the distance to the surface of the liquid) installed in an 8,000 litre mixer vessel.

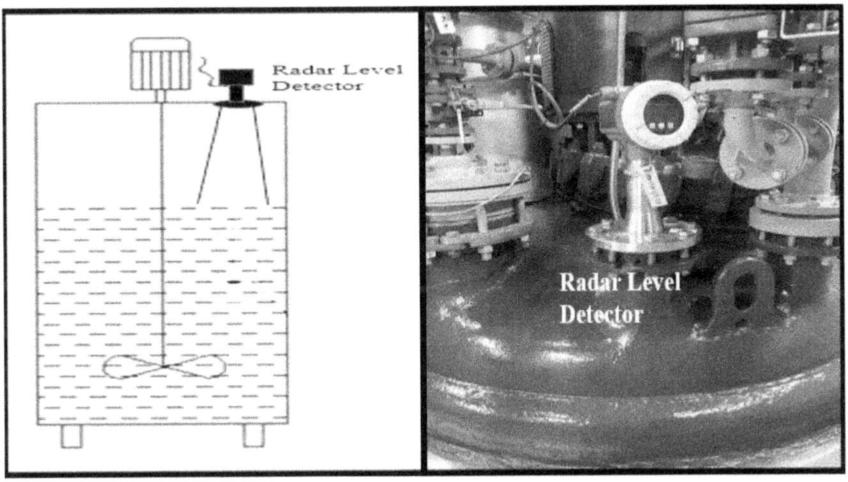

Tip: Ultrasonic waves detect an object in the same way as Radar does it. Ultrasonic uses sound waves, and Radar uses radio waves.

How often should I calibrate? Regulatory bodies including the Medicines and Healthcare products Regulatory Agency (MHRA) define the need for process instrumentation to be periodically calibrated using equipment that is traceable to national or international standards; but it is the **end user's responsibility** to determine the correct frequency for calibrating each instrument. Deciding when to calibrate a device can be difficult, as it requires striking a balance between saving operational costs (by extending intervals between calibrations) and ensuring the reliability of the process.

To date, if an end user decides to extend the period between calibration cycles, they have to do so in the knowledge that the condition of the device in between calibrations can not be easily verified. To overcome such issues, specialist vendors have now introduced level, flow, analytics, and temperature devices with on-board verification technology which provides traceable evidence that the instrument is still operating in accordance to the original manufacturer's specification. Any deviation from factory reference values will be displayed in easy-to-understand diagnostic data, highlighting that maintenance is required or a failure has occurred, in accordance with NAMUR NE 107 and NE 44.

Fully documented, properly indexed, stored calibration reports and calibration certificates with individual serial numbers for all relevant instruments, are essential in ensuring the smooth running of a plant in a safe and cost-effective manner. All instruments must be factory calibrated and certificates of calibration provided with each device. This usually involves the instrument vendor commissioning the installed instrument and issuing a field certificate prior to system hand-over.

All calibration certificates must be traceable to the National Certifying body. For proper operation and optimal performance in a process, each element of a measurement system must be carefully chosen. Fluctuating measurements can not only impact on operating costs but also product quality and safety. The first step of a recalibration is an 'as-found check' where it is determined if the instrument was still operating within specifications before it was taken out of service for calibration. Each component in the measuring system must demonstrate that it is functioning to specification and is fully auditable. If any component of the measurement system does not meet this requirement, out of specification product may be produced leading to costly rework or worse, may lead to an accident.

The calibration range of an instrument is defined as the region between the limits within which a quantity is measured, received, or transmitted, expressed by stating the Lower (LRV) and Upper (URV) range values. These limits are defined by the **Zero** and **Span** values. The **Zero** value is the lower end of the range or LRV and the upper range value is the URV.

For example, if an instrument is being calibrated to measure temperature in the range of **0°C to 100°C,** then LRV = 0°C and the URV = 100°C. The calibration range is therefore 0°C to 100°C. The **Span** is defined as the algebraic difference between the upper and lower range values:

Span = URV – LRV

There are two adjustments in the temperature transmitter: ZERO and SPAN.

Method: Temperature transmitter to be calibrated for a 4 to 20 mA signal with a range of 0°C to 100°C.

- For 0°C, you must get 4 mA and if it is not reading 4 mA (say 4.64 mA), you start to adjust the zero adjustment to get 4 mA.
- For 100°C, you must get 20 mA and if it is not reading 20 mA (say 19.71 mA), you start to adjust the span to 20 mA.

Use of certified, traceable electrical simulators and standards to verify an instrument is processing input signals correctly is mandatory.

A **temperature sensor** must generate the correct output signal so that a Temperature Transmitter/PLC/HMI system can interpret this signal and ensure that it corresponds accurately to a temperature reading.

In **Fig. 1** As part of a **'Process'**, the temperature sensor's transmitter on the top of the mixer vessel is programmed and set up to read as follows:

Temperature range: **(0°C to 100°C)**

Standard corresponding resistance input: **(100 ohms to 138.51 ohms)**

Standard corresponding current output: **(4 mA to 20 mA)**

Fig. 1. - PRT (Platinum Resistance Thermometer)

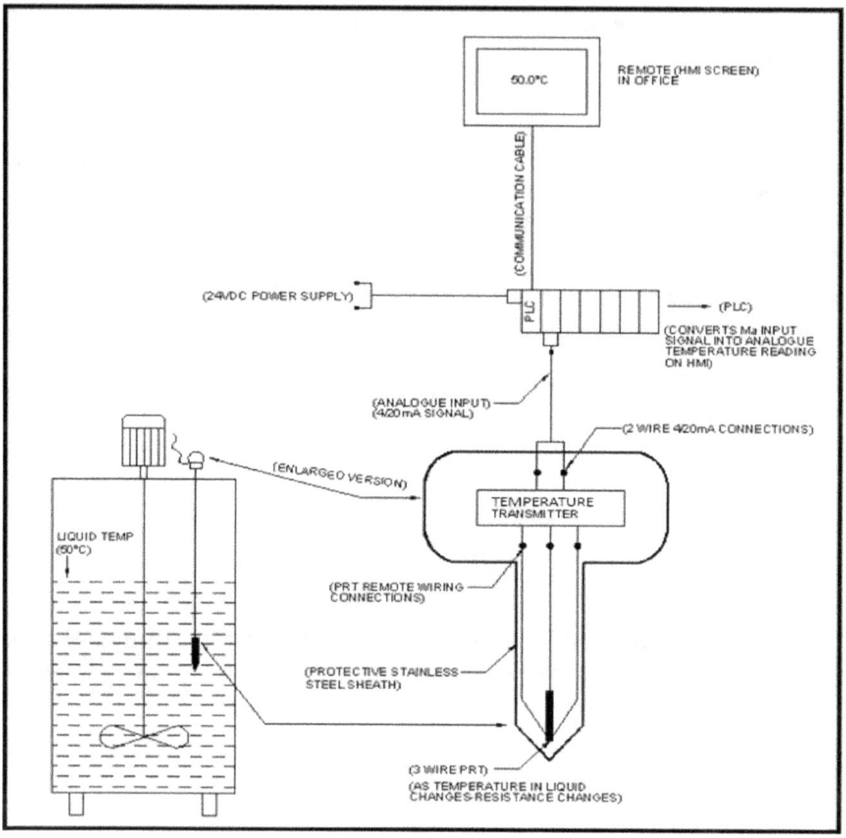

In Fig. 1, the **PRT (Platinum Resistance Thermometer)** and its associated wiring are normally encapsulated and protected by a stainless-steel covering. The PRT (sensing element) is positioned at the tip of the measuring probe, is about 25 mm in length and is connected to the head of the probe with either 2 or 3 wires (lengths of the probes can vary from 10 cm to > 3 metres). This type of temperature sensor exploits the predictable change in electrical resistance in the platinum with changing temperature. Electrical resistance is a measure of a material's ability to conduct electrical current. The PRT provides a known resistance to the **Temperature Transmitter** (TT) as the temperature of the liquid fluctuates in the mixer vessel. The electronic unit within the head of the TT itself is programmed to understand this fluctuating resistance and subsequently converts it to a milliamp signal which is sent to the PLC and is converted to a temperature reading on the HMI screen in the office.

Test procedure: by using a **calibrated resistance decade box** and injecting a known resistance into the temperature transmitter, the pre-set resistance should then correspond precisely to a correct output signal.

To test functionality:

1. Disconnect the wiring from the **PRT input connections** of the temperature transmitter electronic unit, push them to one side.
2. Proceed to connect the wires from the resistance decade box directly into the input connections of the temperature transmitter electronic unit.
3. Inject a resistance of 119.40 Ohms. A corresponding reading of 12 mA should now be measured on the output connections of the instrument using an ammeter connected in series with the output wiring going to the PLC. The PLC then converts this 12 mA signal to read 50°C on the HMI screen via a communication cable.
4. 5 known resistances of between 100 Ohms – 138.51 Ohms are inputted into the temperature transmitter electronic unit which gives an equivalent electrical output signal of 4 - 20 mA which should then show a temperature range of 0 – 100°C, where:

Input: Resistance 100.00 Ohms = Output of 4 mA = 0°C

Input: Resistance 109.73 Ohms = Output of 8 mA = 25°C

Input: Resistance 119.40 Ohms = Output of 12 mA = 50°C

Input: Resistance 128.99 Ohms = Output of 16 mA = 75°C

Input: Resistance 138.51 Ohms = Output of 20 mA = 100°C

N.B. Once the test is complete, reconnect all wiring exactly the way it was disconnected it, observing colour coding and polarity at all times.

A comparison of a TT's performance can also be made with a certified reference thermometer. The certified reference thermometer is put into the tank of liquid and the reading is then cross referenced with the digital temperature readout on the HMI screen.

A TT is normally calibrated using a certified temperature-controlled heating/cooling bath. This proves the entire instrument itself, its associated cabling, hardware, and software without physically disconnecting any wiring **i.e.** from tip of the measuring probe to actual reading on the HMI screen with the aid of FMEDA (Failure Modes Effects and Diagnostics Analysis). The TT and associated measuring probe is removed from the mixing vessel and put directly into a calibrated temperature-controlled heating/cooling bath where known temperatures are set. Checks are then made to see are the readings set on the calibrated bath (e.g. 50°C) corresponding directly to what the digital temperature readout is reading on the HMI screen itself. Auditors can request to see the 'Digital trail' left after a calibration of an instrument has been carried out, the associated paper work alone may not be acceptable, they may ask for the HMI temperature data trend and actually check that the listed temperatures on the calibration sheet match the temperature trend on the HMI screen to the date and time the calibration was carried out.

A **thermocouple** may also be used instead of a PRT. A thermocouple is two wires of dissimilar metals fused (joined) together at the point (measuring junction) where the temperature is to be measured. A thermocouple produces a temperature-dependent voltage as a result of the thermoelectric effect, and this voltage can be interpreted to measure temperature i.e. it produces a minute voltage in proportion to its temperature. Increasing the junction (point of contact) temperature increases the millivoltage level generated:

E.g. Type J Thermocouple:

- 0°C = 0.000 millivolts
- 25°C = 1.277 millivolts
- 50°C = 2.585 millivolts
- 75°C = 3.918 millivolts
- 100°C = 5.269 millivolts

The electronic unit within the head of the TT itself is programmed to understand this fluctuating millivolt signal and subsequently converts it to a 4/20 mA signal which is sent to the PLC and is converted to a temperature reading on the HMI screen in the office. Thermocouples for practical measurement of temperature are junctions of specific alloys which have a predictable and repeatable relationship between temperature and voltage. They are a widely used type of temperature sensor for measurement and control.

Ii is the end user's responsibility to determine the correct frequency for calibrating each instrument. Calibrations intervals should be performed at least once a year or as specified by vendor recommendation, but for **critical applications** a greater calibration frequency may be required to provide the required confidence in the process instrument.

Tip: There are temperature monitoring systems available (e.g. for your Refrigeration or HVAC systems) with high accuracy digital sensors which transmit temperature information wirelessly to the cloud via a battery-powered sensor hub. The hub does NOT require WiFi or cellular connectivity, instead it uses radio network technology developed specifically to transfer small packets of sensor data to the cloud. Cloud-based temperature data is then available anywhere via the Manufacturer's Remote Signals mobile phone app.

Instrument Validation means establishing documented evidence which provides a high degree of assurance that an instrument will consistently produce accurate readings, meeting its predetermined specifications, and ensures that the instrument is working correctly and fit for purpose.

CAT (Criticality Assessment Team) Assessments should also be carried out on instrumentation and equipment that monitors and controls equipment on plant. A critical assessment of all equipment should be made, based on the criticality of the piece of equipment. Regular maintenance and calibration intervals must be maintained and logged to ensure equipment remains safe, compliant, and fully operational.

(CAT) Process

This is conducted, system by system for each plant or area. A complete instrument list is generated by reviewing approved P&ID's and/or **walking down** each instrument in the field. A cross functional team comprising of the System Owners, Engineering, Quality and EHS, categorise each instrument as being one of the following three categories:

- GMP Critical
- EHS Critical
- Non-Critical

N.B. Failure of a critical instrument or late execution of a critical calibration may require a deviation or safety report to be raised and an investigation to be carried out as per a company's SOP system.

GMP (Good Manufacturing Practice) Assessment

GMP critical instruments are assessed as per the criteria defined in an SOP e.g. Quality Impact Assessment Procedure. Instruments are defined as being **GMP Critical** if they were judged to be **Operationally Critical**.

To be operationally critical, each instrument is critiqued against the following questions:

1. Is the instrument used to demonstrate compliance with a registered process?
2. Has the normal operation or control of the instrument a direct effect on product quality?
3. Does failure or alarm of the component have a direct effect on product quality or efficacy?
4. Is information from the instrument recorded as part of the batch record, lot release date or other GMP related documentation?

5. Does the instrument control a critical process element that may affect product quality without independent verification of the control system performance?
6. Is the instrument used to create or preserve a critical status of a system?

If the answer to any of the above questions is 'Yes' and there are no other parallel or downstream components that would detect failure of the instrument, the instrument is deemed GMP Critical.

Safety and Environmental Assessment

EHS instruments are assessed as per the criteria defined below. To be EHS critical, each instrument is critiqued against the following questions:

1. Would failure or inaccuracy of the instrument result in a major accident or environmental reportable non-compliance event?
2. Would instrument failure result in a possible unsafe situation or environmental control or escalation of consequences?
3. Does the instrument mitigate the consequences of an emergency or environmental event?

If the answer to any of the above questions is 'Yes' and there are no other parallel or downstream components that would detect failure of the instrument, the instrument is deemed to be EHS Critical. If responsible for safety instrumented systems on site, it is imperative to understand the responsibilities under the applicable safety standards. Adequately testing safety instrumented functions (interlocks) so they can function properly when required is a must. Thorough maintenance, inspection and testing procedures ensure safety systems maintenance is consistent.

INSPECTION & TESTING

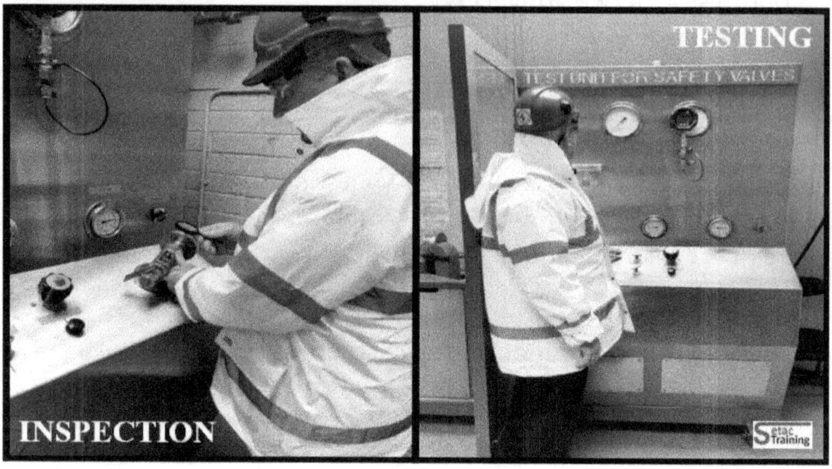

INSPECTION: Any working part has a finite operating lifecycle. As its age and use increase, so does the probability that it will fail. Inspection forms the foundation from which good maintenance practice is developed. Its purpose is to anticipate problems and faults arising, and to initiate preventive maintenance before they develop into potential breakdowns.

TESTING: Pressure Relief Valves (PRVs) serve a critical role in protecting organisations and employees from explosions. Knowing how and when to test and repair or replace them is essential. The valve is designed to limit system pressure, and it is CRITICAL that they remain in working order. When you have equipment pressure relief valves in your facility e.g. storage tanks and pressure vessels, or along your pipeline system, you need to know they are functional and when to repair or replace them.

Non-Critical Assessment - Not using critical standards or methods, as in evaluation. Instrument where failure or inaccuracy will not result in a GMP, Safety or Environmental non-compliance. Instruments are typically operational instruments used for commissioning or engineering purposes. **Process monitoring**

devices along with plant parameters and activities are critical for both operational reasons and in assisting with fault finding should a problem occur. It is important that these devices are fitted on plant equipment in strategic parts of the process to guarantee optimum process monitoring at all times.

In an automated product mixing system, which may have very short cycle times in some of its processes, a company must be able to control and analyse product quality on-line in real time. Any delay in measurement can easily lead to excursions in concentrations of mixtures before being detected and ultimately may lead to inferior product being made which may have to be destroyed or reworked. The key is to optimise the company's processes through increased access to better relevant data by using an instruments advanced diagnostic and not just the primary measured value it gives out to a DCS system. During the initial design phase of a new project, it is important that the best available equipment is selected, the initial cost may be high, but the benefits of virtually trouble-free operation will far out way the initial equipment installation costs overall. The key to efficiently lowering your expenses without compromising the quality of your products is to derive accurate process data from reliable instruments.

Tip: Apply **Zeroth's Law** when dealing with temperature issues on any piece of equipment or process, it basically states: 'If two thermodynamic systems are in thermal equilibrium with a third, they are also in thermal equilibrium with each other'. If two temperature probes (electrical or mechanical) are monitoring a piece of equipment or process and are both reading the same temperature, insert a third calibrated probe into the process to verify actual temperature, if all 3 probes are reading the same temperature, then they are in equilibrium with each other and the temperature readings can be deemed to be correct. Don't naturally assume that because two probes are reading the same temperature that the process is at this temperature especially if the process is **temperature critical**, always verify with a third calibrated probe.

Tip: Avoid the **error of parallax**, this may occur when reading any mechanical instrument from an angle, it is important that a person's head is held directly level with the point being measured to ensure an exact reading is being observed.

N.B. The importance of **'strategically placed' instrumentation devices** with redundancy monitoring/measuring pressures, temperatures, flows .... cannot be stressed enough. A plant's **safe and smooth operation** will be assured if these instruments are suitable for purpose, accurately calibrated, and online at all times.

- Calibration of your temperature instruments are costly in terms of time and money. **Tip:** There are sensors with 'embedded verification capabilities' available that can carry out their own traceable calibration e.g. Trustsens self-calibrating sensor. Human intervention is only necessary if the sensor indicates an error. Calibration certificates can be provided automatically via asset management software. A valid calibration certificate is always available.

6. GOOD ALARM MANAGEMENT

All modern manufacturing and production process control systems provide alarm systems to assist operators in managing abnormal situations. Nevertheless, the integrity and effectiveness of alarm systems can either aid or be a hindrance in responding to these situations.

Alarm Management

Alarm Management is imperative to assessing, improving and optimising plant alarms, thereby increasing the effectiveness of plant operators by only notifying them of a need for their intervention. An appropriate technical means of supervision must be in place.

An alarm management system is crucial to safe and productive operations:

– Reduces unplanned downtime

– Increases safety

– Improves operator effectiveness

– Better process performance/yields

What is an Alarm Management System?

Alarm Management is about safety, the environment, optimising operations and increasing corporate profits. Most plant personnel equate alarm management with reducing alarms; this is not the case, it involves providing operators with enough information to try to prevent abnormal situations from occurring and to prevent any further escalation of those abnormal situations. A poor alarm system results in very large monetary losses every year due to accidents, equipment damage, unplanned plant or unit outages, out of spec manufacturing, regulatory fines, and huge intangible costs related to environmental and safety infractions.

It is important that the operator have some action for any specific alarm, if the action is not required, the alarm should be removed. Reducing or eliminating **alarm floods** liberates an operator to respond to plant demands, enabling them to avoid shutdowns and keep the plant running at optimal performance. In addition, the advanced alarming …. like **alarm shelving**, can dramatically reduce alarms temporarily during a specific period so the operators can focus on important alarms and reduce startup/shutdown time.

The fundamental purpose of alarm annunciation is to alert personnel to abnormal operating conditions. An effective **Alarm Philosophy** outlines key concepts and governing rules for alarm strategy, e.g. what constitutes an alarm and what risk categories pertain to your site operations.

Most control systems that monitor, control and operate equipment are equipped with a series of monitoring devices to ensure reliable operation. If any irregularities occur, an alarm is triggered, and the equipment carries out the necessary switching steps automatically in order to avoid hazardous situations. The alarms associated with these monitoring devices usually range from having no direct influence on the system to its complete shutdown. These alarm set points are normally sub grouped in order of their criticality and indicated in the alarm text with colour coding on a PC/HMI screen e.g.

- Alarm Group 1 – Pre-alarm without direct influence on the system.
- Alarm Group 2 – Partial shutdown of the system. (Hold mode)
- Alarm Group 3 - Complete shutdown of the system.

Alarm grouping, and filtering facilities must be provided to allow for alarm segregation between plant processing trains and equipment.

The ultimate objective is to prevent, or at least minimise, physical and economic loss through automatic or manual intervention in response to the condition that was alarmed.

N.B. A company must revise its procedures for periodic testing and verification of its **back-up control systems** to assure they will function properly at all times and alarm accordingly.

Good Alarm Management principles must be employed and implemented in every industry; the procedures put in place must be adhered to, practiced and properly escalated according to the criticality of the activated alarm. There are sometimes too many alarms annunciated in a plant because of poorly designed alarm strategies (possibly back at the URS design stage) e.g. improperly set alarm points, ineffective annunciation, unclear alarm messages. Design teams must differentiate between alarm management and alarm flooding and try to avoid this scenario where possible:

1. Massive quantities of data arriving on the PC screen alarm banner that are meaningless and without proper context or explanation.
2. Alarm floods makes it difficult to analyse which alarms need immediate attention.
3. Alarm floods makes it difficult to make important informed decisions and take swift action.

Alarm fatigue must be reduced so personnel know that any alarms generated by the system are serious enough to demand their intervention. If the resultant flood of alarms becomes too much to comprehend, then the basic alarm management has failed as a system.

The speed and accuracy at which critical alarms can be identified and require immediate attention will determine how effective an alarm management system is and how quickly it can be responded to.

Operations/Engineering (mainly Operations) must decide which alarms need more attention than others. In addition to the PC screen alarm icon flashing with a drop-down box with written possible causes and associated actions to take, an audible siren must also activate to alert personnel of the severity of such a critical problem. Eliminating extraneous alarms provides better recognition of critical problems resulting in a faster, more accurate response time. Alarm suppression facilities should also be provided for, and prioritised accordingly.

Insist during the installation of equipment that as much information as possible about each system can be put up on the PC screen to assist personnel in diagnosing a fault. Install **shortcut icons/hyperlinks** in a corner of the PC screen which will take personnel directly to view the piece of faulty equipment's documents, e.g.

1. Fault Finding Flow chart
2. Training ppt. presentation.
3. Electrical Drawings.
4. Mechanical P&ID.
5. Parameters e.g. set points, volumes.
6. SDS (Software Design Specification).

These helpful **shortcut icons** and the stored relevant information attached to them, will save a lot of plant downtime as the information required is easily accessible.

Proper Alarm Management and Annunciation alerts 'all concerned' to abnormal operating situations in areas such as:

- Personnel.
- Properly escalating a fault/issue and the response to same.
- Environmental safety.
- Equipment integrity.
- Product quality control.

Nuisance alarms, alarm floods and improperly prioritised alarms all contribute to operator confusion, and thus increase accident frequency. A good alarm management system must be kept in a healthy and helpful state by eliminating nuisance alarms, reducing alarm floods, and ensuring that the necessary alarms are properly prioritised and documented.

7. TROUBLE SHOOTING AND FAULT FINDING

Apollo 13 … Houston, we have a problem!!

One of the greatest examples of **Teamwork** by NASA on how to troubleshoot, fault find, and problem solve by adapting, improvising, and overcoming, they even made feature films about it! Despite great hardship caused by limited power, loss of cabin heat, shortage of potable water, lack of sleep, and the critical need to jury-rig the carbon dioxide removal system, the crew managed to aviate, navigate, and communicate, returning safely to Earth.

Troubleshooting is the process of diagnosing, locating, and correcting malfunctions.

Being a good troubleshooter, problem solver, and decision maker is highly prized in industry and will be a prerequisite on most job applications. Disciplined problem-solving and long standing proven fault finding methods and techniques, teach you to never rush to a conclusion – examine the facts first, determine causality and judge accordingly, then determine a course of action. **Creative problem solving** derives from your ability to ask the right questions, by listening without judging, and then adopting multiple viewpoints. It requires brain storming, solution seeking, and proper implementation. It is the ability to identify and define problems as well as the generation of effective solutions.

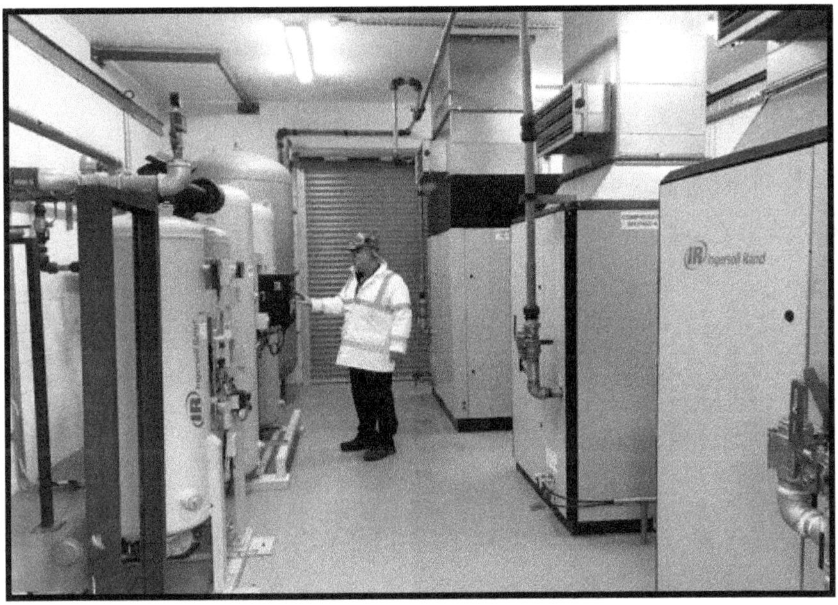

Guided by theory, learn from experience

For effective problem solving, regardless of how difficult a technical engineering problem may be, you **must work through it** using a logical and systematic approach.

The problem must first be defined, cause identified, and the correction properly made in a **step by step** methodical manner. 'Do the basics very well, differentiate between what you know and what you don't'.

Facts and Assumptions

Purpose is to really understand:

- What you know (the HARD FACTS) which are supported by accurate and reliable original data.

What you've assumed:

- Little or no data
- Peoples views and opinions
- What you don't know.

Your investigation must remove as much of your uncertainty as possible, Challenge your assumptions by getting facts (the data). You must use facts to drive down your level of uncertainty. Make sure to base decisions on facts, not emotions.

In the absence of **total knowledge,** you have no choice but to make assumptions.

N.B. However, decisions based purely on assumption are usually wrong.

One of the reasons why so many decisions are based on **incorrect assumptions** is they are often confused with the facts.

Having good problem-solving skills is of real value because you may have to work alone at times, isolated from your peers or manager. Even the most technical positions require the ability to think in different ways. So never underestimate the power of innovative problem solving.

Try not to get caught in the trap of blindly following instructions without first understanding them. Leave your **Macho** decision making at the door. You must realise and recognise the limits of your own capabilities before carrying out any technically challenging work.

Without the understanding or **know how** of where the source of a problem is, it will be difficult trying to resolve it. If you don't anticipate or don't understand potential or real issues, you can't come up with solutions. One thing that sets technicians apart is an understanding of the process, particularly how the instruments monitor and control the process. For example, knowing when a controller can be placed in 'manual mode' without affecting the process and knowing what to do while that controller is in **manual**, requires an understanding of the process.

Without in-depth knowledge and understanding of how a piece of equipment operates, there will be little chance of fixing a fault, especially, if a complex one has developed. A company must always try to ensure that they have a technically trained person in place, who has the right knowledge, relevant information, with the right tools (Hardware – hand tools or software – PC) and at the right time.

E.g. the same analogy applies if sitting for a written or oral test, if you have done your proper preparation study, it is a great feeling, sitting there, knowing you are well prepared. The same goes for a control system that you are responsible for maintaining.

N.B. The information structure that is in place on site will determine how easy or hard a job is going to be. There is no point in a company having all the relevant information if it is very difficult to find.

A piece of equipment's operation, capability and associated written instructions must be understood clearly by going through the instructions, **line by line**, until you have fully understood them before carrying out any maintenance. Seek clarification of instructions from the equipment manufacturer if necessary. The key is to be able to manage the equipments' vulnerabilities and constraints and be able to verify its continued safe operational performance, and how to recover from possible total failure.

Avoid **knee jerk** problem solving, as much as you would like to impress your peers or manager on your fault finding capabilities, you must not be tempted into taking chances. You must **keep your discipline**; stay focussed on proper troubleshooting techniques and adopt the right approach every time using 'evidence based' diagnostics. Ensure the process of solving the problem is managed.

To carry out proper methodical fault finding on a bespoke control system, you must have easy access to, and understand the following:

1. Actual Design Specifications. (Design Philosophy, **URS** – User Requirement Specification, **FDS** – Functional Design Specification, **SDS** – Software Design Specification).
2. O&M manuals (Principle of Operation, Conditions of Start, Process Flow Diagrams, Block diagrams, Decision trees, PPT presentations, Fault Finding flow charts)
3. Data Analytics.
4. 'As Built' P&ID's (Process and Instrument Drawings).
5. 'As Built' Electrical circuit diagram drawings.
6. 'As Built' Mechanical drawings.

All of the above **original** (hard & soft copies) documentation must be kept in a waterproof, flame retardant, controlled cabinet or area. Only signed, controlled, up to date copies of the original drawings and associated documentation can be allowed to be used by personnel on plant and must have the signature of the systems administrator who documents all activities surrounding these essential documents.

Drawings issued by the Engineering department will be stamped 'For Information Use Only'; this stamp should have the signature of the issuer, the name of the recipient of the drawing, and the date of issue. It is the responsibility of the holder of the drawings to ensure that the drawings held are the most recent revision. This is important in ensuring any piece of equipment that has been modified is reflected in all documents via a Change Control procedure to ensure safety and compliance.

Change Control (CC) Procedure definition:

A change control procedure is the process of handling proposed alterations to systems that have been previously designated as fixed, whereby any planned change to a particular system is identified, evaluated, approved, implemented, and documented.

A company's Change Control procedure must be followed; the most important step in a CC, is recognising a change. If the change is not recognised, the CC process will never be initiated! The purpose of change control is to ensure that change occurs in a controlled and safe manner, and also to maintain the validation status of existing systems and to initiate a control process for new systems. It also gives all affected stakeholders the opportunity to review and control any subsequent changes to their systems, and ultimately, to **sign off** before the change is carried out.

This may seem a total waste of time, however, if a change a person has made hasn't been documented, 6 months from now, one of their colleagues might have to fault find the same piece of equipment and realise a change has been made but they do not know the reason why. This may lead to hours of downtime, trying to decipher why the change was made in the first place. The initial person's actions could lead to an accident because other people, unaware of the changes made and the reasons why, caused an unexpected action when the undocumented change was put back to its original state.

Sometimes situations and circumstances dictate that all the above cannot be in place before making the change. Personnel must always get management approval before any change is made and ensure it is documented retrospectively.

You must familiarise yourself with the plants' equipment, be comfortable with individual pieces of equipment operations, know the instrumentation design specifications, capabilities, sounds, smells, temperatures, pressures and flows. You must try to gather as much knowledge as possible of the 'As Built' process and the associated instrumentation, electrical, and mechanical drawings. Physically go to the plant machinery systems in your facility and **walk down** the relevant drawings. Get to know where the AHU's, pumps, control valves, hand valves, pipe work (pipe work colour coding), individual electrical control panels, and the instrumentation such as temperature transmitters and pressure transmitters are positioned. Get to know why they are there, what purpose they serve, and how they work. **The importance of the above cannot be stressed enough**. You must dedicate time to this crucial familiarisation process whether you are on a large site covering several square kilometres with individual processing plants scattered throughout the area, or a commercial building/office block under one roof.

Labelling

Proper, well maintained, visible, legible **labelling** (in all the relevant languages) of site equipment, must be standard across any plant. It is essential not only for operational reasons but also when trying to resolve a fault which must be diagnosed quickly. The labelling should include pumps, control valves, hand valves, pipe work, electrical panels, and any other ancillary pieces of equipment. The associated individual electrical cables, compressed air lines etc. must be numbered and labelled according to the 'As built' drawings. If parts of equipment on plant are not labelled, you must make it your business to do so, and get it done.

Every company has its own set of superheroes; people who are working 'day in and day out' to help a business become the best. Try to avoid, if possible, having heroic maintenance engineers who handle the exceptions and come up with all the answers yet provide no accurate information of how issues were resolved and do not generate corrective actions (CAPA's) to prevent future reoccurrences.

Corrective And Preventative Action (CAPA): Defines the method by which CAPA's are raised, monitored and closed out with the intention of preventing a reoccurrence and ensuring the improvement of a company's Operational, Safety, and Environmental system.

N.B. If you are retiring or leaving your company permanently for another role, your successor should be able to pick up where you left off relatively easily. There will always be a transition phase of learning. If the accurate, relevant information is already in place and easily accessible, the transition period should not be difficult. Your reputation based on how well the systems you left after you were managed and maintained will follow you wherever you go.

FAULT FINDING

Remember the old adage; '***Be prepared and always be proactive in failure prevention***'. You must know the plant you work in, its bespoke systems, know what to do and how to adapt should a control system (hardware or software or both) start to shut down intermittently or fail completely.

You must be familiar with the underpinning fundamental physical principles and properties as are applied within a piece of equipment.

E.g. Control loop: The liquid level in the bulk storage tank is being monitored by the tank level transmitter which automatically turns the pump on/off via SCADA to maintain the liquid level set point. It is important to understand how the entire control loop works. Take the inductive flow switch for example - the inductive flow switch is fitted in the pipeline to detect liquid flow and to protect the pump against dry running. Do you understand its **Principle of Operation?** *i.e. how it works*

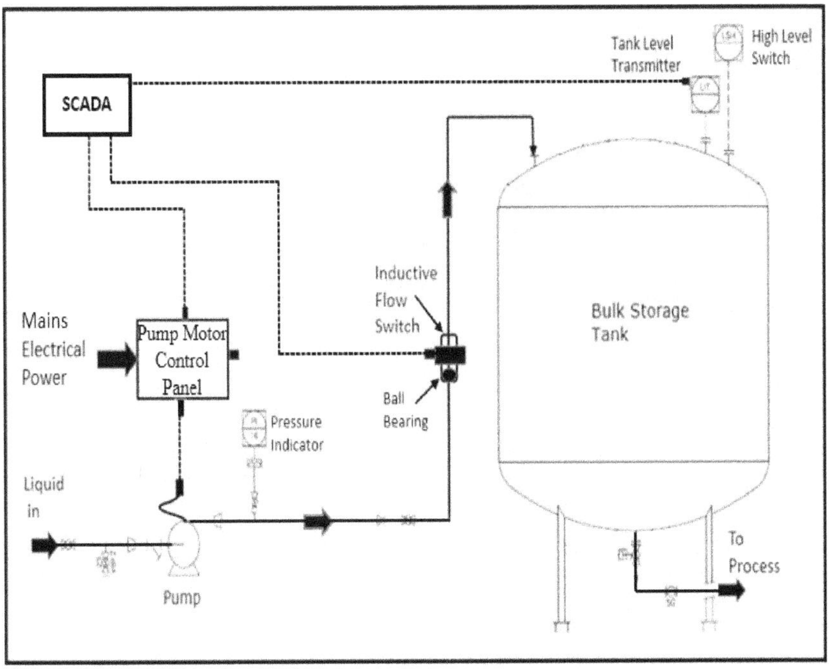

An inductive flow switch, normally fitted on a vertical pipe has a **ball bearing** fitted within the confines of the housing of the inductive switch, this ball bearing moves upwards/downwards when liquid flow is established/not established **i.e.** the ball bearing moves in or out of the induced field of the flow switch signalling to SCADA it has detected flow/no flow in the pipe. *This may seem like common sense to an experienced fault finder but to an inexperienced one, it may be very difficult to decipher.*

You may have studied all the background literature and fault finding methods about a piece of equipment but sometimes it is possible that two obvious problems combine to provide a set of symptoms that can be misleading and send you in the wrong fault finding direction. Being aware of this possibility and avoid solving the **wrong fault** is what must be recognised and understood.

There are many different types of faults, from minor to major, ensure any action that is carried out does not exacerbate the fault. Can the piece of equipment run safely, even if not at 100 % efficiency until you feel confident in carrying out the repair?

Don't turn a minor fault into a major issue by your actions. **Resist the temptation to rush an investigation.** This usually results in misdiagnosis, misunderstanding, considerable time wasting, and potentially disastrous consequences.

'Understand first, then Investigate'.

- Take your time and always be disciplined, logical, and systematic
- Gather the evidence
- Analyse the evidence
- Look at the symptoms, and consider their significance and understand this first
- Review the anecdotal evidence; analyse, and understand
- Consult others
- Write a PROBLEM STATEMENT

- Investigate and risk assess based on good quality factual data
- Decisions are to be based on best available information
- Plan carefully
- Execute with caution
- Test and prove thoroughly
- Document and archive.

N.B. Plan the work and work the plan, those who fail to plan, plan to fail.

Forensically analyse all information available using a firm foundation of core skills which should be constantly refined as more experience is gained.

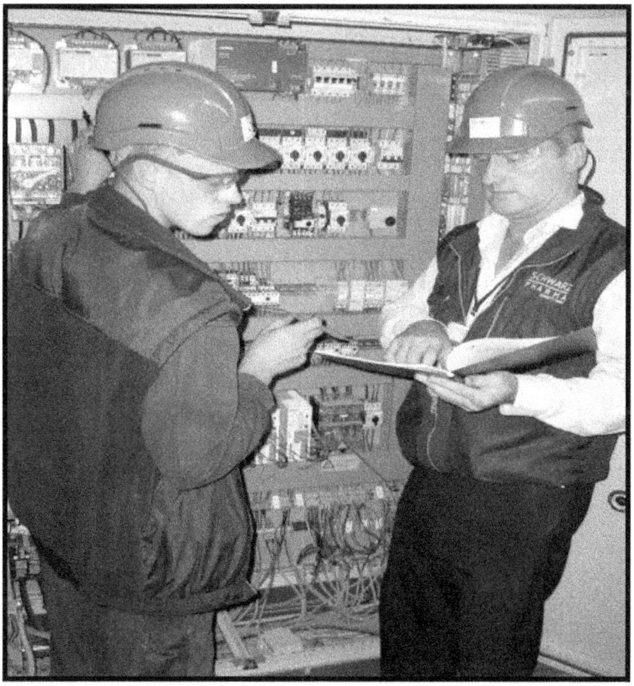

Apply a logical **'Gather the Evidence'** approach to finding and rectifying a fault.

Investigation leaders and 'Peeling the Onion'!

Wanting to get to the root cause of a problem as quickly as possible is the ultimate goal, but sometimes, impatience leads to mistakes being made. 'Don't be too eager to find the answer quickly, instead, try to understand the question better.' A good root cause analysis procedure must be applied for ascertaining and analysing the causes of problems. Being able to differentiate and recognise 'what is a symptom and what is a root cause' (not just causal factors) using proven techniques is key.

What does this mean in practice?

- Keep the investigation 'broad'
- Keep asking open ended questions and listen
- Gather and document as much factual data as possible
- Gather and document as much anecdotal data as possible
- 'Paint a Picture' - don't get specific
- BE PATIENT!

Peeling the Onion is a metaphor used to encourage you to 'undress' the problem. Fault finding is like peeling an onion, you may have to peel away many layers before finding 'root cause'.

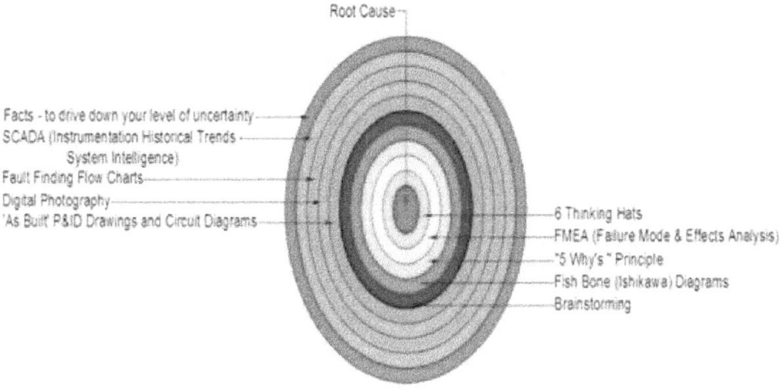

The 10 most useful aids synonymous with 'Trouble Shooting, Fault Finding, & Problem Solving' are:

1. Facts and Anecdotal evidence
2. BMS/SCADA/Data trends
3. O&M manuals and Flow charts
4. Digital Photographs
5. 'As Built' P&ID's, Electrical and Mechanical Drawings
6. Brainstorming and Consult with the Vendor/System Designer
7. Fish Bone (Ishikawa) diagrams
8. 5 Why's principle
9. FMEA (Failure Mode & Effects Analysis)
10. Six Thinking Hats

1. Facts – to drive down your level of uncertainty. Establish what you know and what you don't know. Take notes and review. **Anecdotal evidence:** is evidence collected in a casual or informal manner and relying heavily or entirely on personal testimony. Thus, even when accurate, anecdotal evidence is not necessarily representative of a typical experience, **accept it with caution.**

The objective of any investigation is to make any remedial

decisions based on the FACTS. The more facts you have, the better! Make sure facts are facts… and not assumptions!

Guide to making good Decisions:

- Don't be in a rush to find the answer, understand the question first
- Figure out what you know and what you don't know
- Gather as much accurate data and analysis as is necessary to resolve the dilemma
- Acquire the informed opinion of trusted SMEs
- THEN assess and manage risk

2. **BMS** (Building Management System) - BMS systems are 'intelligent' microprocessor-based controller networks using communication protocols such as BACnet, Modbus, and LonWorks,

installed to monitor and control a buildings technical systems and services such as lighting, power systems, fire detection systems, security systems, hydraulics and especially HVAC systems. Intelligent Feedback is essential for asset utilisation and availability. It also simplifies the operational and decision-making processes. Intelligent feedback is also key in ensuring comfort and safety functions in buildings.

Buildings and Machinery are constantly communicating but are you listening? Invest in technology, digitalisation enables you to listen and react. The technology will be doing all the work, the BMS operator must ensure the technology is working. A Building Management System should be maintained with an appropriate level of servicing. As with any software driven system, data and files should be backed up on a regular basis.

SCADA (Supervisory Control And Data Acquisition) – is a control system architecture that uses computers, networked data communications (e.g. Fieldbus, Profibus, Industrial Ethernet) and graphical user interfaces (GUI) for high-level production process supervisory management but uses other peripheral devices such as Programmable Logic Controllers (PLC's), Actuator Sensor Interface (ASI), and discrete PID controllers to interface to the process plant or bespoke machinery. This 'System Intelligence' allows personnel via a PC screen GUI mimic diagram in a control room to remotely control, make parameter adjustments, and monitor in 'real time' a production process on the factory floor and react accordingly to associated alarms. Historical **Data Trending** is one of the many key features of this system and also when determining the stability of control algorithms; tuning the system; reviewing instrumentation historical trends; and analysing process and failure data.

Network Topology – from sensors to digital services

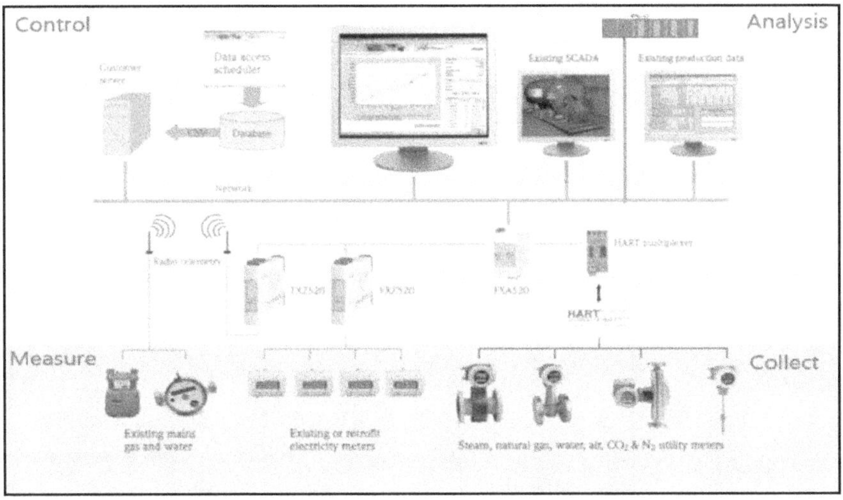

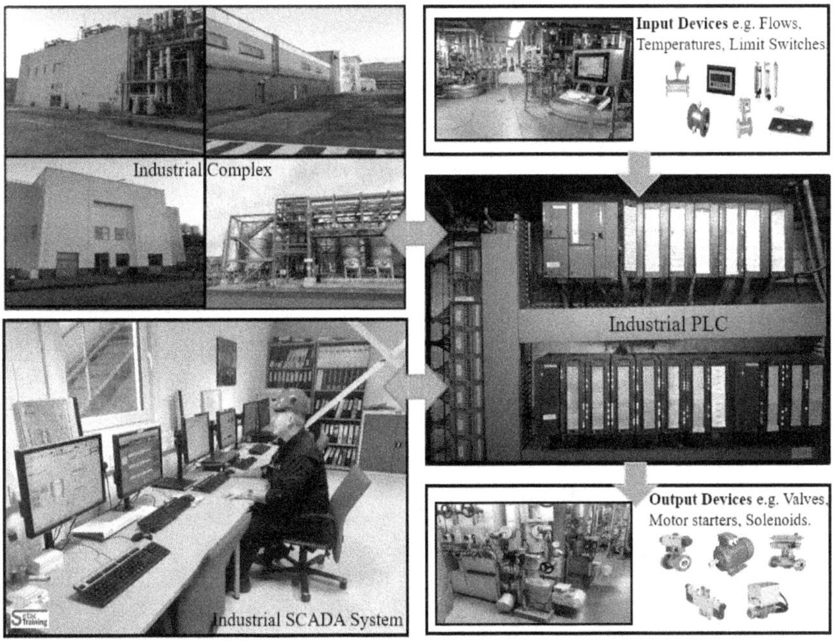

3. **O&M** (Operations & Maintenance) manuals/ **Flow charts** - Review and Generate a picture of the 'Who, What, Why, Where, When and How'.

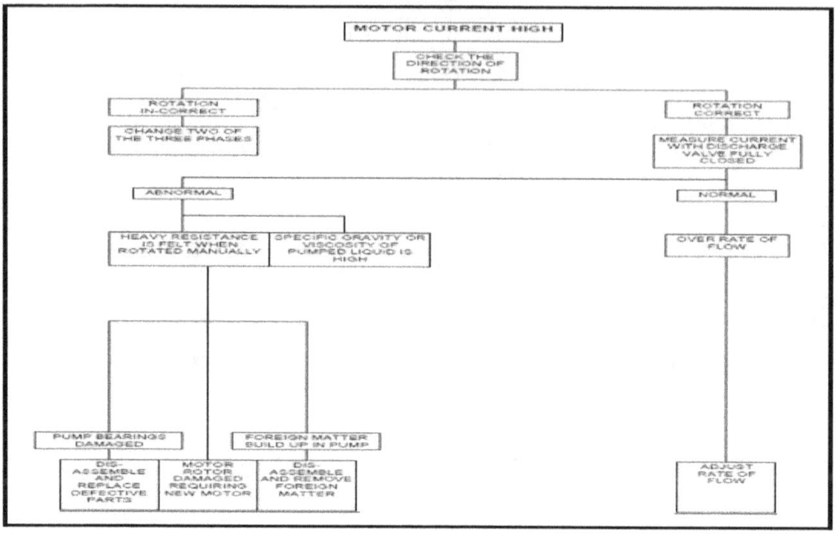

4. **Digital Photographs** - Become an **'Engineering Detective'** - Review the digital pictures/videos of when a piece of equipment was operating smoothly (all mechanical & electrical settings and digital HMI faceplates/screenshots – especially during commissioning). Digital photographs and videos make great reference material should a problem occur. If a setting was changed accidentally or a part has become worn, it can be checked against the photo. Make sure they are clearly identified and stored on a shared hard drive that is secure and easily accessible.

5. **'As Built' Mechanical P&ID** - all equipment and associated pipework clearly identified, and everything labelled.

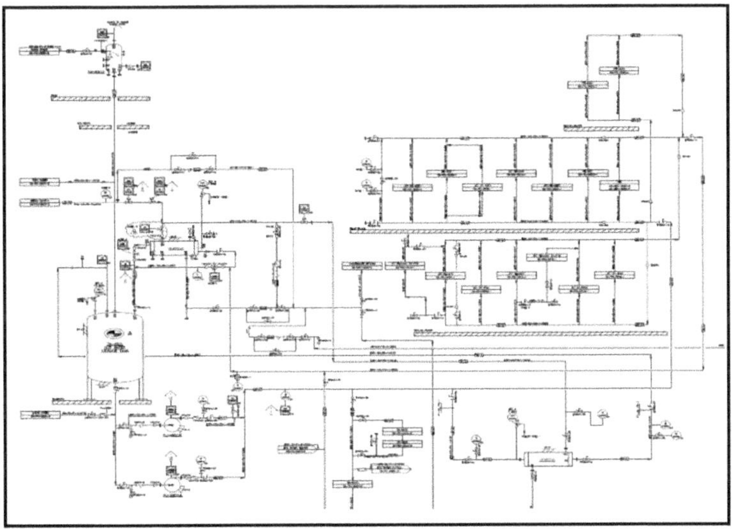

'As Built' Electrical Drawing - all equipment and associated cabling clearly identified, and everything labelled.

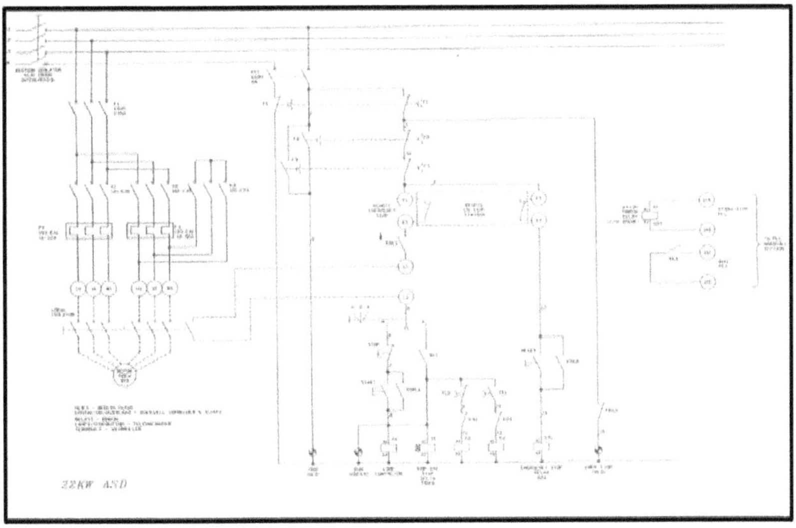

6. **Brainstorming**: is a group creativity technique designed to generate many ideas for the solution of a problem. How? By having a relaxed meeting (3-4 people). Seat the participants *side by side* in a semicircle of chairs facing a flip chart or whiteboard. Select a leader. Write up a Problem definition. Generate ideas. No discussion or evaluation. Reduce list (agree criteria). Consult with the system designer if available. Consult with the **Vendor** in leveraging equipment supplier expertise and knowledge.

7. **Fish Bone (Ishikawa) diagrams.** To examine effects or problems to find out the possible causes and to point out possible areas where data can be collected. A **Cause and Effect Diagram** is a graphic tool that helps identify, sort, and display possible causes of a problem or quality characteristic. A Cause and Effect Diagram helps determine root causes; encourages group participation; uses an orderly, easy-to-read format; indicates possible causes of variation; increases process knowledge and identifies areas for collecting data.

There are four steps to constructing a cause and effect diagram:

1. Brainstorm all possible causes of the problem or effect selected for analysis.
2. What has changed and when? Classify the major causes under the headings:
   - Equipment
   - Process
   - People
   - Materials
   - Environment
   - Management or add your own heading for your bespoke process
3. Draw a cause and effect diagram.
4. Write the effects on the diagram under the classifications chosen.

Cause and Effect diagram:

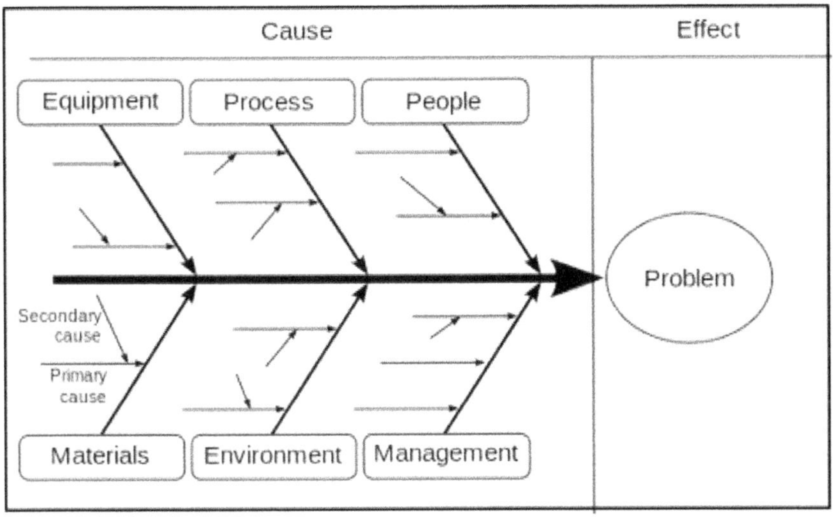

5. Why's principle - to get below the surface of the problem. You may have to ask 'Why' five times before getting to the root cause of the problem.

1. We didn't make the production schedule - **Why?**
2. The machine stopped - **Why?**
3. The machine's main motor MCB tripped - **Why?**
4. A motor winding burned out due to excessive temperatures - **Why?**
5. Inadequate cooling on the motor's cooling ribs due to dirt build up - **Why?**

Answer: We have no Preventative Maintenance Programme in place.

8. **FMEA (Failure Mode & Effects Analysis)** – uses methodologies in a step-by-step approach for identifying all possible failures in a design, a manufacturing or assembly process, or a product or service. It assesses the risk associated with those

failure modes, it ranks the issues in terms of importance, and identifies & carries out corrective actions to address the most serious concerns.

9. **Six Thinking Hats:** helps people see a situation from the same perspective when a problem develops.

E.g. Solving Problems - Blue, White, Red, Green, Yellow, Black.

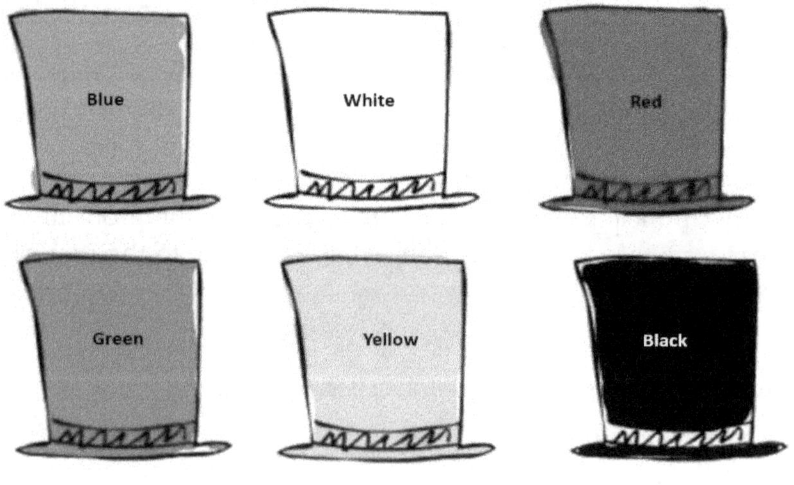

The meeting may start with everyone assuming the **Blue hat** to discuss 'Process Control' and how the meeting will be conducted and to develop the goals and objectives (the Blue hat can also be used at the end of the meeting where it asks for summaries, conclusions, and decisions). Next the discussion moves towards **White hat** thinking which is about data and information. It is used to record information that is currently available and to identify further information that may be needed. The discussion then moves on to **Red hat** thinking to collect opinions, intuitions, reactions, and emotions to the problem. This phase may also be used to develop constraints for the actual solution such as who will be affected by the problem and/or solutions. Next the discussion moves on to the **Green hat** which involves creative thinking and the generation of new ideas and possible solutions. The **Yellow hat** now takes centre stage where it looks for the benefits in a situation. This hat encourages a positive view even in people who are always critical. **Black hat** thinking is used for critical judgment of the solution set, sometimes it is easy to overuse the Black hat.

Because everyone is focused on a particular approach at any one time, the group tends to be more collaborative than if one person is reacting emotionally (Red hat) while another person is trying to be objective (White hat) and still another person is being critical of the points which emerge from the discussion (Black hat). The hats aid individuals in addressing problems from a variety of angles and focus individuals on deficiencies in the way that they approach problem solving. This tool enables us to look at things in a collaborative way, beyond our normal perspective to see new opportunities. It provides a means for groups to plan thinking processes in a detailed and cohesive way, and in doing so, to think together more effectively. "Be 'Open Minded', your brain isn't going to fall out."

The extent of preparation should be directly proportional to the size of the problem to be rectified – (replacing a light bulb versus replacing a piston in a 10,000 HP engine).

Tip: You must remind yourself to make a **conscious decision** about your intended actions and how you are going to go about them. Replay proposed actions in your mind and plan strategies regardless of the type of work that is to be carried out. Be prepared for the inevitable obstacles as they arise. Try to avoid, where possible, making big decisions regarding the dismantling of a machine if under stress, physically or mentally tired, suffering from sleep deprivation, feeling unwell or in a hurry. If this situation arises, take a **time out,** think about what must be done and the possible consequences (if any). Decisions made under these conditions can prove very costly to both you and the equipment if a mistake is made.

Fatigue can kill: we are not good judges of our own impairment. Fatigue can affect performance in several ways:

1. It can degrade **decision making** - short-term memory problems and an inability to concentrate.
2. It can slow **reaction times** - reduced hand-eye coordination or slow reflexes.
3. Noticeably **reduced capacity** to engage in effective interpersonal communication.

Ensure decisions are based on accurate information.

E.g. ***a fault has occurred in a machine:***

What are the **Conditions of Start**? *Do not* start changing parameters (PID settings) in controllers, making unauthorised modifications or adjustments, pushing buttons, turning knobs, closing/opening hand valves .... if you are not positive you know what you are doing.

Your fault-finding discipline must always be maintained, no matter how much you would like to press a button or open a valve. You must **keep your hands to yourself** and ensure any action taken is based on competent engineering fact. Once you have collected the evidence, you must think about what it tells you.

Separate the various symptoms, use your process of elimination skills and try to work out the significance of each one, try not to suffer from **Analysis Paralysis** either (thinking gets in the way of doing or trying to fully understand the problem) where you may be receiving multiple pieces of information from all quarters possibly leading to information overload.

Do not make whole scale changes, take a structured approach, and take one step at a time. 10 to 15 minutes of thought can save hours of downtime and personal anguish. Always act with conscious thought; actions should be taken based on evidence, not feeling.

E.g. replacing a faulty PCB (Printed Circuit Board) in a machine:

Before discarding the faulty unit, ensure to observe/jot down notes/take a photo of any jumper connection links or any switch set up arrangement that may be in a bank of switches on the faulty PCB itself and replicate exactly both links and settings on the new PCB prior to fitting.

Before you start changing anything or moving anything, train your brain to scan, stop, look, listen, and learn. **It's easy to miss something you are not looking for.** Use your eyes to look at the symptoms.

Use your **'Natural Senses'**:

- **Do visual checks**, look out for anything unusual, oil stains, leaks, burn marks.
- **Use your sense** of smell to detect anything out of the ordinary, burning oil, rubber.
- **Use your sense of taste** to detect anything out of the ordinary e.g. Dry mouth.
- **Listen out** for anything that doesn't sound right e.g. knocking/rubbing noise.
- Get to know the noises a piece of equipment generates, check for any excessive vibrations by **actual touch**, this could be first sign of wear and tear of a mechanical bearing (remember, there are approximately five million sensory nerve receptors in your skin alone).

Key points to remember when Fault Finding

- Always exercise caution and systematic thought; always use logical and deductive reasoning.
- When fixing the fault: never underestimate the difficulties of any repairs to be carried out.

Questions to ask yourself before you start the work:

- Have I got correct technical manuals/information required to carry out such a repair?
- If heavy lifting is required, is the equipment (e.g. hydraulic lifting jacks, portable power tools) readily available to carry out such a repair?
- Have I the specialist equipment and bespoke tools needed to carry out the work? Machines will normally have complex pieces of individual components with tight tolerances attached to them e.g. a gearbox.
- Are all the spare parts available as required if needed?
- Is the vendor/manufacturer who supplied the equipment readily available? Ensure to have the emergency mobile, landline phone numbers and email addresses of the vendors and manufacturers service people for expert advice should anything go wrong.

N.B. When using tools and equipment, ensure they are safe to use and fit for purpose. Ensure they can always be accounted for and know their location, for the following reasons:

1. If tools are left loosely on a pipe or steel girder, they can fall and hit someone, causing serious injury.
2. Trying to find for example, a wrench that has fallen into a myriad of hydraulic pipes, hoses or cables can be very difficult and time consuming.
3. If a screwdriver falls into a rotating mechanism accidentally and is not noticed, there could be serious consequences should the machine be turned on, the mechanism could be jammed or damaged. The screwdriver can also become a missile if ejected from the mechanism at a fast speed which could cause serious injury.
4. A spanner/wrench that has been mislaid and fallen into electrical switchgear unnoticed, could have disastrous consequences if the electrical equipment was switched on, possibly causing short circuits across **'Live'** contacts resulting in a large flash/electric arc

which could cause serious burns to a person or be possibly fatal and may also cause severe damage to the electrical equipment itself or cause a fire.

5. **Portable equipment,** by its nature, is more susceptible to damage than fixed electrical equipment. It is also more likely to be used in different environments and is often directly in contact with the user. If using this type of equipment, ensure it is **PAT (Portable Appliance Tested)** tested. Ensure the AC transformers, power leads, power tools and associated plug tops and sockets are all in good condition and fit for purpose before using them e.g. RCD protected, properly fused, no cuts, **no damaged or non - standard joints** (including taped joints) are evident in the power leads. Ensure also that power tools (drills, jigsaws, grinders) whether they are AC powered or DC powered have no sign of physical damage to the outer cover of the equipment or obvious loose parts or screws. Ensure the outer covering (sheath) of the cable is being properly gripped where it enters the plug or equipment and no coloured individual wires are visible.
6. When using long power cables, it is important to consider the amount of **voltage drop** that can occur, as it can cause the cable to become hot and unsafe. As the current passes through a longer and longer conductor, increasingly more of the voltage is 'lost' (unavailable to the load), due to the voltage drop developed across the resistance of the conductor.
7. Ensure any lifting equipment (hydraulic jacks, winches, block and tackle) used, is properly serviced, certified, has no apparent physical damage present and is fit for purpose.
8. Tools and equipment can be very expensive to replace should they be abused, mishandled, mislaid or stolen.

Abnormal Operation Analysis and Correction:

Four logical steps are required to effectively analyse a problem and make the necessary corrections:

1. Define the problem and its limits.

2. Identify all possible causes.
3. Test each cause until the source of the problem is found.
4. What was the initiating factor? - try to establish what caused the fault in the first place and then make the necessary corrections.

N.B. Ensure any work carried out does not cause loss of control or escalation of consequences in the event of equipment failure.

The First Step: is to define the limits of the problem:

- When a problem develops, compare all information with normal conditions. Up to date P&ID's, SCADA historical trends, logbooks, or PDA's with daily recordings of key parameters, defining expected acceptable limits of the equipment's operation will be invaluable here. Well maintained PM records are a very useful source of information which should also be referenced when solving an intermittent problem/fault. Intermittent faults are regarded as the most difficult class of faults to diagnose and are cited as one of the main root causes of 'No Fault Found'. There are a variety of technical issues relating to the nature of the fault which make identifying intermittent.

- **Team Resource Management** plays a key role when trying to resolve a problem. This is where your people skills come in. The best option, if available, is to speak to the person who reported the issue. The information gleaned may assist your fault-finding fact mission, especially, if the person is very experienced in the equipment's operation. Ask machine operators, other maintenance personnel or instrument technicians for input in resolving the problem. Two heads are better than one. Ask questions to help gain clarity on the details of the problem. **Communicate with others as clearly as possible to avoid misunderstandings.** Do not dismiss a person's comments or observations out of hand, respect the opinions of others, it may have no relevance, or it may be the exact piece of information you are looking for.

- Knowledge, experience, up to date O&M manuals, engineering drawings, digital data recording and consistent records are the basis for avoiding the unusual especially when applying **RCA (Root Cause Analysis)** techniques. Problem Solving & Root Cause Analysis is about avoiding costs and minimising losses by getting to the root causes of problems and failures and solving them. Unexpected failures are not a normal way of life and they can be avoided. Losses are not inevitable - they too can be avoided by adopting a structured problem-solving approach. Conducting an effective root cause analysis can be a daunting task. And without a good root cause analysis, your CAPA will not result in the desired improvement.

Personnel must learn to get to the root of a failure, incident, or non-conformance in their area of work:

1. Analyse problems thoroughly and effectively.
2. Identify the connections between different effects.
3. Prioritise problems quicker and more objectively.
4. By being more effective in a group or with a team when solving problems.
5. Communicate complex issues, visually and verbally.
6. Take the emotion and speculation out of problem solving.
7. Take a process/systems view of managing your operations.
8. Anticipate problems to prevent them from occurring.

- When the **direct cause** of a problem is not immediately evident e.g. of one thousand machine parts being made per hour, one hundred of them are undersized and are rejected. Use of the **5 Why's** principle, for example, could now be implemented. As well as looking for root cause, you must also take into consideration the contributing causes. Probably they would not have caused the problem on their own, but, increased the likelihood of the problem occurring. Lack of **follow through** on RCA lessons is a common weakness and is often due to an ineffective logging or a progress tracking system.

- Hearsay, opinions, beliefs, speculation, or anecdotal information based on or consisting of reports or observations of unskilled observers must be received with caution as these observations and reports can lead the fault finder in the wrong direction. Challenge them to find out the factual basis of their opinion or decision as information is only as reliable as the source. Ask if they are basing what they've said on gut feeling or engineering fact? It is important the latter takes precedence when considering the best way to tackle the problem. Remember the adage 'Facts not Opinions'.

- Gut feeling is one thing but causing **€100,000 worth of damage** to a piece of equipment because of 'Gut feeling' fault finding techniques is another. Just because you believe or want something to be true, doesn't make it true; this type of thinking is simple, logical, and WRONG. Reactive, emotion based, habitual and non-cognitive behaviours must be replaced with analytical, reasoned, intentional and reflective behaviours instead.

*Hard data + Soft Data + Wisdom **(knowledge, experience, and sound judgement)** = Good Decision Making.*

- If you work a shift pattern and come across day to day engineering issues that need addressing, you must not leave it to others; pass on all relevant information to your direct supervisor and to the oncoming shift. Include all relevant personnel who may be directly affected by these issues. Inform verbally, by written shift report or via email. **The most important point is the communication and follow up**, even if this means contacting the plant after personnel have gone home to ensure the relevant information has been received and understood by the on-site shift personnel. Never underestimate how essential this follow up is.

- ***"Simple misunderstandings can lead to huge and costly delays."***

- Ensure whatever message is passed on is clear and unambiguous, whether to internal shift personnel, other departments, or vendors. Do not leave an email, verbal or written message open to interpretation. It may seem obvious to you, because, you are familiar with the system, to others it may be very difficult to decipher. In other words, **'Add the detail'** this will make everyone's jobs easier.

- When you have been given an important message verbally, ensure you repeat it to the person and how you interpreted what they just said to remove any misunderstandings. Ensure your mind is in 'thinking mode'; removing any other distractions you may be thinking about; your brain works sequentially and can only do one thing at a time. Concentrate on what the person is saying, write it down if necessary, if something doesn't sound right or doesn't make sense, make sure it is clarified. Be exact in all communication and correspondence to all, leave no room for guesswork. Ensure to follow up with the other members of the team on the successful resolution to a problem regardless of who resolved the issue. The plant must be handed over in a safe, clean, and fully operational state to the incoming shift. Go the extra yard to make this happen and leave the plant the way you would like it to be left when you are coming on shift.

- **Tip:** During and at the end of a working day, take ten minutes to recap what work you carried out. Has the plant been left in a fully operational and safe state? Did you:

1. Leave anything running which should now be switched off?
2. Take any system or individual automatic valve/motor out of **automatic mode** to facilitate maintenance works and did not put the piece of equipment back to its original state and test it once the work was completed?
3. Remove the **lock and tag** from a mechanical isolation valve?
4. Remove the **lock and tag** from an electrical isolator?
5. Check and double check your work before you **signed off** on the permit to work?

- A small pocket note pad should always be kept and little reminders written in during the day to jog your memory, at the end of a shift of what actions were carried out and what was 'turned off' or 'turned on' so you could carry out your duties. You should then go through the actions **one by one**, ensure everything is back on line and **fit for purpose** before leaving site.

- If you are having difficulty trying to fix a fault, **say so**. You must realise and recognise the limits of your own technical capabilities, it's important to be aware of your own limitations. If you've been working on a piece of equipment for hours and cannot resolve the fault, **Stop,** and **Reflect** on your current predicament. Don't waste anymore time. Recognise you have exhausted your fault-finding capabilities and need help. Use the time instead to escalate the issue. Inform the relevant people or department(s) of the issues you are encountering, then start looking for advice from other colleagues or vendors. Research more information and ultimately learn from the experience.

- "Transcend old habits, **never assume**, when the word 'assume' is broken up, it can be interpreted as making an 'ass' out of 'u' and 'me', Don't **fall into the trap of making assumptions**." Assumptions most of the time lead to incorrect decisions and can prove very costly. Decisions based purely on assumption are usually wrong. One of the reasons why so many decisions are based on incorrect assumptions is that we often confuse them with the facts. Always approach the fault with the actual manufacturers' operational manual of how the piece of equipment works, **not** use the most dangerous phrase in the English language **'this is how we always operated it'**. The O&M manual is an essential part of the machine and must accompany it during its entire lifetime. Just because a piece of equipment has been operated in a certain way over a period, doesn't mean that this is the proper **mode of operation**. There may have been an undocumented modification carried out. Reference to the manufacturers' manual will help identify any modifications done to the original design.

- **Maintenance procedures** are put in place to ensure any work carried out is completed according to the manufacturer's instructions and always comply with defined work procedures. In that way, repeatability and accountability is always maintained, the intent is to remove bad work behaviours as well as human error**.** If there aren't proper Preventive Maintenance schedules in place over the lifecycle of a piece of equipment, complacency can set in. Safety interlocks might **not** be working as they should and a **work around** may have been put in place. The piece of equipment could then be operated in this incorrect manner, with the potential for an accident or incident to occur.

- The key to minimising equipment problems are scheduled **service intervals and routine inspection**. Written records indicating date, items inspected/replaced, service performed, and machine condition are important to an effective routine maintenance programme. Standardise maintenance schedules and tasks; have pictures, checklists and steps printed on each work order, this will reduce variance between work done by different maintenance team members. From such records, problems can be identified and solved routinely. To avoid breakdowns, a piece of equipment will have operational parameters **e.g.** temperature or pressure 'low and high' set points within which the machine can work safely.

- If a piece of equipment has been operating per the manufacturers instruction over several years suddenly stops for no obvious reason, you must proceed to make a list of all deviations from normal operating conditions.

- ***Do not rush in and make wild guesses.*** Think logically before tackling the issue head on. Time spent on preparatory work to refresh your memory on how a system operates, can save valuable time. It is easier to do a job right than to explain why you didn't. Remove any items not relating to the symptom and separately list those items that might. Use this list as a guide to further investigate the problem.

The Second Step: is to decide which items on the list are possible causes and which items are additional symptoms. Again, the extent of preparation should be directly proportional to the size of the problem to be rectified. Use the systematic approach; has a decision tree or a fish bone diagram been developed based on the piece of equipments' operation and real time recordings of relevant parameters? Using these types of diagrams and writing down all the possible conditions and potential problems with other members of the team can help build up a picture or pattern that can be used to assist in finding the fault more quickly.

This type of **brain storming** is only as good as the people who are carrying it out. Try to have the right mix of Subject Matter Experts (SMEs) who have a good understanding of the actual problem and the affected equipment. The team should comprise of both engineers and operational personnel and involve any other department who they think may help them in resolving the problem.

The Third Step:

- **Do not** alter several things at once.
- Identify the most likely cause and take action to correct it. **Think** before any action is taken. **Do not** change a mechanical setting or electrical parameter if you're not sure of the ramifications of such a change. **Do not** change anything unless it can be returned to its original position.
- **In solving a wrong problem, a new problem can be created.**

Do not withdraw a piece of equipment from a unit unless it can be put back to its original state. Observe the orientation of how each component is interlinked with other components. Be extremely careful when removing multiple individual pieces of equipment to get at a faulty part, as any breakages can prove very costly and increase downtime dramatically. The importance of making sure to retain every nut, bolt, anti-vibration washer, screw …. in a safe place during disassembly cannot be stressed enough. Lay them out neatly in a structured sequence on a bench and ensure they all can be accounted for.

Tip: Use a piece of cardboard and some masking tape, as you remove the screws, nuts and bolts …. tape them to the cardboard in the sequence you took them out and replace them in the reverse sequence after repair is complete. If after reassembly, there are some bolts and nuts left over, it will be clear, reassembly hasn't been successful.

- Trying to find or replace a special length bolt that has gone missing can be very time consuming and frustrating. Also consider that some bolts may be longer than others for a specific purpose and must be replaced exactly as they were removed.

Tip: Locate another piece of cardboard and use it as a template.

Draw a rough sketch of the part that is being dismantled and as the bolts are being taken out, push them through the cardboard as a reminder how they should be replaced. Digital cameras or mobile camera phones can prove very useful in having actual real time pictures of the original installation to assist when reassembling.

- If a digital camera/mobile phone is not available, always have a **black permanent marker** to mark the pieces of equipment that are being removed.

E.g.: a mechanical seal on an agitator shaft in a mixing vessel must be replaced. Draw a straight line through all the individual sections before disassembly and use the line as a reference when reassembling.

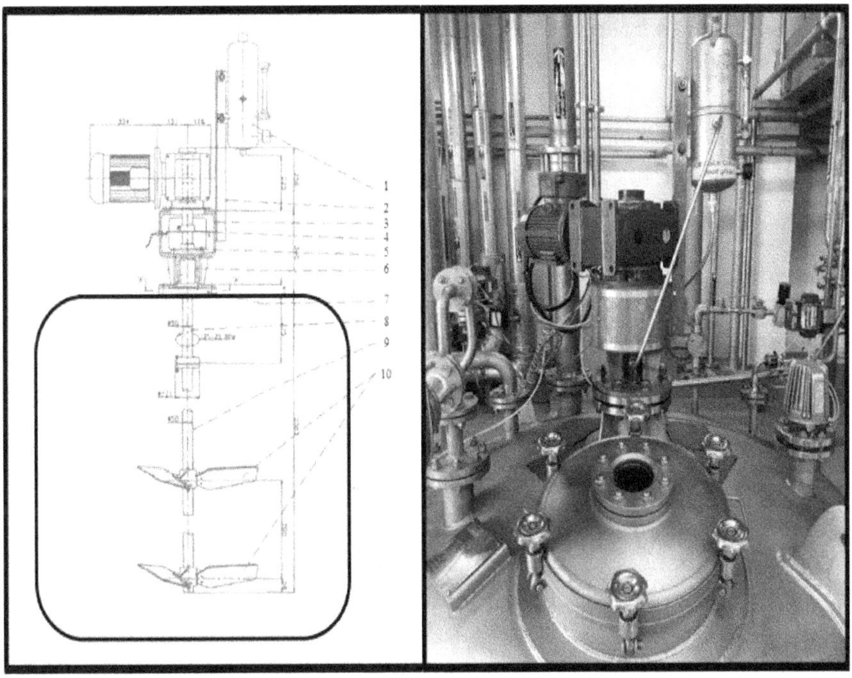

- When reassembly has finished, mark each bolt, nut, screw with a little yellow paint/permanent marker after they been tightened or torqued to a manufacturer's recommended specification. This gives added assurance that it can be visibly seen if any bolt etc. without a yellow mark has not been properly secured.

- If a fault has been misdiagnosed and the piece of equipment that was thought to be the cause of the fault turned out to be functioning properly, replace the original part, as electrical and mechanical parts can be very expensive and sometimes very hard to replace.

  If the symptoms are not relieved, move to the next item on the list and repeat the procedure until the cause of the problem has been identified.

- Take as many digital photos and videos of a smoothly operating piece of equipment as you can and 'time and date' them. Make sure they are stored in a place that is secure and easily accessible.

The Fourth Step:

- Once the cause has been identified and confirmed, analyse the fault or faulty equipment. Investigate why it failed. Remove the cause of the failure and generate a CAPA (if required) to prevent the fault reoccurring. Proceed to make the necessary corrections. Keep an open mind when implementing a CAPA. Formulate new questions to provide a different direction if need be**, it doesn't make sense to do things the same way and expect a different result.**

- Test the equipment to ensure it is working properly and per the manufacturers' instructions, specifications, and guidelines before putting it back into service.

- **Keep incident reports,** learn from them, and decide how the incidents can be prevented in the future. If a fault rectifies itself, always remain suspicious and carry out as many investigations as possible to determine what caused the fault in the first place. Most faults that self-rectify, normally fail when the piece of equipment is needed the most. **Be aware**, an undetermined cause to a problem leaves a situation that can happen again as no preventative action can be put in place to mitigate against the problem reoccurring.

There are 3 areas that must be clarified when dealing with total failure of equipment:

1. **Facts** – Deal with the facts that are evident. They must be written down and clarified. Focus and rely on facts and accurate data before dealing with anecdotal information.
2. **Conditions** – Have any environmental conditions changed dramatically? Has an unexpected instance occurred? E.g. lightning strike, excessive ambient temperature, freezing temperatures.
3. **Circumstances** – Has something happened or has someone done something to the piece of equipment that would not be normal and may have been missed during the fact-finding mission? **i.e.** a change of a parameter or mechanical setting could have been made during the previous days' preventative maintenance schedule. Some faults may take months to manifest before failure occurs, **e.g.** Metal fatigue on a structure may start with a single tiny crack. When mechanical stresses are physically put on the structure, the crack may get bigger and longer and may ultimately lead to breaking point and catastrophic failure.

It is much easier to establish facts when the equipment is running, even if not properly: noisy; excessive vibrations; and intermittent stoppages. When the piece of equipment has completely shut down and with no indication of what is wrong (internal failure), the problem becomes a lot more difficult to diagnose **e.g.** motor seizure in a screw air compressor - it could be a motor winding failure, motor mechanical bearing failure or a mechanical screw failure.

What you learn from one control system (hardware, software, process, and instrument drawings) can help when called onto another similar, but not identical control system where a problem has occurred. You must write good reports on how and what you did to remedy a fault (inserts photos, P&ID's, electrical drawings, relevant emails). File them in a manner which can assist you or others if the same or similar problem arises in the future. These archived reports can prove invaluable in saving time and effort should this fault reoccur. **There must be a record**, whether it is a hand-written document or electronically recorded of all activities carried out on a piece of equipment.

If a control system is running perfectly, you must take the time to either **write down or electronically record** all parameters: P.I.D. (Proportional, Integral, Derivative) control instrument loops settings; temperatures; pressures; flows …. for all associated equipment within the control system. Keep these parameter lists in a safe fire proof place (not in the same building) where they can be retrieved should a PLC control system crash occur and does not recover properly. This recorded information can be used as a checklist for the smooth restart of the system.

Safety Management of control systems must be employed to ensure safe operation of a machine. The control system must be integrated with safety systems which are installed in the field in order to partially or completely isolate/shutdown equipment in case of maintenance or extraordinary operations. Guards and interlocks are designed and fitted in the interests of safety and UNDER NO CIRCUMSTANCES should the equipment be operated with guards removed or interlock switches bypassed.

Control and safety interlocks are put in place to ensure proper safe operation of equipment; upon detection of abnormal operating conditions, a safety circuit or hardwired interlock uses mechanical means to place equipment and processes in their defined fail-safe state thus ensuring the safety of personnel, plant equipment and the environment. When fault finding, your job is to find out what the interlocks are and in what sequence they operate. If you can gain some understanding of the **Principle of Operation**, it will be easier to solve the problem.

Most machines built today have PLC's (Programmable Logic Controller - is an industrial computer control system that continuously monitors the state of input devices and makes decisions based upon a custom programme to control the state of output devices). Normally hard wire safety interlocks which work independently of the PLC are also part of the entire control system. If it has a HMI (Human Machine Interface) attached, try to navigate its matrix and find its graphical alarm screen for text that may tell you what the problem is. Most HMI's have mimic graphics of the machine itself and the piece of faulty equipment will flash **Red** if it has failed to start. If a malfunction is suspected but there is no diagnostic message on the HMI display, the **conventional way** of fault finding must be implemented where you physically go inside the equipments main control panel. Use the standard procedures described next as a guide to verify that the equipment hardware and process connections are in good working order.

Tip: Always approach the most likely and easiest-to-check conditions first.

Before tackling the entire control system, **wear the appropriate PPE**. Then try not to overlook the obvious:

1. **Electrical power** - is the correct mains voltage present? **e.g.** 400 VAC @ 50Hz? Be aware of back voltages especially if there are step down transformers and power supplies providing specific voltages (110 VAC, 24VAC, 24Vdc, 12Vdc) needed for the HMI's, PLC, control panels and instrumentation. These can give misleading 'back feed' voltages and cause confusion. Establish correct voltages are present, starting with the machine's main electrical isolator and then methodically isolate each individual breaker or control fuse and all the while establishing that the correct voltages are present according to the manufacturer's specifications.
2. **Compressed air** - is the correct main air pressure present? e.g. 7 Barg and where PRV's (pressure reducing valves) are used, are they set at there respective correct pressures? e.g. 1.4 Barg, 3 Barg.
3. Other utility supplies are available and online.

If it's a piece of equipment which uses a variety of utility systems such as refrigerant cooling systems and Nitrogen for example, check that all are present and at the correct pressures required by the equipment. Pressure and flow switches are normally fitted to these utility services to ensure smooth operation, if not energised (fail safe), the machine will stop and in doing so, protects both personnel and the machine itself.

Check the safety interlocks:

- Emergency stop buttons are not pressed.
- All electrical isolators to internal motors are switched on.
- Key switches are properly set (Hand, Off, Automatic).
- Micro switches/sensors on doors, guards .... are working and engaged.
- Check for any physical damage done to the machines ancillary equipment (solenoids, automatic valves). Check electrical cables for damage, hydraulic and pneumatic hoses for leak or rupture.

The electrical isolator for the machine will be mechanically interlocked with the main electrical panel door. The panel must be electrically isolated before the panel door will open. Once inside the control panel, check the electrical drawings, check for tripped circuit breakers, blown fuses, controller failure (no LCD display), any sign of burned contacts and investigate the cause of failure.

Most MCC (Motor Control Centre) electrical panels are now being built with **IT intelligence**. Each motor in the plant will have its own cubicle. It can be completely withdrawn from the main electrical panels' common power bus system and worked on. This motor cubicle has all the associated control equipment needed to operate the motor in 'Automatic' or 'Hand' with relevant run/alarm LED indicators (power on; stopped; running; tripped status, number of starts; running hours). The motors' cubicle operation is monitored by a PLC system with a HMI attached to it. Reports can then be generated, disseminated, and used to focus maintenance resources in the areas that most need them.

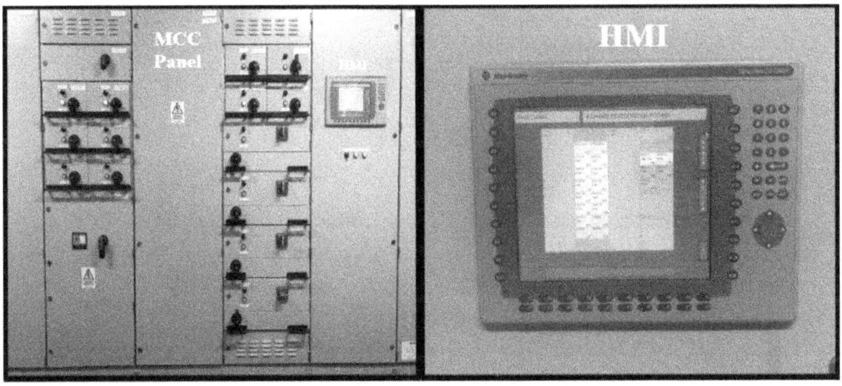

This type of automation infrastructure exhibits much more intelligence than conventional systems and greatly assists fault finding. Diagnostic functionality is improved because programmable systems offer constant test outputs with full diagnosis through the software – something that can't be done with conventional systems. The result is that faults are easier to identify and put right, so that downtime is reduced. Try to familiarise yourself with the internals of the motor cubicle and its associated electrical drawing. There could be anything up to 10 individual main components involved. It is important to identify each component and where it is located on the 'As Built' electrical drawing. It is also important to understand **their functions and all the interconnecting component interlocks**: electrical isolator; MCB's; control fuses; contactors; digital timers; relays; current monitoring devices; overload protection devices; device cards to interface with the PLC/HMI system (which will monitor all these components in real time for tripped/failure status) + all associated cabling and connectors. If any one of these components fail, is there another readily available?

N.B. One of the advantages of having a HMI screen is that if all the **Ready to Start** conditions/interlocks are not met before starting a piece of equipment (e.g. Overload tripped) the system will not make the **Start command** available. The HMI alarm screen can then be interrogated to find out immediately what individual **component** has tripped. Investigation into the root cause of overload trip can start immediately.

Once you have figured out the machine's **Principle of Operation** and have discovered the fault, you must proceed with caution and always be vigilant.

The following would be a typical example:

A 24Vdc electro pneumatic 5/2 way double acting solenoid valve, when electrically activated, extends, and retracts a pneumatic piston:

Principle of Operation:

- Fig. 1 (a) At rest, compressed air comes out of port 2 and into port B of the piston, air in the piston is vented to atmosphere via port A into port 4 and out of port 5 of the 5/2-way valve, the piston is in the retracted position.
- Fig. 1 (b) When y1 energises, and y2 is deenergised, the 5/2-way valve changes state, compressed air comes out of port 4 and into port A of the piston, air in the piston is vented to atmosphere via port B into port 2 and out of port 3 of the 5/2-way valve, the piston extends.
- When y1 is de-energised and y2 energises, the 5/2-way valve changes state to its original position and compressed air comes out of port 2 and into port B of the piston, air in the piston is vented to atmosphere via port A into port 4 and out of port 5 of the 5/2-way valve, the piston retracts.
- Proximity switches detect the position of the piston (extended or retracted).

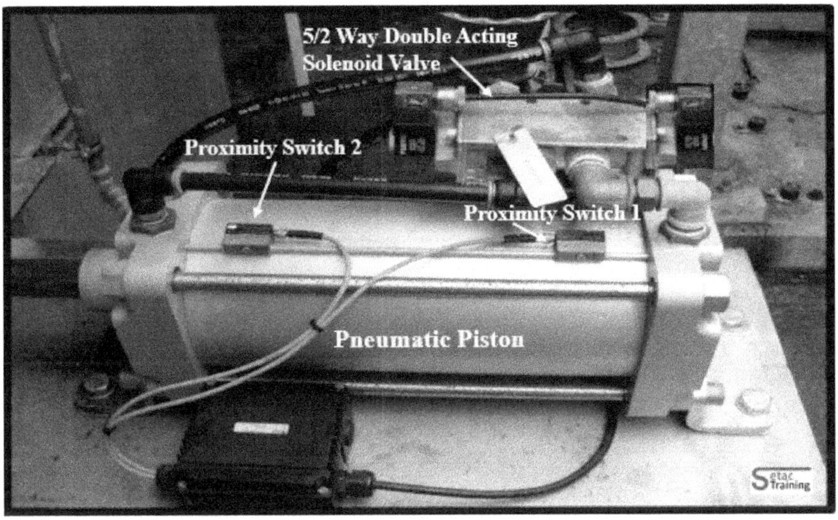

Fig.1 (a) Retracted Position:

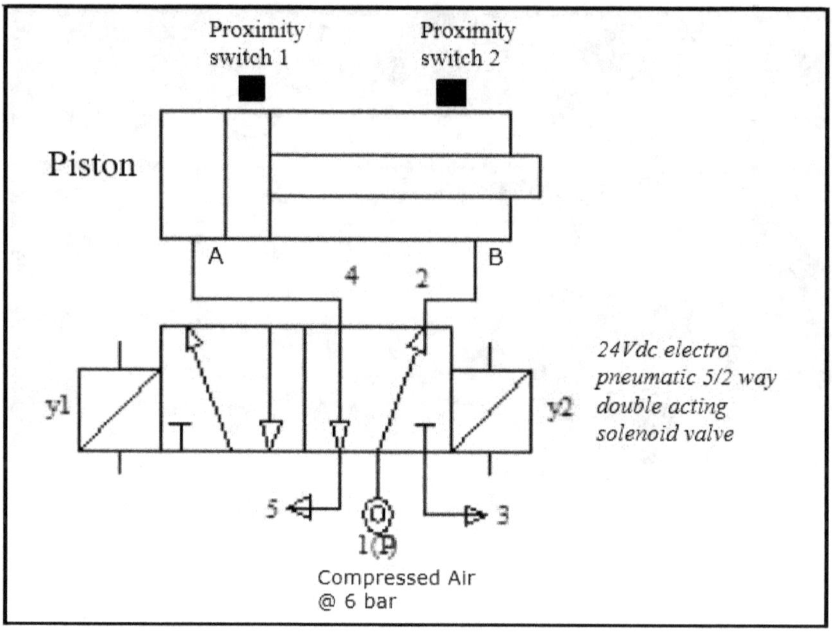

Fig.1 (b) Extended Position:

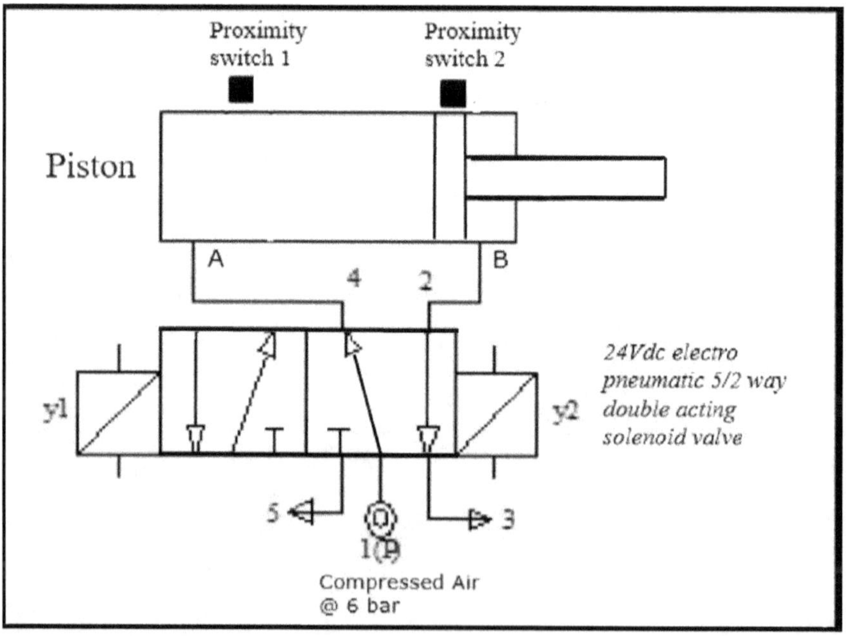

Fault: *the pneumatically operated piston* **has failed to extend;** *the*

following procedure should now occur:

1. Check for any obvious compressed air leaks and check that the correct compressed air supply to the solenoid valve is present (check local compressed air pressure gauge).
2. Check to see is there a manual over ride button available on the 5/2-way valve mechanism itself, if there is, activate it, if the piston extends, check the electrical supply.
3. Check for any loose electrical connections and cables to the solenoids.
4. Check that the correct voltage (24 Vdc) to y1 solenoid coil is present. Check that there is zero voltage to y2 solenoid coil.
5. Check the valve mechanism (listen for clicking sound, when solenoid coil is energised). No clicking (either solenoid coil is faulty or the mechanical actuator in the pneumatic valve is mechanically jammed).

It is discovered that the 24Vdc coil on y1 solenoid which activates the valve's mechanical mechanism is faulty and needs replacing.

Tip: If you want to learn more about electro pneumatic solenoids (or other similar type equipment), access the internet, and do a search e.g. 'Videos on how an electro pneumatic 5/2-way solenoid valve works'. Educational videos will become available for your viewing and learning.

Hard copy or electronically written corrective procedures are necessary and very useful when these types of faults occur. A trouble shooting guide (with photos) should look something like Table.1, with individual possible faults and actions listed per instrument, valves etc. insist on it from the vendor:

Table.1

Procedure in case of fault: Immediate check and/or repair.

Possible cause **Corrective measures**

Compressed air supply. **Check:**

- is compressed air present and at the correct pressure?
- compressed air connections/air leak?

Electrical or cables defective. **Check:**

- electrical connections and cables to solenoids.
- is correct supply voltage present at the coil of the selected solenoid
- solenoid valve coil e.g. possibly burned out?

Proximity switches defective. **Check:**

- switches; replace as necessary.

5/2-way valve defective. **Check:**

- mechanical movement of valve, is it jammed?
- Seals
- Strip valve and rebuild if necessary.

Piston **Check:**

- mechanical movement of piston, is it jammed?
- Seals
- Strip piston and rebuild if necessary.

Create a troubleshooting checklist (with photos) for each piece of equipment on site. Acquire as much knowledge as you can from the vendor who supplied the equipment, peers, or colleague's; then collate and populate the information into one document that everyone can use.

No trouble shooting document is ever written in stone, consider it to be evolving. As a person comes across unforeseen issues and rectifies them, ensure this new information is captured where it can be referenced so others don't have to cover the same ground as the original fault finder. This saves valuable time and money to the company which could be better spent on more constructive issues.

What might be mind boggling to some, may be blatantly obvious to another. Despite all efforts, sometimes it is easy to get caught up in a fault and get nowhere fast. **Always seek help.** Discuss the fault with other peers/manager. Get their input if they have also been trained on the piece of equipment, **good counsel can be invaluable.**

Call in expert help if available and learn from them. Have in place, a direct contact with the vendor manufacturer if a problem occurs. The problem being encountering may never have happened before and was not foreseen by the manufacturer who designed the equipment.

There may be no magic answer. Knowledge and experience of the equipment will play a key role in diagnosing the problem along with the manufacturer's assistance.

N.B. If a new piece of equipment comes on site and is to be commissioned by the vendor's technical engineering personnel, ensure you are with them every step of the way, no matter how busy you are.

The tacit experience gained from the experts sometimes can not be found in O&M manuals. The same goes for existing pieces of equipment that may need to be given an overhaul by the vendor. Ensure as many relevant key personnel on site are trained. Insist that personnel and time be made available to gain this knowledge and experience.

Quick Test No.1: Complete the Circuit to make it work (Answer at the back of the book).

Test: Cylinder 1A will not extend when push button 1S2 is pressed.

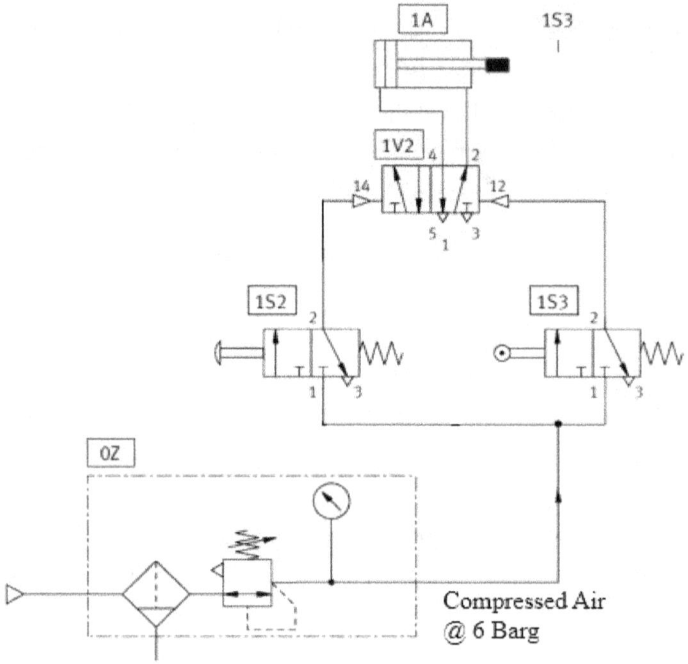

Note: Asset Criticality Ranking (ACR) is a process to evaluate each asset's importance to the business with focus on factors such as safety, regulatory impact, revenue loss, and reduced capacity. It involves the development and implementation of an ACR process to rank each asset based on criticality to the business in order to guide tactical operations and strategic improvement efforts. ACR is a part of the **Reliability Based Maintenance** methodology which is a systematic approach to the proactive preservation of critical asset functionality and operation. Keep an open mind. Try to grade (Scale of 1-10) how important a piece of equipment is and the consequences to the plant if it fails. Try to figure out all possible failure scenarios and put in place measures to mitigate against such scenarios occurring. **This is where risk identification, analysis**

and control comes into play.

Example 1 - *an electric 3 phase motor which turns a mechanical pump and pumps liquid from a bulk storage tank to a processing area starts to fail intermittently (Fig.3):*

Fig. 3:

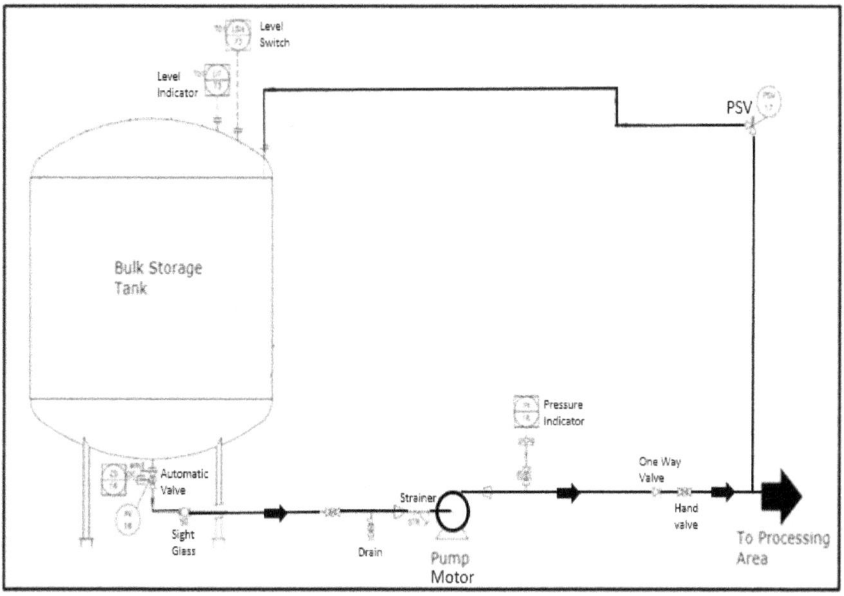

Pump Motor 'Direct On Line' (DOL) Start/Stop Circuit:

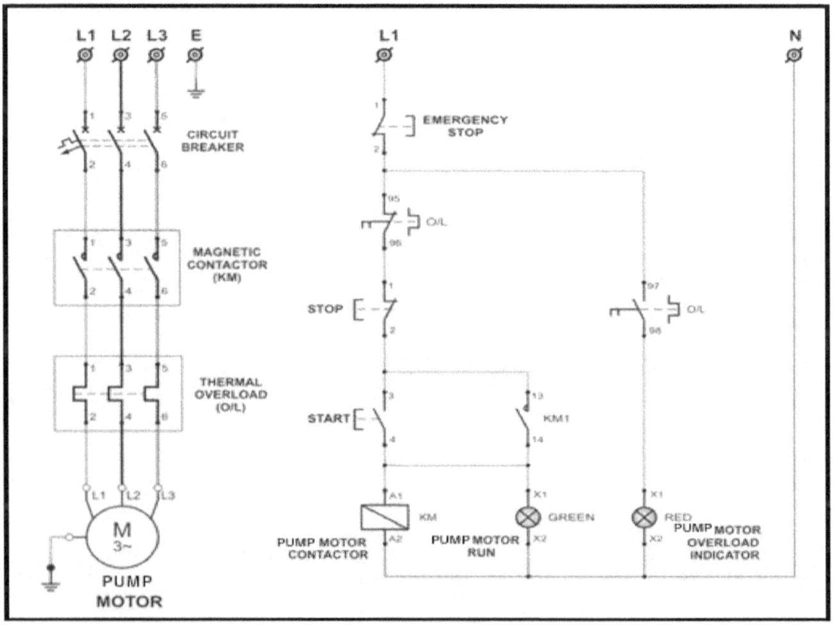

The pump motors' **thermal overload protection device** starts tripping intermittently, stopping the motor. (a thermal overload is designed to open the starting circuit and thus cut the power to the motor in the event of the motor drawing too much current from the supply for an extended time. The overload relay has a normally closed contact which opens due to heat generated by excessive current flowing through the circuit. Thermal overloads have a small heating device that increases in temperature as the motor running current increases). Under no circumstances should a thermal overload be adjusted above the current (amp) rating on the nameplate of the motor.

N.B. Never replace nor choose an automatic-reset thermal overload protected motor for an application where the driven load could cause personal injury if the motor should restart unexpectedly. Only manual-reset thermal overloads should be used in such applications.

Another problem develops; the motors' circuit breaker starts to trip. The problem is believed to be a motor problem. Despite standard electrical checks (correct voltages present on all phases and agree with motor rating plate and load factor, integrity of cable connections checks, winding continuity, resistance checks, and short circuit tests) indicating no issues with the motors windings, the motor is replaced, and the same problem reoccurs. **Considerable downtime has now elapsed, the problem still exists, and the fault finder is still at square one**.

If a good installation qualification (IQ) has been carried out with all the original parameters and recordings of the actual motors' performance is available (benchmark or baseline figures), use this as a starting point also check any PM records that may have been carried out on the motor. If over a period, the motor current or temperature rises inexplicably, investigate immediately before the motor fails completely. If no information is available, read the maximum current rating on the motors nameplate and measure the running current. If the motor is running at its maximum rated current or above, investigate immediately. If it is a new installation, questions must be asked **e.g.** is the motor rated for the mechanical load? Can the mechanical load on the motor be reduced until the problem is found and rectified? Is the motor itself suitable for the environment it was selected to operate in? e.g. poorly ventilated area or excessive environmental temperature conditions.

The actual problem is found to be that the pump that is being driven by the motor is mechanically overloaded due to an ingress of solids (making it harder to pump) thus causing the motor to work beyond its rated capabilities causing it to trip out on overcurrent.

The most common causes of pump failure are due to incorrect installation. The pumps' installation and operating instructions must be read thoroughly but with particular attention to the following points:

- Correct baseplate installation
- Pipework connection
- Pump/Motor alignment
- Impeller clearance/Imbalance
- Coupling fitting
- Direction of rotation
- Lubrication

Other causes of pump failure:

- **Pressure** - a pump will not operate properly without sufficient inlet pressure, the pump will cavitate.
- **Cavitation** - is caused by the rapid formation of vapor pockets (bubbles) in a flowing liquid in regions of very low pressure and collapsing in higher pressure regions, often a frequent cause of structural damage to the propellers or other parts of the pump.
- **Leakage** - leaks caused by mechanical failures can be catastrophic. Early detection of abnormal conditions such as

cavitation, pressure imbalance or excess vibration can help avoid leaks and their consequences.

- **Vibration** - there are several causes of vibration which can damage seals or internals and cause pump failure e.g. Pump bearings can fail due to overload, excessive wear, weather or substance related corrosion, failure of the lubricant, overheating, or contamination.

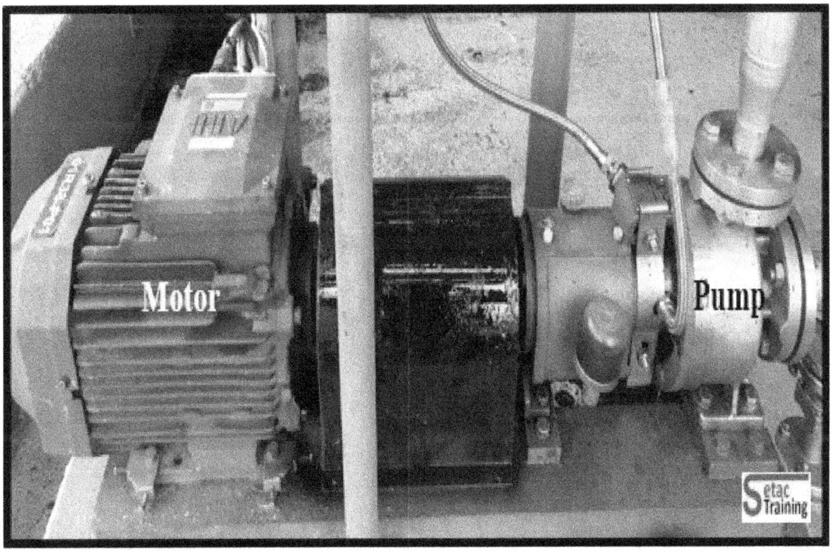

Example 2 – An 'AHU supply fan' feeds fresh filtered air to office areas.

Problem: *The 3 phase electric motor which rotates its 'fresh air intake' supply fan via belts and pulleys in the air handling unit (AHU) enclosure in* ***Fig.4*** *is experiencing high running currents and high temperatures.*

Fig. 4:

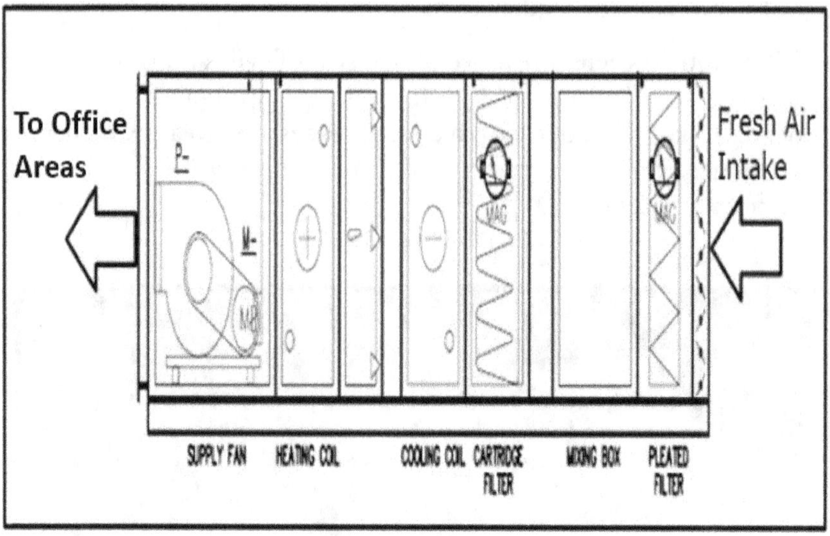
To Office Areas
Fresh Air Intake
SUPPLY FAN
HEATING COIL
COOLING COIL
CARTRIDGE FILTER
MIXING BOX
PLEATED FILTER

AHU internal 'Supply Air' motor/fan arrangement:

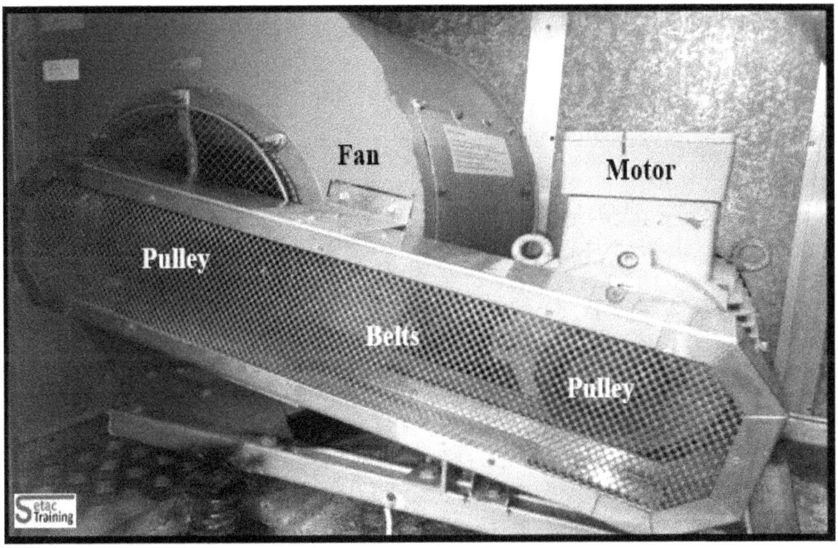

Isometric – Duct work layout:

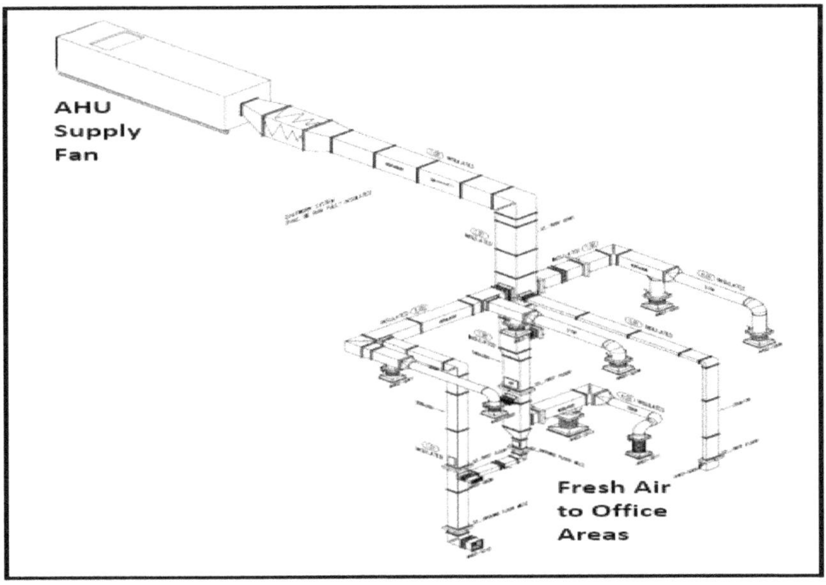

Check:

(i) Electrical current load against maximum load current rated on the name plate of the motor (a PM record of the motor must be kept and maintained). This log can help in determining what constitutes normal versus abnormal operations.

(ii) If the motor is running at its maximum rated electrical load current capacity or over its maximum permissible running temperature, check:

What protection relays are monitoring the motor's operation and if activated on a trip condition, will stop the motor:

(a) A voltage monitoring relay which guards the motor against the damaging effects of phase loss, under and over voltage, phase imbalance, phase reversal and voltage quality of incoming power line.

(b) A current monitoring relay which provides the motor with under and over current detection.

(c) The motors thermistor or thermal cut out over temperature protection device (if fitted, will be embedded in the windings).

The protection device (temperature transmitter) on the mechanical bearings (if fitted).

The motor is either overloaded or under rated. More than likely it will be the former, especially if the motor has been running normally and within operational parameters over a long period. There are many factors that can cause the motor to draw excessive current, over heat and cause electrical windings to burn out (insulation failure). Check the motor's resistance across its windings, the ohms reading for each winding must be the same (or nearly the same).

N.B. The common rule states that insulation life is cut in half for every 10°C of additional heat to the windings.

As an example, if a motor that would normally last 20 years in regular service is running 40°C above rated temperature, the motor would have a life of about 1 year. The following are some examples that may lead to premature aging of a motor and ultimately failure:

(a) Over voltage.

(b) Mechanical overload.

(c) Mechanical bearing failure.

(d) Inadequate cooling on the motor frame itself (poor cooling due to high ambient temperature, clogged ducts .... are typical examples of nonelectrically induced temperature stress on both the motor and insulation system).

It can also be on the fan side:

(a) Mechanical overload.

(b) Mechanical bearing failure.

(c) Foreign bodies on the fan.

(d) Belts (worn or broken) and pulleys (improperly aligned) which are driven by the motor.

(e) The fan blade itself may be imbalanced or loose on the shaft.

(f) Intake filters blockage.

Remove the belts between the motor and the fan. Start the motor on its own, if the motor is still running hot it will help exonerate the fan as being the problem.

The main questions to ask are: Is the motor circuit **tripping**? (MCB, thermal overload, motor protection relay or VSD tripped). Is the control circuit to the control equipment of the motor itself **interlocked** with a **safety device?**

E.g. A differential air pressure switch is fitted across the fan which measures the pressure differential between the suction and delivery side of the fan itself to prove the fan is spinning. If there is no pressure difference detected on the pressure switch after starting up the fan motor, normally after a period of 10 seconds, an alarm will be activated to shut down the AHU fans' control circuit which will in turn, stop the fan motor. All these **safety/trip monitoring devices** must have at minimum, alarm indicator lights with audible alarms attached to ensure the alarm is properly identified, recorded, and should remain active until accepted and reset by the maintenance personnel.

In other words, is it a **motor control circuit issue** or **a safety device issue?**

If it's a motor issue:

1. The supply voltage (400V) and the full load current (FLC) of the fan motor must first be identified from the motor nameplate – e.g. 400V/45 Amps.

2. The voltage must be measured with a Volt meter, a reading of 400V should normally be expected across all 3 phases. The current must be measured on all 3 phases going to the motor with a clip on ammeter. Normally, if all is OK with the windings of the motor and associated connections, the current readings across all 3 phases should be evenly balanced. The actual running current of the motor will be dictated by the mechanical load it is driving. A higher current usage could be indicative of a defective motor bearing and is **why it is important to record and document the base current load of a new motor installation for reference in the future.**

3. Heat and poor ventilation can have a very detrimental effect on a motor and may not manifest itself for hours after the motor is operational. Has the motor got PTC thermistor protection? - as the temperature in the motor windings rise and reaches the PTC sensor's temperature rating (which is embedded in the motor windings), the PTC sensor's resistance transitions from a low to a high value which triggers an alarm. The set point level which is set on a thermistor monitoring controller whose relay contact is wired in series with the

motor's starter control circuit, if activated, will shut down the motor itself. The thermistor controller galvanically separates the thermistors in the windings from the motor's starter control circuit (no physical contact between the 2 circuits).

4. Remove any accumulated dirt from the motor frame by wiping, vacuuming, or brushing and ensure the motor's air passages are clear. If a lot of dirt has accumulated on the motor frame and is not cleaned, this will lead to the motor running hot because the air passages are clogged up reducing cooling air flow. The heat will reduce the insulation life of the windings and will cause motor failure. **If the dust is conductive, it may cause a fire.**

5. If it's a Variable Speed Driven (VSD) motor, its associated VSD controller will show the actual fault on its HMI screen if any has occurred. VSD's normally have a thermistor protection circuit built into them where there is no need of a separate thermistor controller. The thermistor in the windings are wired directly into the VSD itself.

6. **No alarm condition should be automatically reset.** It must be a manual operation and the fault/trip must first be identified, recorded, and rectified to prevent reoccurrence before the motor can be brought back on line.

7. The manufacturers' motor and fan construction arrangement drawing and the **as built** electrical and mechanical drawings must be available. This information could help identify what could possibly be shutting down, interlocking, or tripping the fan motor.

A standard hot water duty/standby pump distribution loop system:

Fig.5 – Overview

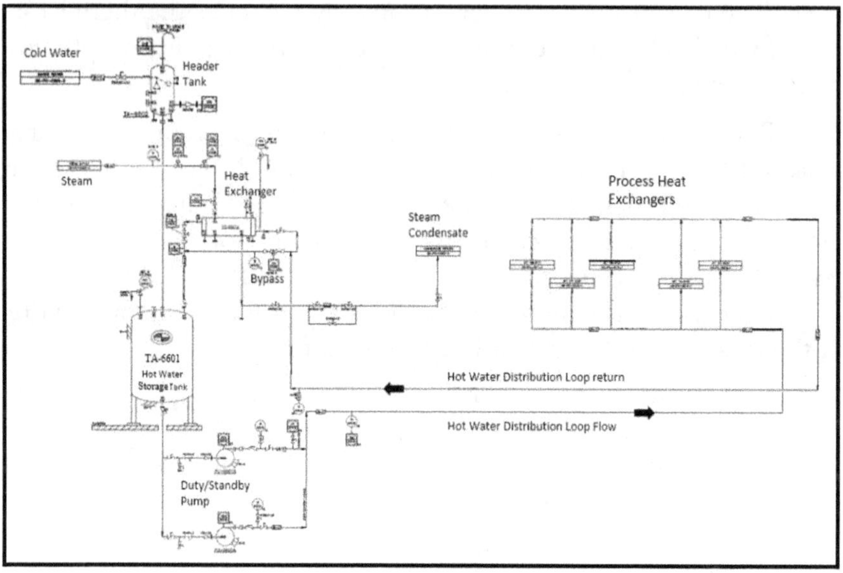

Fig.6 – Expanded "Duty/Standby Pump" view

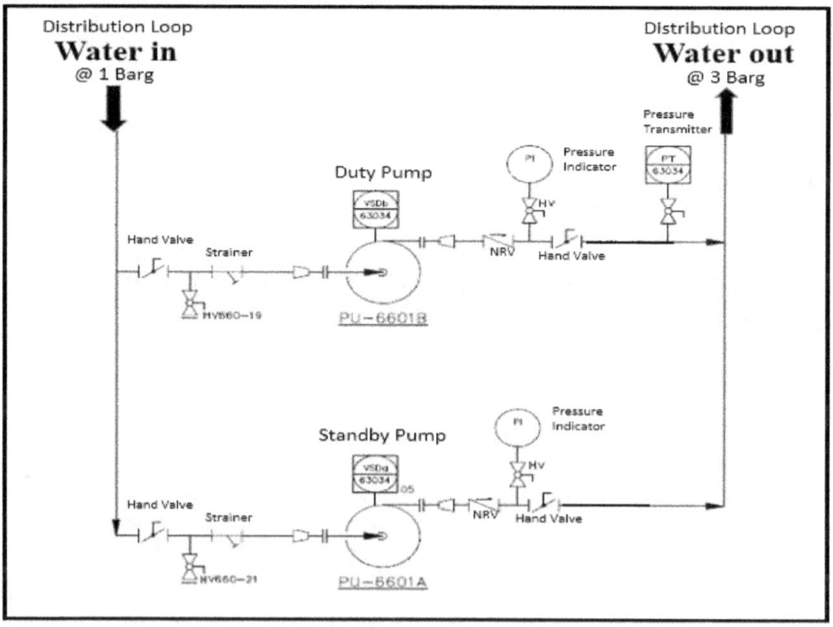

Photo of the hot water duty/standby pump distribution loop system arrangement:

Issues that may cause **'flow problems'** on a standard hot water duty/standby pump distribution loop system arrangement:

- Electrical isolator for the selected pump may be turned off or motor **Emergency Stop** pressed.
- The selected pump may be turning the wrong way.
- Local isolation hand valves either side of pump could be closed.
- If the water system pressure is not being maintained, check the speed of the **online** motor on either PU 6601A/B to see if it is running at 100%/50 Hz on the VSD.
- The **strainers** fitted on the suction side of the pumps PU 6601A/B may be blocked – isolate/remove and clean.
- Check that both motors are not spinning at the same time (look at the motor fans) even though power is only being delivered to one motor. The **Non-Return Valve** fitted on the delivery side of either of PU 6601A/B can stick open and cause possible **water short circuiting**. Water is just circulating within the duty/standby

pump arrangement (causing the non-powered motor to turn via its pump impeller) with very little hot water being delivered to the distribution loop and the units that require heating.

- Pump impeller could be damaged/clogged with debris/pitted.

Quick Test No.2: Complete the Circuit to make it work (Answer at the back of the book)

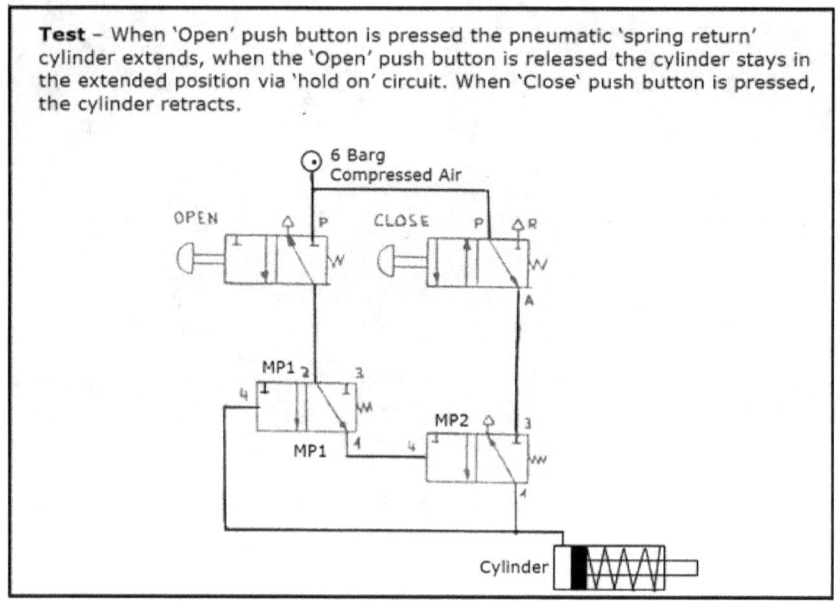

Test - When 'Open' push button is pressed the pneumatic 'spring return' cylinder extends, when the 'Open' push button is released the cylinder stays in the extended position via 'hold on' circuit. When 'Close' push button is pressed, the cylinder retracts.

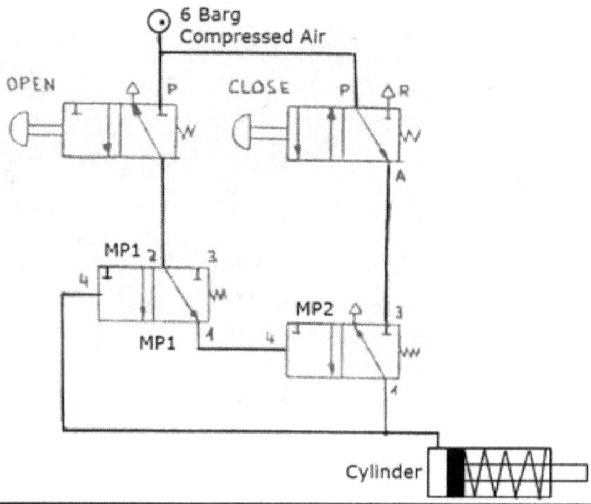

Tip: Avoid poor greasing practices. Bearings must be lubricated/replaced when scheduled on a PM or if they are noisy or running hot. **Do not over-lubricate**, excessive grease and oil creates dirt and can damage bearings.

N.B: A motor breakdown can cost up to **€10,000** per hour in downtime if it happens in a paper mill or another continuous process. Try to install a high-quality energy efficient motor where possible.

E.g. a high-quality energy efficient 55 kW motor costs in the region of €3,000. If it's a low-quality motor, it will cost a bit less. In a 24/7/365 continuous process, the high-quality energy efficient motor will use up its own cost in electricity during the first 30 days of operation. The low-quality motor will use more energy. Having been built from lesser quality materials; it is less able to convert electricity to mechanical energy and will instead produce more heat. Not only does it cost more to run, the heat will eventually start to break down the insulation materials inside the motor. This in turn may lead to a motor failure, with the ensuing downtime costs as well as the costs for a new or rewound motor. Meanwhile, the high-quality energy efficient motor keeps on running.

Think before making a purchase, do not compromise on motor quality, top of the range process performance motors use the best materials and manufacturing practices to minimise energy consumption and are designed to achieve a possible service life of up to 30 years.

You must be prepared to accept that there will be **Highs and Lows** during your career.

The '**High**' of overcoming and rectifying a serious fault, putting the plant back online, saving the company vast sums of money is very gratifying and also gives a sense of deep confidence in your abilities.

The **'Low'** will also be experienced. There will be a time when you believe you have all the bases covered, but an unforeseen problem arises which you can neither diagnose nor rectify. You may feel you have lost **face** in front of your colleagues. Doubts may start building up on your abilities and fault-finding methods. Learn from it, put it down to experience, and move on. If dedicated to a discipline, your professionalism will make sure these 'lows' are kept to a minimum.

Hindsight is a great thing, having diagnosed a fault which may have taken you hours to pinpoint and rectify. Then having to listen to your peers or the **Monday morning quarter back** telling you how they would have resolved the fault in minutes.

It is easy in the cold light of day, under normal circumstances, to discuss what you would or wouldn't do when a fault develops unexpectedly, rational thinking will prevail.

Words are easy to say where a list of possible causes might be rolled out by several people. It is a lot harder in the real world to prove each suggestion. Ultimately, one of the suggestions might be right and then having to listen to 'I told you that was the issue'. This type of fault finding is too **Hit and Miss**. Stick to your proper fault-finding methods no matter how tempting it is to 'take a chance and see what happens'.

Check, double check, and treble check any work you carry out. Be satisfied that all work completed was done correctly without compromising safety and to the best of your abilities, regardless, of the time pressures you may have been under to get the piece of equipment back on line. Try to 'audit' work done on a regular basis. Not only review actual work done, also look at subsequent failure & repair history reports.

One of the tests of being a good engineering practitioner is the ability to recognise a problem before it becomes an emergency. As the qualified person, it's your responsibility to make sure everything is running as intended. In doing so, you must keep a look out for any potential threats. **Think on your feet and adapt to the situation.** Following standard procedures is extremely important when rectifying an issue, but if an unusual one should arise, free thinking must play its part. Determine what the best solutions are, and act quickly and efficiently to resolve.

In extreme circumstances, if under pressure, tired, exhausted and time is critical in rectifying a fault, sometimes rational thinking may not be employed. There may be only a limited amount of time available, possibly minutes, **for critical decision making to be carried out** and to ensure the decisions made, are the right ones.

E.g. You may find yourself in a highly charged atmosphere (you are the ships' engineer, out at sea, the engines have failed, the ship is listing and dead in the water and there is a violent storm coming in, 20 lives plus the ship and its cargo are in your hands). Taking the wrong course of action, could possibly exacerbate the problem.

Running around throwing your hands in the air and shouting obscenities is not going to help. This isn't a time for covering your back, finger pointing or the blame game. Remember, apportioning blame is not constructive, look for ways to resolving the problem instead.

You must remember your training, remove outside influences and distractions from your mind, stay calm and focused. Investigative 'Lessons Learned' meetings of what went wrong can be held later once the issue has been dealt with. Preventative measures can be agreed to avoid reoccurrences.

Crisis Management is a situation-based management system that includes clear roles and responsibilities and process related organisational requirements company-wide. The response shall include action in the following areas:

- Crisis prevention
- Crisis assessment
- Crisis handling
- Crisis termination.

The aim of crisis management is to be well prepared for crisis, ensure a rapid and adequate response to the crisis, maintaining clear lines of reporting and communication in the event of a crisis and agreeing rules for crisis termination. A crisis mindset requires the ability to think of the worst-case scenario while simultaneously suggesting numerous solutions. Trial and error is an accepted discipline as the first line of defence might not work. It is necessary to maintain a list of contingency plans and to be always on alert. Organisations and individuals should always be prepared with a rapid response plan to emergencies which would require analysis, drills and exercises. 'Immersive Learning' will play a key role here.

TV documentary channels and internet videos of industrial explosions, fires, plane crashes and the associated human fatalities are very informative where you might think that this will never happen to me. **Think Again -** 70% of reported incidents in the oil and gas industry worldwide are attributable to human error, accounting for more than 90% of the financial loss to the industry. You may work in the Nuclear, Aeronautical, Chemical, Petrochemical industry where exothermic (process that generates its own heat) reactions take place.

E.g. As the shift maintenance engineer, you may be faced with a situation where the safety equipment designed to mitigate the consequences of an emergency has failed. The entire cooling system (including back up cooling system) fails during an exothermic chemical process (Reference example: Fukushima Daiichi nuclear disaster - Japan - 11 March 2011). A critical situation has now arisen, the process temperature starts to rise exponentially with a risk of explosion and the cooling systems must be put back on line as soon as possible. This is your **'Call to Action'**.

Delegate**,** if there are other personnel on site, ask them for assistance, whether they are technical or not. Let them take the phone calls you may be receiving from other site personnel warning of impending dangers. They must be told to write down all information accurately, who the person is and what phone number they can be contacted at. If the calls been received are only symptoms of the cause been worked on, stay focused on the task. Being bombarded with non-essential information can be very distracting. Deal with the **symptomatic issues** systematically according to their criticality once the actual **cause issue** has been rectified and the cooling is back **on line.**

N.B. Never put yourself or other staff at risk by going into a building where there may be a possibility of an explosion, gas leak, flooding .... actually occurring. Get out of the area immediately and alert the emergency response team (you maybe a member) who are professionally trained to deal with such critical issues. Remember, in the space of seconds you can be overcome by fumes or electrocuted should there be water streaming down into an electrical panel. You may be about to enter an area where you see another member of your team lying on the ground. It is a human urge to help but you must **'Stop and think before you enter'**. Always be aware of the dangers around you.

You may enter a control room where there are numerous flashing lights and alarm sirens ringing in your ears, warning of some impending danger unknown to you.

This can add complexity to an already highly charged situation. Be aware of **'Automation Confusion'** - you might be receiving misinformation from the instruments that control the process which feed into the control system or from the control system readouts themselves. There may be multiple erratic alarms going on and off. Can you trust them? First try to analyse conflicting information and try not to become 'fixated' on just one issue. You must try to decipher which alarm is **critical** and needs your immediate attention and which alarm is non-critical but needs you to recognise and be aware of it.

Don't panic, panicking will blur your thinking. This is where your 'Fluid Intelligence' must kick in. Ensure not to misread the information being received from the current environment. Review the standard operating procedures; ensure all settings on the control panel are as they should be.

Has a valve been opened or closed in error by personnel during operational checks? If one of a multitude of instruments on a control panel is malfunctioning or giving questionable readings, perform a primary scan of the entire control panel, and check all instruments and their respective readings. Are any other units not working properly or giving the same type of erratic readings? Go through them one by one, starting with the most critical instrumentation readings.

Flexible behaviour will play an important role in this situation. You must have the ability to adjust your thoughts to changing situations and conditions. Adapt and adjust your thinking to new information. Flexibility involves being able to train yourself to reinterpret unexpected situations that may at first give a sense of doom or alarm. Flexible people have the capacity to smoothly handle multiple demands, shifting priorities and rapid change. Be ever mindful of 'task saturation' **e.g.** trying to do too many things at one time. Some critical action on a check list might be missed, which could make the situation worse. If a person can't assess what's going on in their environment, they'll have difficulty adapting their responses to this type of **real time** information.

If there are others also trying to solve the problem, don't suffer from **'collective brain thinking' -** everyone focussing on one issue, which may be only a symptom of the actual cause. Don't get caught in this trap, all involved must take in as much information from the environment as possible. Disseminate it and make decisions based only on accumulated **accurate** information.

Mental Gymnastics may start to kick in. There may be a 'total disconnect' between what you know should be happening, and yet the exact opposite is occuring. Your **Mental Flexibility** enables you to adapt to unfamiliar, unpredictable, and fluid circumstances. Try not to become confused and make rash decisions. Remember your fault-finding discipline must be adhered to. ***Manage the problem.*** Watch out for patterns and try to logically understand exactly what is to be done and the consequences of those same decisions before any actions are taken and ensure all decisions made and actions taken are recorded to be reviewed after the crisis has passed. The first objective is to get the piece of equipment stabilised and in a safe operational mode. Figure out what is controllable, what is operational and not operational. Don't take short cuts or bypass a safety system without fully understanding the ramifications of such activities.

N.B. Only propose solutions when the **cause & effect** of any actions carried out are fully understood.

You must stay focused, calm, and remember your training and above all, give answers, based only on accurate engineering fact.

It is inevitable that sometimes people will be stressed and tired and will have to make key decisions! So, remember to:

1. Design processes and procedures that are brain friendly.
2. Don't over load the logical brain, we're not very good at multi-tasking, it increases stress which in turn kills your brain cells and your thinking.
3. When your stress levels have been exceeded you can't think rationally and perform your task right. Be wary of the effects of stress on your performance.
4. Use pictures, diagrams, and schematics, in fact anything else other than words.

These cases may sound extreme, but, all events happened. Hopefully, you will never be confronted with any similar type events, but having been trained properly and to avert disaster, the feeling alone from this will be very satisfying.

The above examples stress the importance of knowing the plant's systems inside out, proper training, work practice and preparation on how to handle critical situations. Imagine yourself being in this serious situation and not having the proper training and not knowing what to do, lives, including your own, could be lost as well as severe damage to the plant. The costs associated with such an accident can be astronomical. The environment and the company's reputation could also be affected. **Adequate expert training must be given.**

A person cannot be expected to trouble shoot a problem on a piece of equipment, if they have not been taught or trained how to deal with the problem in the first place or if inadequate alarm systems warning of impending danger or systems operation failure are not installed.

Insist on proper training. Asking a person to self train on a complex piece of equipment is **not recommended** as this person cannot train properly and be competent to work with this equipment when **they don't know what they don't know.**

In any industry, whether it is a manually intensive production line or a highly automated production plant where product output is paramount and making money may be the dominant factor, **'Safety'** sometimes, can be over looked. Watch out for hazards, identify, report and rectify them.

Don't wait for an accident to happen. Even if the piece of equipment is running without any operational problems, it doesn't mean it is running safely.

Regardless of the time constraints or how much pressure you are put under by superiors to get a piece of equipment operational, **you must not compromise on your experience, knowledge and know how**. <u>If you know it is not safe</u> to put a piece of equipment back into service, you **must not** let a superior over-ride your decision. Ensure any concerns are documented and followed up if decisions are over-ridden.

N.B. Use Best Engineering Practices across all disciplines. It should become the exception rather than the rule **not** to use these practices.

Part of maintaining a plant will be written checks (hard copy or electronically). These will be carried out on pieces of equipment throughout the site using GDP (Good Documentation Practices). You must clear your mind. Make a conscious decision to take your time. Write neatly, legibly, and fill in the correct information. Leave no boxes unfilled and ensure to sign and date all written information carried out. **E.g.** When writing a cheque, the currency amount is filled in, crossed, signed, and dated, the same goes for a standard maintenance check.

These checks can prove very useful should a problem occur on a piece of equipment. The archiving of this information can assist in fault finding if a subtle but changing pattern manifests itself on the equipment over a period of time e.g. **contemporaneous notes** (are notes made at the time or shortly after an event occurs. They represent the best recollection of what you witnessed).

N.B. These records will be crucial should an accident or incident occur. This is one of the first documents (maintenance records) along with training records that will be reviewed by incident or accident investigators.

Your signature is your bond and by signing a maintenance check and approving it, you are signing that everything is in order, the equipment is within acceptable running parameters, is fit for purpose and suitable for operation. Signatures are incomplete without documentation of the date of the signature.

N.B. Never sign your name to a check that you didn't witness or carry out.

In the technological world we live in with computer screens and their associated control systems and instruments now operating and monitoring industrial plants and office buildings. It is easy to sit in a control room and assume that everything is under control. 99.9% of the time, these systems work flawlessly, but, if erratic readings are showing up on the computer screens (unexplained high or low temperatures, pressures, flows) .... go on plant if possible and physically check the reason for the erratic readings. **Also watch out for flat lines on data trends on PC screens where there would normally be slight oscillations.** Flat lines can be indicative of a faulty reading (possible faulty instrument issue, open or short circuit on the control cable between the instrument and control panel, blown control fuse or faulty PLC input card).

Don't fall into the trap of always taking for granted and assuming that the computer read outs are correct. Automation systems are programmed and can only react to known operational and safety conditions that have been envisioned by the engineers during the initial design. The automation system then acts appropriately to deal with the issue. It can do nothing about an unforeseen condition that was not anticipated by the designers. A condition such as this can have a devastating effect on a piece of equipment's safety and operational system and will not be detected by the automation system, as it is not programmed to do so. Don't let flippant remarks from others such as, "It always does that …" satisfy your curiosity, **investigate and report.**

Modern control systems have fail-safe systems including, duplicate watchdog systems, duty/standby, redundancy and self monitoring hard wired critical variables of the equipment which are flagged, alarmed and actioned by the control system that monitors them. Never underestimate the human factor in any control system, i.e. **to physically check any fault on plant regardless of how inaccessible a piece of apparatus may be** (CCTV equipment may be used to observe parts of a process from a central control room, for example, when the environment is not suitable for humans).

Always insist on the **best turnkey project** installation possible **e.g.** the more 'data trended' instruments giving real time intelligent feedback information that are fitted on installations, the easier it will be to keep the plant running efficiently and if a fault develops, these instruments can help the fault finder to diagnose the fault quicker. Remember, a control system is only as good as the instruments that input information to it, so it can monitor and control the outputs accordingly. Data acquisition is essential. The key is to maximise uptime and this is best achieved if you are armed with pertinent machine data.

Summary

- Always be proactive in failure prevention.
- It is critical you understand a piece of equipment's **Principle of Operation**. When you know how it works, it is easier to trouble shoot a fault should it appear.
- There are many different types of faults, from minor to major. Ensure any action you carry out does not exacerbate the fault.
- The extent of preparation should be directly proportional to the size of the problem to be rectified.
- **Do not** make whole scale changes, take a structured approach, and take one step at a time.

Don't start changing parameters (P.I.D settings) in controllers, pushing buttons, turning knobs, closing/opening hand valves if you are not positive you know what you are doing.

- Exercise caution. Use logical and deductive reasoning always.
- An intermittent failure if left unaddressed is likely to crop up repeatedly until finally, corrective action is taken to rectify.

8. SPARE PARTS

A properly managed supply of spare parts and materials can make a significant contribution to profitability, not only through its role in the minimisation of downtime losses but also (and not unimportantly) through the minimisation of the costs of acquiring and holding the inventory of spares itself. **Spare parts form the bedrock and are the lifeblood on which plant capacity and operational reliability is built.** A fully automated machine will normally have all the latest technologies attached to it, if any part fails and stops it, a full inventory of critical spare parts must be available.

Technical training and knowledge are a huge advantage, being able to diagnose a fault quickly is key, but, if in the event of failure, the spare parts are not available to perform a quick turn around, the machine will remain inoperable. The lack of spares can be at the heart of a company's down time problem. The impact of not holding a required spare part can be massive. Downtime costs often far outweigh the purchase and holding costs of spares.

Spare parts and operational reliability are intrinsically linked, this requires appropriate storage, treatment of, and timely access to the required parts. Yet, spare parts are also the most overlooked contributor to reliability outcomes.

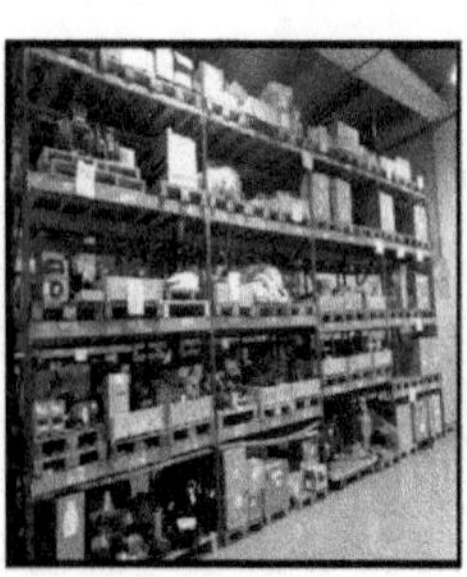

Tip: Try to standardise on pieces of equipment, this will keep spare parts to a minimum and avoid keeping costly pieces of equipment sitting on shelves in the engineering store area. Can you use some other company's stock (share/pooling)?

Maintenance inventory must be managed in a way which adds value to the business and its bottom line. Ensure properly trained experienced **Stores and Procurement** personnel can manage and supervise the process of setting and managing maintenance inventory from stock creation, through stock control and onto stock disposal.

Personnel must understand:

- Why companies must carry Inventory & Spare Parts
- The costs associated with holding Inventory
- Equipment Criticality & Failure Modes
- Risk of Plant Failure and how to assess
- How to develop a Maintenance Bill of Materials (BOM), Manufacturers recommended spares, Design Engineers recommended spares, Risk based approach to spares holding
- The various Inventory Models, Types & Systems
- Inventory Reduction Techniques
- The Inventory Management process to enhance Maintenance and company performance.
- Inventory Management Control (Processes, Technology – systems, bar codes & RFI (Request for Information), Economics, KPI's)
- Warehouse Design & How to Store Stock.

No company can operate at a high level of output without a reliable supply of functional spare parts. A set quantity of all critical spare parts (with dedicated individual part numbers) must be kept on engineering stores shelves. When the minimum amount of any spare part has been reached an automatic reordering of the spare part should be initiated and restocked if possible via **a dedicated computer-based engineering stores management software system.**

An engineering stores management software system allows the protection of a company's investment in plant equipment. This type of system will ensure a company manages its engineering inventory. Downtime will be reduced; engineering costs will be kept under control and investment in engineering spare parts will be minimised by having such a system.

Key Benefits

- Gain a clear visibility on spare parts and consumables usage via electronic stock management system bar-coding and automatic email communication process guarantees stock levels are upheld.
- Facilitates the processing of all transactions accurately and efficiently - less paperwork.
- All purchasing and associated costs of spare parts ordered and managed via PO - no PO/no spare part.
- Analyse engineering stores effectiveness through KPI Reporting.
- Improved system information quality with key information available to other key users - finance department/ external auditors.

Tip: Pay sufficient attention to storeroom security, do not leave your inventory data management in hands of non-stores personnel who may be more concerned with other issues rather than storeroom documentation.

If asked to maintain a piece of equipment or a vendor supplies a new piece of equipment, ask about the spare parts and possible asset obsolescence. Do they keep a minimum stock on their shelves? How quickly can they arrive on site if needed? This could save the company having to keep expensive parts on plant **just in case**. Ask about the after sales service they provide.

Vendor ***General Terms of Guarantee*** *example:*

We hereby guarantee that all our products are manufactured using superior quality raw material and components and that our manufacturing processes are controlled by extremely accurate quality assurance; from the material choices until the product is ready for use. All our products and accessories are designed and manufactured according to the EN standards and therefore flawless operation is guaranteed under appropriate conditions specified in the EN standards. Furthermore, we guarantee that all our products are carefully tested in our own factories before delivered to the customers. We kindly ask you to read and follow carefully the use and maintenance instructions, enclosed with the product. Correct installation, maintenance, and use ensure long-term durability of the product.

Generally, a company tends to get what it pays for, so when choosing equipment which will last for 5, 10 maybe 20 years, quite often it is not the cheapest supplier that should be considered but the supplier that provides the best overall value when machine specifications and support are considered.

Replacing components with non authorised parts could lead to a sharp reduction in performance, a risk of premature failure and could possibly compromise safety and quality. Always work, where possible, with **best in class** suppliers whose products, technical innovation and support can be fully utilised when needed.

Tip: Vendors will expedite **spare parts orders** much quicker if a customer specifies the order number of the machine part or plant. The following guidelines should be observed:

- Spare parts should only be ordered in accordance with the vendors' recommended spare parts list that accompanies the equipment
- Orders placed by telephone to the vendor must always be confirmed in writing with a valid purchase order (PO) supplied by the customer before delivery.
- Ensure to include cost of **delivering** the actual spare part to site in the purchase order, as the associated costs can be substantial

- If a vendor receives orders without the above specified information, they will not normally guarantee correct delivery. They will not bear any expenses which are incurred due to the need to **exchange parts** which can be very expensive especially if it involves large items coming from overseas
- Delivery of parts ordered in accordance with the spare parts list shall be made based on the vendors **General Terms of Sales and Warranty**.

By observing these points, it makes it easier for the vendor to process orders effectively and to provide fast delivery.

9. FACILITIES MANAGEMENT

ISO 41001:2018 focusses on building management and was implemented to establish the requirements a facility management system must meet in organisations that:

1. Want to demonstrate the efficiency and effectiveness of facility management, and what it contributes to the organisation's goals.
2. Pursue consistency in defining the requirements and needs of all the parties involved in the facility management process.
3. Aim to be sustainable in a highly competitive environment.

Facilities Management represents a continuous process of service provision to support a company's core business. Facilities Managers are now facing more challenges than ever. Chief among these are:

- The need to optimise the performance of assets to ensure business continuity
- Extend equipment lifecycles without unacceptable loss of equipment performance
- Develop strategies to reduce operating costs and generally, contribute to the company's revenue flow
- Efficiently collect, store, secure, analyse and operationalise data to generate value and maximum benefit for the company
- Workplace Wellness. The overarching aim of an FM must be to enhance and provide an optimal work environment that meets the needs of employees and visitors through smart technology e.g.

Buildings should be assessed and designed to promote more active, comfortable, and productive lifestyles.

- It's an FM's job to ensure a building and its services function reliably, efficiently, and effectively.

A Facilities Manager wears many hats:

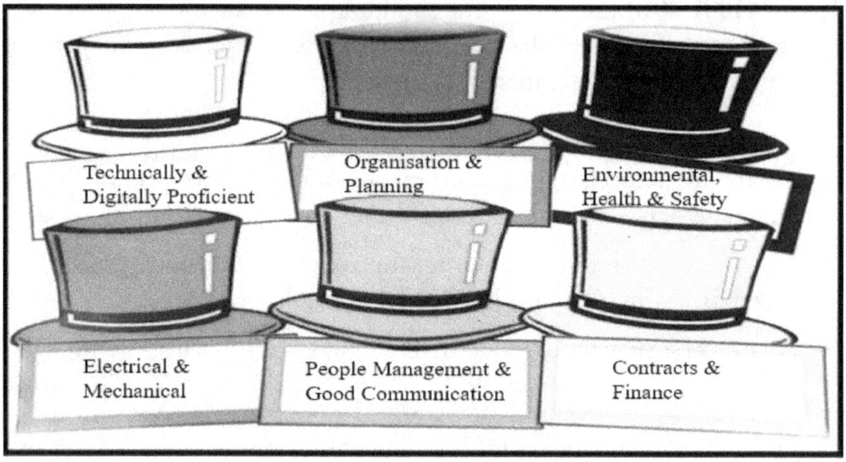

On top of this, technologies are shifting and there's an ever increasing body of regulation to comply with, covering areas from health and safety to environmental protection. The Facility Manager must be focused on the End User/Occupier's workplace demands while optimising operational (OPEX) and capital expenditures (CAPEX). A company must try to create a new mindset among employees to take a new look at processes based on real time information and business knowledge. Implementing **Continuous Improvement** (CI) projects, involving onsite personnel, who know the associated systems and processes, will play a key role in its future success. Your industry must be as safe and efficient as possible to compete in the global market place .... key factors to ensuring these requirements are the high efficiency, maximum reliability and optimised performance of your plants' assets. As maintenance techniques become increasingly sophisticated, Building Managers are presented with an opportunity to take a more positive and proactive approach to **asset life-cycle management**, viewing it as a profit centre, rather than an expense. The total life cycle of any engineering system is the most important factor to consider. This is the overall cost of the system over its entire lifetime and includes not only the initial capital investment but also the:

- Cost of installation
- Maintenance
- Servicing
- Energy consumption.

Only by taking all these factors into consideration and adopting a comprehensive approach to a new system installation, is it possible to benefit from all the potential savings.

N.B. You must monitor assets closely to understand what normal performance looks like and ensure it is digitally captured and archived for future reference. Looking at degradation, system alerts and **Mean Time Between Failures (**MTBF), you can start to predict through learnt performance data when assets might falter and service them in advance of that point. The benefits to this approach are numerous and include cost savings, better asset performance and reduced downtime.

Smart technology and IoT are transforming the future of industry, energy, mobility, cities and work. For example, the majority of today's assets in processing plants using the Asset Management model already deliver much more information than just a single process value. This additional information can range from more process values to self-diagnosis about the asset's health or even the prediction of potential problems that might occur in the near future. Fully networked "things" will simplify business life in the future. Wi-Fi has removed the human-factor and has given assets the power to talk to each other, resulting in the joined-up capturing and sharing of information. Products become assets that can be connected to the internet, which means they can be remotely monitored **E.g.** IIoT provides FM's with the ability to virtually monitor and optimise the condition of their entire water network using environmental sensors, energy meters and building automation technology. These devices, in real time, can relay information on their performance, condition, and environment. This data helps companies to better understand how assets are used, to predict when they need maintenance, and to sharpen the total cost of ownership models. By knowing when faults can be avoided or their impact minimised, and with a better understanding of risk and financial exposure, companies can establish service-based contracts with service-level agreements based on reliable data.

Building Management Systems (BMS)/Building Automation Systems (BAS) and HVAC systems have an intuitive interface that allows a wide range of monitoring, reporting and diagnostic tools across multiple sites in a single integrated system. Current generation BMS systems are now based on 'open communication protocols' and are 'WEB enabled' allowing integration of systems from 'multiple systems vendors' and access from anywhere in the world. Ubiquitous connectivity allows you to visualise and manage data, creating detailed analyses and predictions for Condition based Maintenance (CbM), which allow for a flexible maintenance model that can be modified for individual assets.

Example:

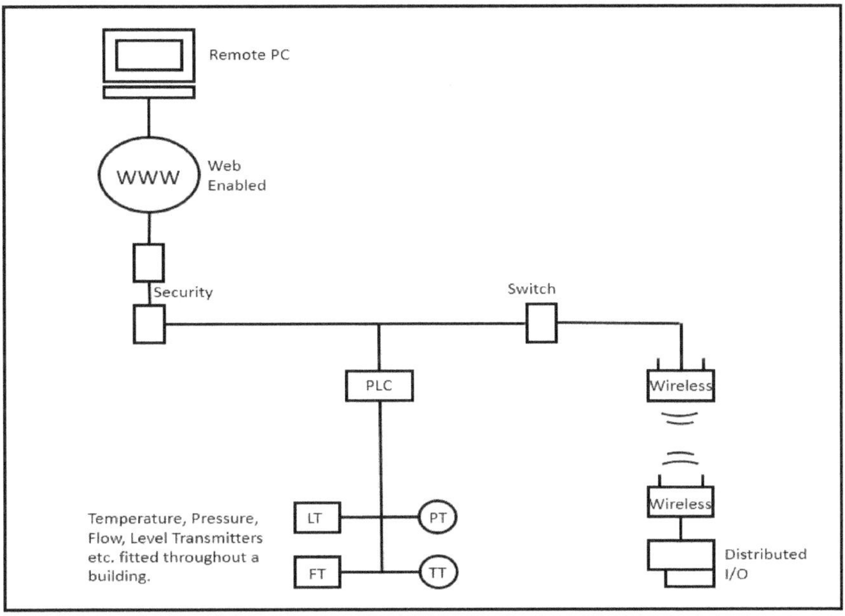

Invest in Technology for Asset Utilisation and System Safeguards

1. Progressive companies realise how **Automation** can:

- Help increase asset utilisation and availability
- Deliver Smarter Utilities
- Improve business performance.

2. Technology assists **Operation Personnel** to learn new ways to help increase plant uptime and to safely optimise production via operator interface technology.

3. Technology helps **Maintenance & Reliability professionals** to streamline work procedures and become more predictive thereby increasing their asset utilisation.

4. **Intelligent Feedback** is essential for asset utilisation and availability. It also simplifies the operational and decision-making processes. Intelligent feedback is also key in ensuring comfort and safety functions in buildings.

Soft FM Services

All businesses require a range of services in order to operate smoothly. Efficient **Soft Facilities Management** is important for a thriving business. It involves delivering such utilities, ranging from waste management and cleaning services to pest control or office moves and redecoration. Soft FM not only provides a cleaner environment to work in or visit but helps to make a far more productive working environment (e.g. ergonomic solutions to employee's workstations and equipment tailored to their needs and comfort) whether in an office, cultural building, or any other sort of property. It deals with:

- **Waste:** Effective waste management includes dealing with sanitary waste, recycling, and disposing of confidential waste efficiently.
- **Cleaning:** Whether general or industrial cleaning is needed, cleaning services must be completed to the best standard.
- **Security:** Protection of employees and the business normally comes under the control of the FM department, in particular, the maintenance of security hardware. Regular maintenance of security systems is necessary to ensure that the equipment operates at optimum performance. It is necessary to re-evaluate the effectiveness of individual security measures and the complete security system periodically.

The FM department security staff must have received market sector specific training so that they can deliver a full range of security services from security surveillance and protection of personnel and property, to the provision of bespoke, integrated solutions combining manned and systems-based security to maximise protection for the business.

FM security services include:

- Manned Guarding
- Mobile Patrols
- Key Holding
- Alarm Monitoring and Response
- CCTV and Intruder Alarm Systems
- Access Control
- Remote Monitoring
- Security Screening
- Car Park Management

- **Aesthetics:** Involves repairing small scale damage to external and internal areas or arrange for full landscaping or in-house decoration. It's important to keep the inside and outside of your building looking its best.
- **Office moves and set up:** can be very disruptive, it's the FM's job to ensure it is carried out smoothly and quickly, including dealing with waste management or recycling old furnishings.
- **Catering:** This includes kitchen equipment, vending or full menus and eating areas for clients, staff, meetings, visitors, special events or hospitality functions, quality food solutions and professional catering staff.

Pest Control: Any premises, residential or commercial, can experience a pest problem or infestation at some stage. Among some of the more common pest problems are Rats, Mice, Woodworm, Flies, Cockroaches, Pigeons, Wasps. A good service provider will provide solutions for every pest problem and will use only safe, registered, and user-friendly products. They will also use the latest technology such as Electronic Reporting and eMitter Permanent Pest Control Monitoring to ensure pest control is managed in the most efficient and up-to-date manner possible.

Hard FM Services

Involves **Reactive and Preventive** (short and long term) **Maintenance** services revolving around short term reactive fixing or mechanical and electrical planned Preventive Maintenance (PM). The latter services are longer term and include management of power distribution systems, lighting systems, lift servicing, fixed-wire testing, HVAC etc.

Reactive/short term management services:

- **Power issues:** unexpected power issues must be dealt with efficiently e.g. tripped individual circuit breakers or RCD's. Finding, investigating and resolving the fault and knowing where to reset the tripped breakers quickly
- **Lighting issues:** unlit or badly lit areas must be addressed to ensure adequately lighting is provided e.g. replacing flickering lights (incandescent bulbs or fluorescent tubes)
- **Lift and door maintenance:** something so simple can have such an effect if broken. Entrance doors or lifts must be quickly repaired so that there is as little effect on your workforce or visitors as possible
- **Cold and hot water services:** ensure correct water temperatures and water pressures are maintained at their required set points throughout the building(s) e.g. wash hand basins and showers
- **Heating systems:** Boiler Maintenance - a reputable firm must regularly service your boiler(s). Gas-fired boilers should be serviced once a year; oil boilers twice a year. A regularly serviced boiler can **save as much as 10%** on annual heating costs
- **Cooling systems: e.g.** Air Conditioning Units Maintenance – your service provider must regularly service your air conditioning unit(s) at least once or twice a year
- **Drainage fixes:** issues with partial or total blockages must be resolved quickly e.g. sinks, urinals, toilets

- **Painting and Decorating:** regular upkeep is what it takes to preserve, beautify, and extend the life of your property
- **Building fabric and joinery repair:** scuffed carpets that pose a safety hazard or broken drawers must be repaired or replaced
- **IT/communication/audio visual issues**: quickly deal with and reduce interruption to **day to day** operations due to random problems with computers, phones, internet, or other communication devices

Preventive/longer term management services:

- **MV/LV Power distribution network** e.g. transformers, switchgear
- **Backup power generators and UPS systems**
- **Lightning protection:** Future proofing your building and your business equipment from lightning strikes by fitting 'Lightning Protection Systems' e.g. Surge protection and lightning arrestors.

Test points, and earth network Earth resistance testing by qualified personnel.

- **Fire detection and suppression systems:** Future proofing your building and your business from potential fires, including (setting up and regularly testing) smoke/heat detectors, break glass units, water sprinklers, fire extinguishers, assembly points and interior signage
- **Lift Servicing and Inspection:** Lifts (passenger, goods, or both) are subject to a 6-monthly **thorough examination** by a competent lift equipment professional person after which he/she must issue a report of the examination which contains all the information prescribed in the Regulations

- **PAT and fixed-wire testing:** Portable equipment must be periodically assessed, tested and certified under a PAT programme.

The frequency of testing and certification must be in accordance with legislation currently in force. PAT testing ensures that portable and transportable equipment is maintained in a safe condition to avoid any hazard to person's or property.

It involves the periodic testing and visual inspection **(Protection Through Inspection by a competent person)** of any electrical system(s) that conducts electricity around a property or building. This includes any hard wiring and items such as main panels, distribution boards, lighting, and sockets. The risks to employees, employers and businesses are very real – not only from the risk of electrical shock but also the risk of fire caused by faulty appliances.

According to PAT testing regulations, if the certificate of the competent person indicates that the portable equipment tested was 'not safe and without risk' on the day of the test, the employer shall ensure that the equipment is not used until it is made safe and certified. The results of PAT inspections and tests must be recorded and kept available for 5 years from the date of inspection. These must be available for inspection by an inspector and access is made available to users of the equipment upon request.

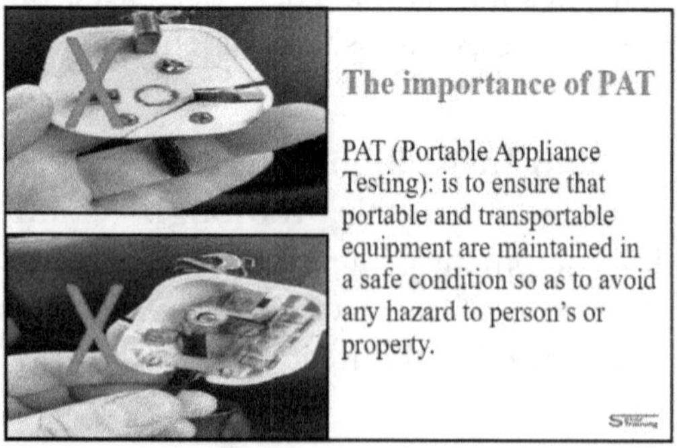

The importance of PAT
PAT (Portable Appliance Testing): is to ensure that portable and transportable equipment are maintained in a safe condition so as to avoid any hazard to person's or property.

- **Heating, Ventilation, and Air Conditioning (HVAC):**

A **Building Management System (BMS)** is based on a network of controllers which offer closer control and monitoring of building services performance, including, heating, ventilation, and air conditioning. This is graphically shown on a mimic computer screen (Fig.1) in real time: AHU's; pressures; temperatures; flow rates; and relative humidity parameters. It also shows AHU fan motor speeds and 'differential pressure indicator status' fitted across the AHU intake and extract fans and associated filters.

Fig.1: Typical BMS HVAC PC Screen

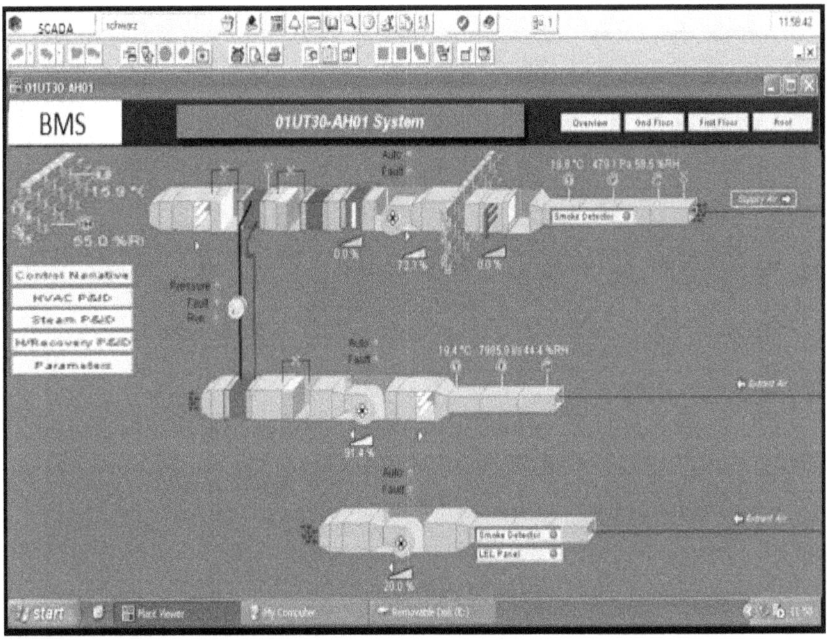

Air Handling Unit's (AHU's) (Fig.2) - heat, cool, humidify, dehumidify, ventilate, filter, and distribute air in a building. Ventilation systems supply air to the space and extract polluted air from it. It involves installing efficient heating, air conditioning and ventilation systems, making your building's environment appealing and comfortable to workers and visitors alike.

Fig.2

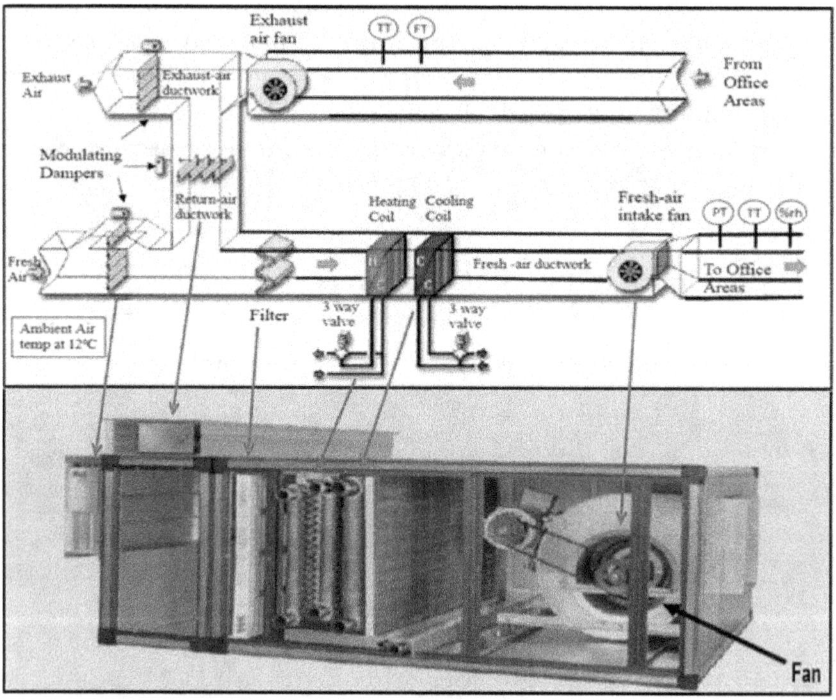

The demand for air conditioning in buildings is growing rapidly in response to more intensive building use.

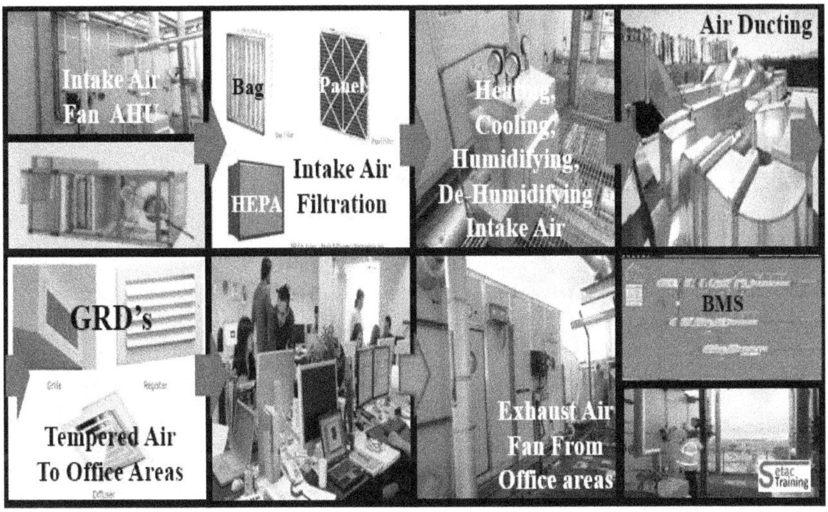

Buildings must be both properly ventilated and energy efficient. Every Euro, Dollar, Yen etc. still counts for most businesses. **HVAC systems by nature are energy-intensive in thermally treating air.** HVAC systems in commercial buildings are responsible for almost a half of total energy use. Unnecessary ventilation can waste energy and cost you a lot of money. For example, ventilation accounts for around 30% of heat loss in most commercial buildings (an estimated 25% in industrial buildings).

Tip: Understanding how air flows in and out of your building and the quality of that air is the first step towards designing systems that help you control improvements in the energy and occupational performance of your building relating to air. The internal fan itself is basically an air pump that creates a pressure difference and causes airflow. An **ambient air survey** will help you assess the Indoor Air Quality (IAQ) of your building and help you understand, control and evaluate the air allowing you to improve energy efficiency and the health of the buildings occupants. Buildings have two lines of defence: its thermal envelope and its air handling units. The thermal envelope is designed to keep thermal comfort heating and cooling in, while simultaneously keeping outside air and contaminates out. The air handling units, meanwhile, introduce the regulatory required amount of outside fresh air into the building.

As well as the need to keep a building's running costs in check, maintain compliance and reduce downtime, government targets and the public consciousness around sustainability are also pushing the envelope to maintain key assets in an intelligent way.

The purpose of a HVAC system is to maintain an environment that ensures the comfort of its occupants. Well ventilated workplaces are more comfortable, healthier and more productive places to work. Poor air quality and incorrect temperatures in buildings can result in loss of productivity, absenteeism, and in some cases, medical problems. Indoor air quality includes temperature, odour, and high or low levels of gasses.

The Goals of a HVAC System:

1. Controls Temperature
2. Fresh air circulation
3. Air Filtration
4. Efficient and Economical
5. Unobtrusive and Quiet
6. Heat Recovery System.

An important factor in fabric design is what occurs inside the building. If the activity and equipment inside the building generate a significant amount of heat, the thermal loads may be primarily internal (from people and equipment) rather than external (from the sun). This affects the rate at which a building gains or loses heat. Building Configuration also has significant impacts upon the efficiency and requirements of the building fabric. Careful study is required to arrive at a building footprint and orientation that work with the building fabric to maximise energy benefit. Higher concentrations of people and equipment will generate more heat making air conditioning or recirculation of air more important than providing heat.

Human Comfort is subjective and can be defined as a state of satisfaction a person has with their environment, however, many factors can contribute to this:

1. Air temperature.
2. Relative Humidity.
3. Air movement.
4. Air quality.
5. Acoustics.
6. Aesthetics.
7. Clothing.
8. Work space dimensions.
9. Lighting.
10. Activities.
11. Clean toilet and washing facilities.
12. Rest and eating facilities.
13. Emotions & Perception.

To have 'thermal comfort' means that a person wearing a normal amount of clothing feels neither too cold nor too warm. Thermal comfort is important both for a person's well-being and for productivity. It can be achieved only when the air temperature, humidity and air movement are within the specified range often referred to as the 'comfort zone'.

Where air movement is virtually absent and when relative humidity can be kept at about 50%, the ambient temperature becomes the most critical factor for maintaining thermal comfort indoors. However, temperature preferences vary greatly among individuals and there is no 'one temperature' that can satisfy everyone. Nevertheless, an area which is too warm makes its occupants feel tired; on the other hand, one that is too cold causes the occupants' attention to drift, making them restless and easily distracted.

For most area types and activities, it is recommended to maintain a minimum fresh air supply rate of **10 litres per second per person** into building through outdoor-air intakes. Typical 6/10 air changes per hour is used for offices.

Even minor deviation from comfort may be stressful and affect performance and safety. Workers already under stress are less tolerant of uncomfortable conditions. Maintaining constant thermal conditions in buildings is important and should be controlled and monitored via a Building Management System.

The optimum 'comfort zone' environmental conditions to provide a healthy and effective environment for people working in office areas are:

- Temperature: 21 - 23°C
- Relative Humidity (RH): 40 – 60%
- Carbon Dioxide (CO_2) Levels: <600 ppm
- Air Changes Per Hour: 6 - 10 ACPH
- Noise Control: 30 - 40 db.
- Lighting: 500 – 700 lux

Car Vs Monitoring display unit of three of the environmental office conditions in 'real time' that are outside the optimum conditions above:

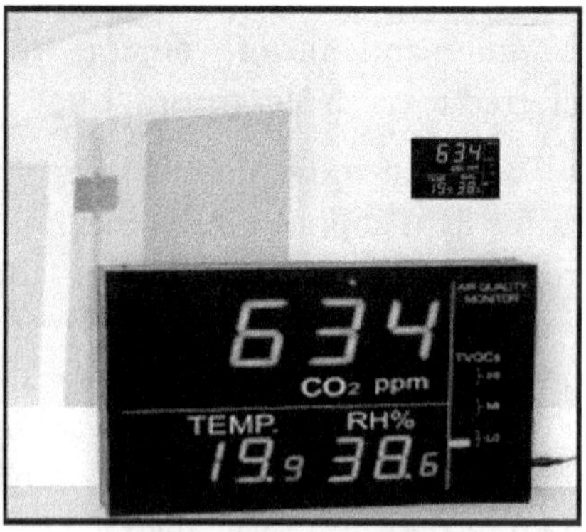

E.g. Controlled Environment - COMFORT is truly what vehicle passengers and workplace personnel want. Automotive HVAC systems have done a great job of adjusting a car cabin's temperature. A passenger sets the climate control to 22°C (72°F), and before long, the cabin's temperature achieves or approximates the desired comfortable air temperature. Building HVAC systems via a BMS should provide the same environmental conditions considering that people may sit in their cars for two hours a day versus their workplace for eight hours or more.

AHU – Filtration: Air filtration is a cost effective and highly efficient way to purify the air entering or exiting your business or facility. Air quality improvements can be made using advanced HEPA (high-efficiency particulate air) filters to screen harmful pollutants, while equipping workplaces with air quality monitors can also warn building managers about tiny foreign particles that could harm lungs or spoil quality product.

Panel, Bag & HEPA Filters (Fig.3) –

Pleated Panel filters (e.g. G4) are suitable for ventilation and air conditioning systems which require a higher efficiency and greater dust holding capacity. A pleated **Panel** filters' main purpose is to act as a prefilter where it is typically installed in front of a more efficient filter to extend the life of the more expensive higher efficiency filter.

Bag filters (e.g. F9) are low to medium efficiency air filters and are used where greater dust holding capacity is required than a disposable Panel Filter can offer. Filters with a large filter surface are more energy economical than filters with a smaller filter surface. This also usually applies to products with a high dust holding capacity. A bag filter will therefore always perform better than a panel filter.

A **HEPA** filter (e.g. H13) is a type of high efficiency air filter. Basically, a HEPA filter can remove and trap up to 99.95% airborne particles. HEPA is the acronym for "High Efficiency Particulate Air".

Fig 3.

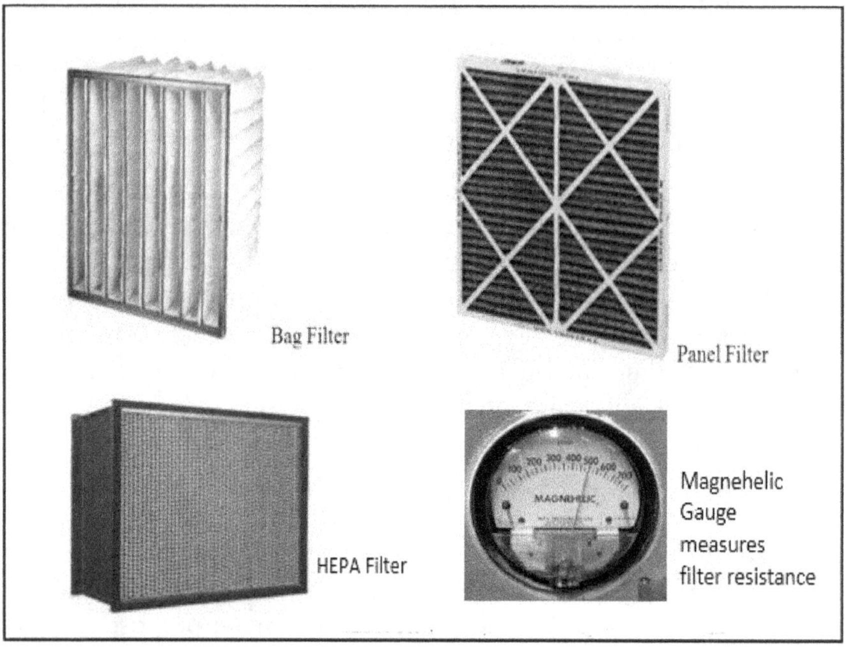

The initial filter air flow resistance and final filter air flow resistance are typically measured as **pressure drop** across the filters.

Filter Energy Savings:

Choosing the right air filter can be a big part of your company's energy saving plan. With soaring power prices and new energy directives, it truly pays to save energy in your air filtration systems. Low cost filters clog quickly, causing a higher resistance to the airflow which results in an energy cost penalty.

Quality air filters capture particles and maintain the proper airflow two to three times longer than low cost filters and requires less frequent filter changes. Fewer filters, less labour, reduced waste AND the biggest savings is energy costs. How? By selecting filters designed for lower average lifetime resistance, the HVAC unit doesn't work as hard to pull air through the system.

E.g. In the average commercial building, 50% of the energy bill is for the HVAC system and 30% of that is directly related to the air filtration, so it always pays for you to choose the best low energy air filter combination for the right filtration application.

Tip: Cleaning fans & air ducts and replacing clogged filters as required can improve system efficiencies by up to 60%. Therefore, regular maintenance is critical in keeping airways clear and to keep systems operating at design efficiency.

AHU Duty/Standby Arrangement Example:

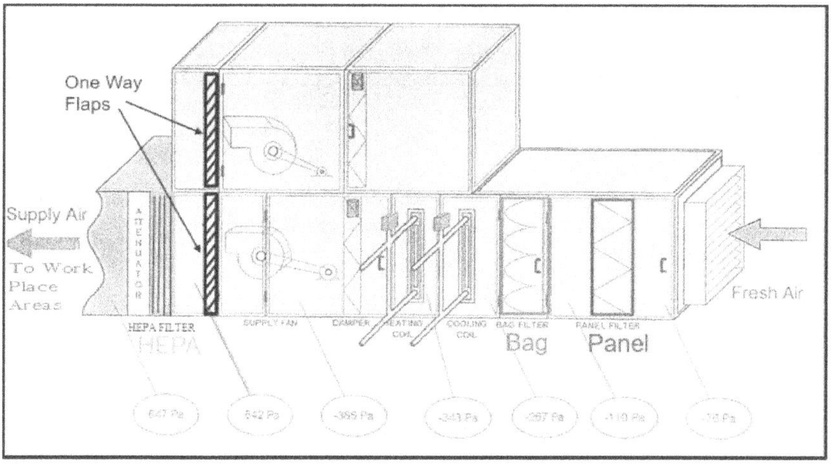

Variable Air Volume (VAV) is a type of heating, ventilating, and/or air-conditioning system. Unlike constant air volume systems (Pictured in Fig.3 above), which supply a constant airflow at a variable temperature to an entire single building zone. VAV systems vary the airflow at a constant temperature, they were developed to meet the varying heating and cooling needs of multiple building zones. You can save as much as 30% in energy costs with a VAV System.

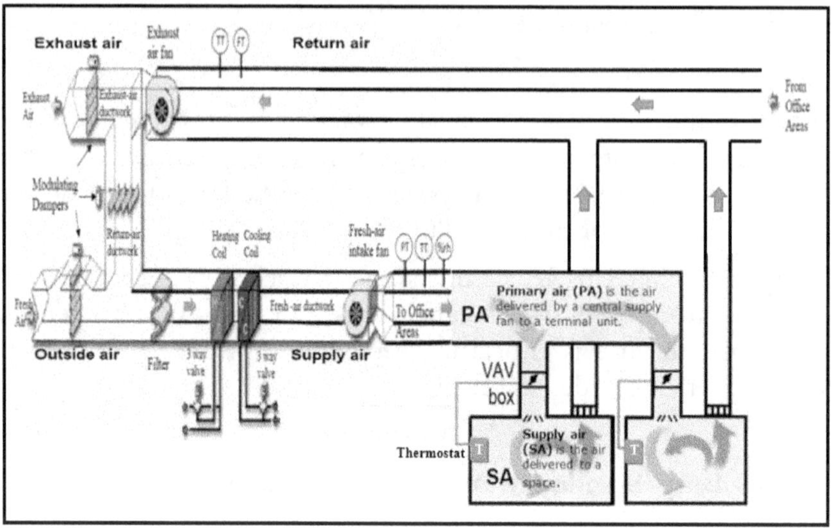

Tip: An **anemometer** is a test instrument that measures air velocity. Air velocity (distance travelled per unit of time) is usually expressed in feet per minute (FPM). By multiplying air velocity by the cross section area of a duct, you can determine the air volume flowing past a point in the duct per unit of time. Volume flow is usually measured in cubic feet per minute (CFM). Handheld anemometers of certified accuracy are an excellent, portable tool for performing tests on HVAC system performance e.g. for optimisation and diagnostics.

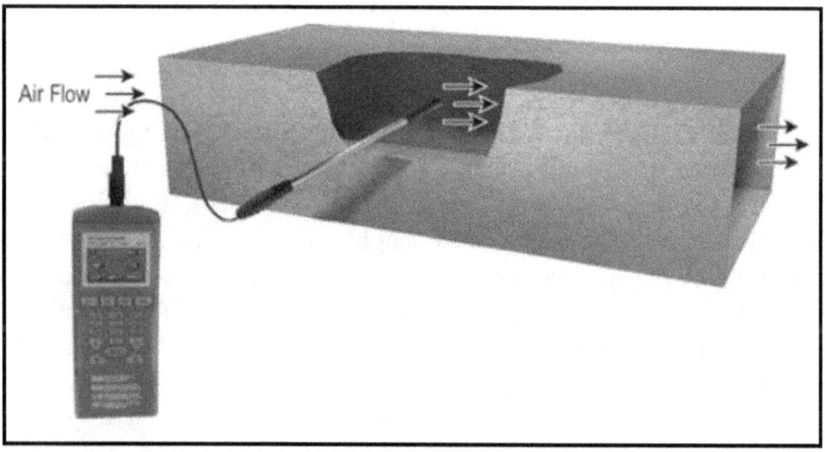

AHU – Heating System:

The heating coil (e.g. using steam) increases the temperature in office areas and other building spaces to compensate for heat losses between the internal space and the outside environment (Fig.4). **N.B.** Heating costs rise by about 8% for each 1°C of overheating. A 2°C temperature difference discrepancy may not seem much but when the system must heat 34,000 m3/hr of intake ambient air from 10°C to 20°C, it will have a big difference in heating costs.

Tip: Ensure the Temperature Transmitters (TT), Pressure Transmitters (PT), Flow Transmitters (FT) and RH probes are calibrated annually.

Keys to obtaining design efficiency of a heating system in the field include:

- Sizing the system for the specific heating load of the building being constructed
- Proper selection and installation of controls
- Sizing and designing the layout of the ductwork, piping, heat exchanger, boiler (steam) and fan motors for maximising efficiency
- Insulating and sealing all ductwork.

Fig 4.

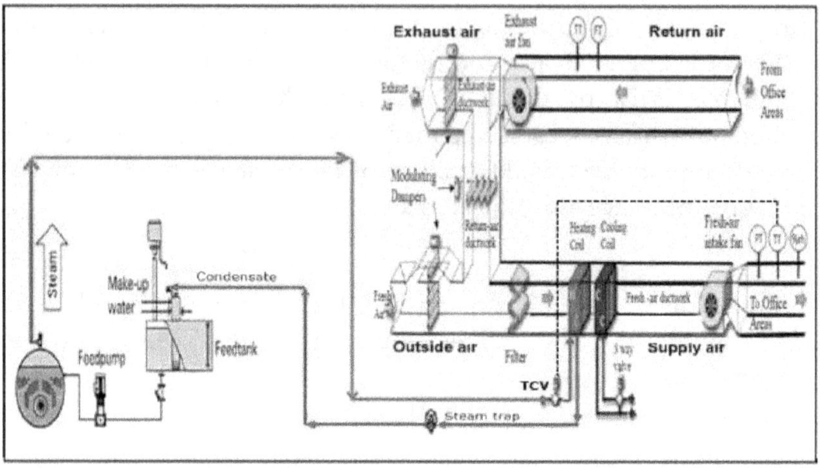

AHU – Cooling System:

The cooling coil decreases the temperature in office areas and other building spaces where heat gains have arisen from people, equipment, or the sun and are causing discomfort (Fig.5).

Keys to obtaining design efficiency of a cooling system in the field include:

- Sizing the system for the specific cooling load of the building being constructed
- Proper selection and installation of controls
- Correctly charging the Refrigeration system with the proper amount of anhydrous (meaning it is without water) Ammonia refrigerant
- Sizing and designing the layout of the ductwork, piping, refrigeration unit, cold heat exchanger, glycol/water mix, balancing valves, and fan motors for maximising efficiency

- Insulating and sealing all ductwork.

Fig 5.

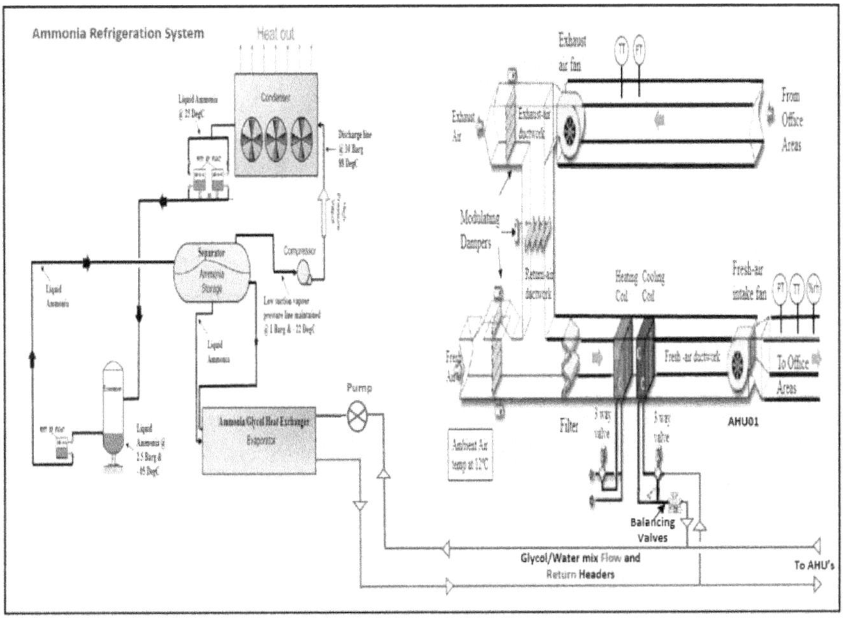

Tip: Cooling systems designed with a constant flow principle circulate large amounts of a glycol/water mix to the AHU cooling coils. During commissioning of the cooling system in Fig.5 above, hydraulic balance is often achieved by using **Balancing Valves**:

Expanded view of Balancing Valves in Fig.5:

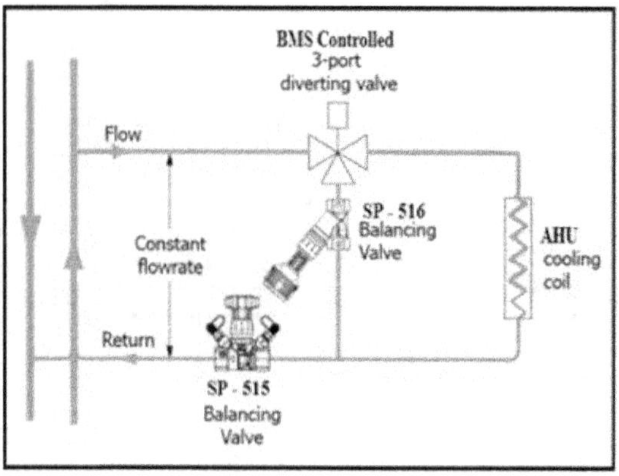

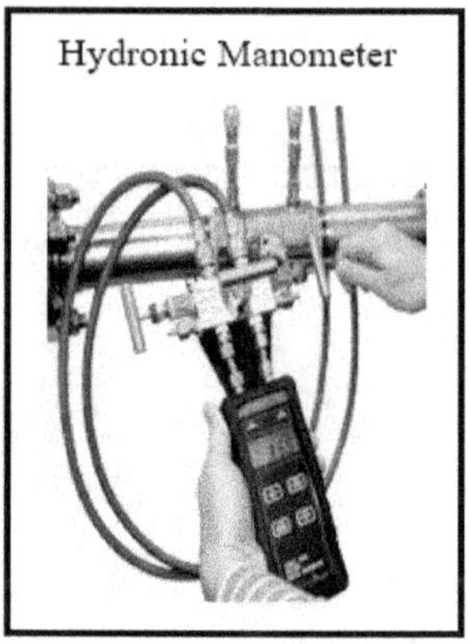

For a proper hydronic balance, a series of manual balancing valves are installed, varying in size according to the required flow for each AHU in the system. They are typically balanced for terminal flow using **differential pressures.** To do that as efficiently as possible, it is important to ensure that the bespoke 'design flows' to each AHU in the system are correct; some cooling coils are bigger than others. Manual balancing valves are the best option in these types of systems. Once installed, they can be commissioned and set to the design flow using a Hydronic Manometer without the risk that flows will change due to partial load conditions.

Tip: Hydronic Manometers are used to balance hydronic heating and cooling systems, check pump performance, and to set balancing valves. Balancing hydronic systems reduces energy consumption, saves money and enhances system performance.

AHU - Heat Recovery System: Heat Recovery Ventilation (HRV) (Fig.6) is an energy recovery ventilation system using recirculated liquid filled (e.g. Glycol/Water mix) heat recovery heat exchangers between the intake and exhaust air flows. The BMS will switch on the heat recovery system pump when the outside ambient air temperature reaches its set point **e.g.** 12°C.

Fig.6:

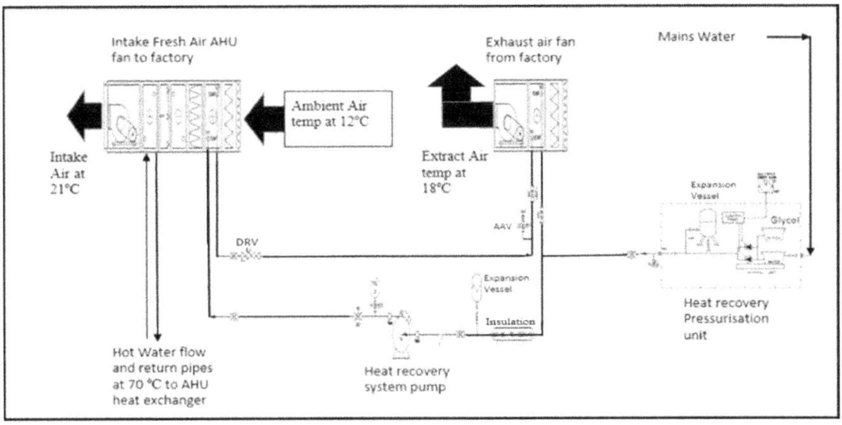

AHU – Economiser: Using ambient air temperature air for heating and cooling rather than by mechanical means. Modulating dampers (Fig.7) on the fresh-air intake, exhaust air and return-air ductwork will enable a BMS to control the mixing ratio of air in order to achieve the optimum condition of air exiting the mixing section. This minimises the heating or cooling load of the unit.

Fig.7:

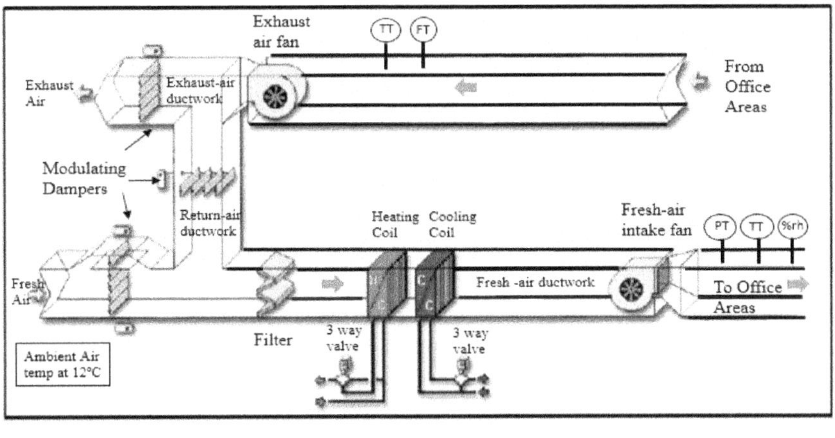

AHU – Humidification is the process in which the moisture or water vapour or humidity is added to the air without changing its dry bulb (DB) temperature. It increases humidity (moisture) in a building by **directly injecting** pressurised steam into the airstream (Fig.8). Relative humidity levels below 20% can cause discomfort through drying of the mucous membranes and skin.

Fig.8:

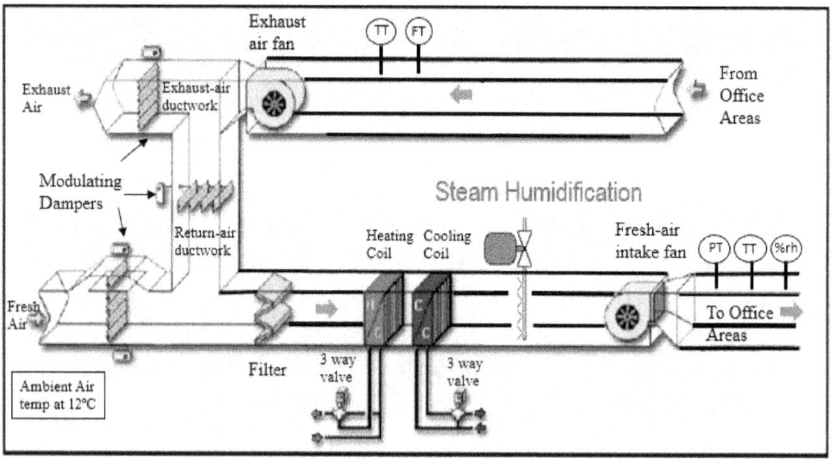

AHU – Dehumidification reduces the level of humidity in the atmospheric air. The cooling coil extracts water from the air which is maintained at a temperature lower than the dew point temperature of the air by the cooler liquid passing through it (Fig.9). Excessively humid air can cause mould and mildew to grow inside buildings, both of which pose numerous health risks. It can also make some people extremely uncomfortable, causing excessive body perspiration that can't evaporate in the already-moisture-saturated air.

Fig.9:

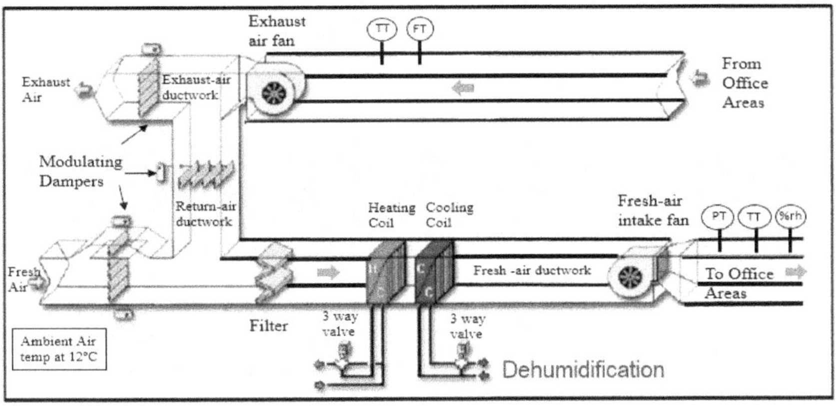

Building & Office Areas – GRD's

GRD's – Grilles, Registers, Diffusers (Fig.10)

A **Grille** is a perforated cover for an air duct (used for heating, cooling, or ventilation, or a combination thereof), they have no damper to close off the flow of air.

A **Register** differs from a grille in that a damper provides control to direct and diffuse the amount of air that enters a room.

A **Diffuser** is an air distribution outlet designed to introduce air into a conditioned space to obtain a desired indoor atmospheric environment. It usually located in the ceiling and consisting of deflecting vanes discharging supply air in various directions and planes and arranged to promote mixing of the air which is supplied to the room with the air already in the room.

Fig.10:

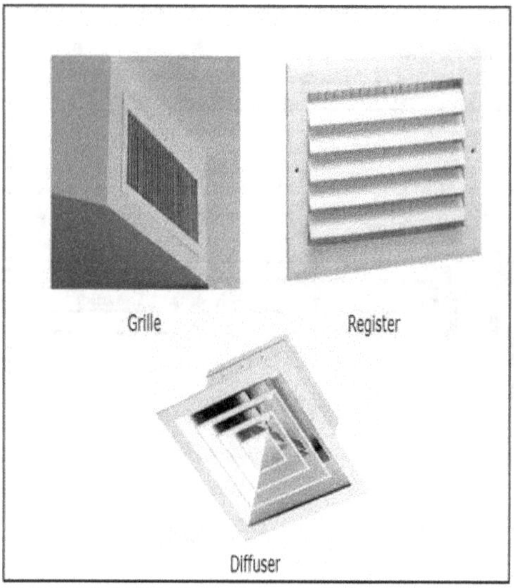

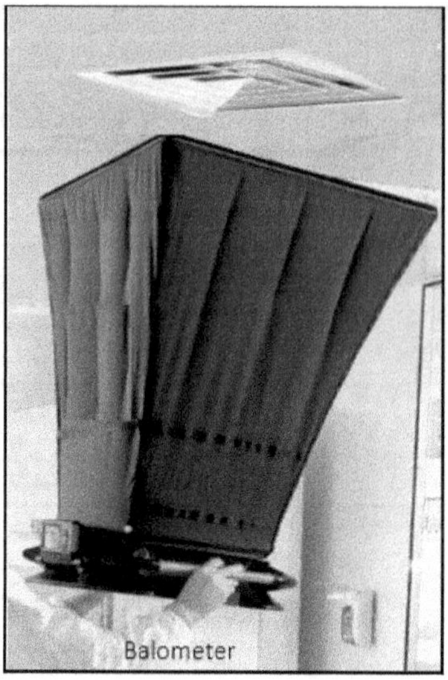

Tip: Airflow from standard GRD's is measured using a calibrated air balancing hood (Balometer). The primary reason to measure airflow is to discover if there are any ventilation issues and the original design and specification of the ventilation system for the building and its occupants is maintained. It must be checked at least annually by the verification step of measuring, adjusting and balancing airflow.

Reference Building Temperatures examples:

Sector	Building/Room type	Temperature (°C)	Temperature (°F)
Office/Service Companies	Computer Rooms	19-21	66-70
	Banks, Building Societies, Post Offices	19-21	66-70
	Offices	21-23	70-73
Hospitality	Restaurants/Dining Rooms	22-24	72-75
	Bars	20-22	68-72
	Hotels	19-21	66-70
Schools/Further and Higher Education	Educational Buildings	19-21	66-70
Industrial /Factories	Heavy Work	11-14	52-57
	Light Work	16-19	61-66
	Sedentary Work	19-21	66-70
Hospitals and Healthcare	Bedheads/Wards	22-24	72-75
	Circulation Spaces/Wards	19-24	66-75
	Consulting/Treatment Rooms	22-24	72-75
	Nurses' Stations	19-22	66-72
	Operating Theatres	17-19	63-66
Public Buildings	General Building Areas	19-21	66-70
	Law Courts	19-21	66-70
	Libraries	19-21	66-70
	Exhibition Halls	19-21	66-70
	Laundries	16-19	61-66
	Churches	19-21	66-70
	Museums and Art Galleries	19-21	66-70
	Prisons	19-21	66-70
Retail	Retails Buildings	19-24	66-75
Sports and Leisure	Changing Rooms	20-25	68-77
	Sports Halls	15	59
	Pool Halls	28-30	82-86

Reference Building Air Changes Per Hour (ACPH) examples:

Sector	Building/Room type	ACPH
Office/Service Companies	Computer Rooms	15-20
	Banks & Building Societies	04-08
	Offices	06-10
Hospitality	Restaurants/Dining Rooms	08-12
	Bars	20-30
Schools/Further and Higher Education	Educational Buildings	04-12
Industrial /Factories	Heavy Work	08-15
	Light Work	08-10
	Welding Workshops	15-30
	Boiler Rooms	15-30
	Laboratories	06-15
Hospitals and Healthcare	Bedheads/Wards	06-08
	Sterilising	15-25
	Operating Theatres	25-26
Public Buildings	General Building Areas	04-08
	Libraries	03-05
	Laundries	10-30
	Toilets	06-10
	Museums and Art Galleries	12-15
	Police Stations	04-10
Retail	Shopping Centres	06-10
Sports and Leisure	Changing Rooms	15-20
	Swimming Pools	10-15

Note: Lung disease accounts for 12,000 work-related deaths each year. Breathing in dust, gases, vapours and fumes at work can cause life-changing lung disease or make existing conditions worse. Companies must do the right thing and protect their workers from work-related lung disease.

De-Stratification Fans

When hot air rises it 'floats' on top of cold air because it is less dense. This creates what's called a 'stack-effect' in a building and as a result, areas end up being heated from the top down. Typically, a temperature difference of up to 10°C can happen within a 10m high indoor space.

Ceiling fans gently induce air movement which ensures that the air is mixed to produce a uniform temperature. Consequently, building HVAC thermostat settings can be reduced with associated reductions in energy costs but with improved occupant comfort. Where cooling is required, ceiling fans provide an alternative to 'energy and maintenance hungry' air-conditioning systems.

They don't cool the air but, set at a higher speed, they use the 'wind chill' effect to provide evaporative cooling for occupants. They can also work in conjunction with air-conditioning where the movement of air improves the effectiveness of the air-conditioning, meaning set-points can be reduced and equipment operation reduced.

De-Stratification fan installation example:

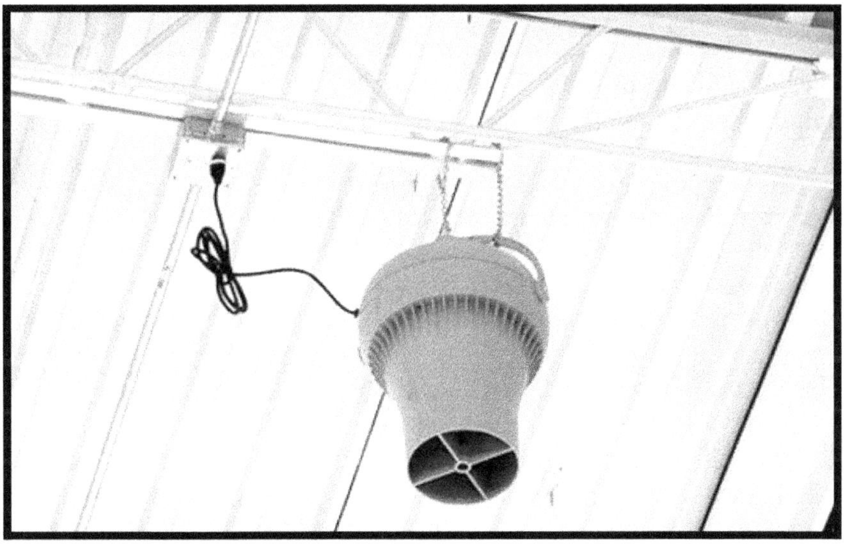

Psychrometrics:

Although the principles of psychrometry apply to any physical system consisting of gas-vapor mixtures, the most common system of interest is the mixture of water vapor and air, because of its application in **heating, ventilating, and air-conditioning**. In human terms, our thermal comfort is in large part a consequence of not just the temperature of the surrounding air, but (because we cool ourselves via perspiration) the extent to which that air is saturated with water vapor. Many substances are hygroscopic, meaning they attract water, usually in proportion to the relative humidity or above a critical relative humidity. Such substances include cotton, paper, cellulose, other wood products, sugar, calcium oxide (burned lime) and many chemicals and fertilizers. Industries that use these materials are concerned with relative humidity control in production and storage of such materials.

In industrial drying applications, such as drying paper, manufacturers usually try to achieve an optimum between low relative humidity, which increases the drying rate, and energy usage, which decreases as exhaust relative humidity increases. **In many industrial applications it is important to avoid condensation that would ruin product or cause corrosion.**

Molds and fungi can be controlled by keeping relative humidity low. Wood destroying fungi generally do not grow at relative humidities below 75%.

Some Psychrometric properties

- **Dry-bulb temperature (DBT):** is the temperature of air measured by a thermometer freely exposed to the air but shielded from radiation and moisture. DBT is the temperature that is usually thought of as air temperature, and it is the true thermodynamic temperature. It indicates the amount of heat in the air and is directly proportional to the mean kinetic energy of the air molecules.
- **Wet-bulb temperature (WBT):** is the temperature a parcel of air would have if it were cooled to saturation (100% relative humidity) by the evaporation of water into it, with the latent heat being supplied by the parcel. A wet-bulb thermometer will indicate a temperature close to the true (thermodynamic) wet-bulb temperature. The wet-bulb temperature is the lowest temperature that can be reached under current ambient conditions by the evaporation of water only. Wet-bulb temperature is largely determined by both actual air temperature (dry-bulb temperature) and the amount of moisture in the air (humidity). At 100% relative humidity, the wet-bulb temperature equals the dry-bulb temperature.
- **Dew point temperature**: is the temperature at which a given concentration of water vapour in air will form dew. More specifically, it is defined as the temperature at which a given volume of air at a certain atmospheric pressure is saturated with water vapour causing condensation and the formation of dew, which is the condensed water that a person often sees on flowers and grass early in the morning. Dew point varies depending on the amount of water vapour present in the air, with more humid air resulting in a higher dew point than dry air. Furthermore, the

higher the relative humidity, the closer the dew point to the current air temperature, with 100% relative humidity meaning that dew point is equivalent to the current temperature. In cases where dew point is below freezing (0°C or 32°F), the water vapour turns directly into frost rather than dew.

- **Humidity:** is the amount of water vapour in the air. Water vapour is the gaseous state of water and is invisible. Humidity indicates the likelihood of precipitation, dew, or fog. Higher humidity reduces the effectiveness of sweating in cooling the body by reducing the rate of evaporation of moisture from the skin.

Air Conditioning (A/C):

E.g. **Standard Cassette** with outdoor **Air cooled Condensing unit**:

An **Air Conditioning unit** is perfect in situations where maximum airflow is required from minimal space, they are popular in offices, shops, hairdressers, and other retail premise's. It works by using refrigeration to chill indoor air, taking advantage of a remarkable physical law: When a liquid converts to a gas (in a process called phase conversion), it absorbs heat. Air conditioners exploit this feature of phase conversion by forcing special chemical compounds (e.g. R-410A refrigerant) to evaporate and condense over and over again in a closed system of coils. (**Cold Side** - Standard Cassette) The Air conditioner contains a fan that moves warm interior air over these cold, refrigerant-filled coils. When hot air flows over the cold low-pressure evaporator coils, the refrigerant inside absorbs heat as it changes from a liquid to a gaseous state. To keep cooling efficiently, the air conditioner has to convert the refrigerant gas back to a liquid again. (**Hot Side** - Air cooled Condensing unit) To do that, a compressor puts the gas under high pressure, a process that creates unwanted heat. When a vapour is compressed both the pressure and temperature of that vapour increases. All the extra heat created by compressing the gas is evacuated to the outdoors to a second set of coils called condenser coils, and a second fan. As the gas cools, it changes back to a liquid, and the process starts all over again. Think of it as an endless, elegant cycle: liquid refrigerant, phase conversion to a gas/ heat absorption, compression, and phase transition back to a liquid again.

The Parts of an Air Conditioner

- Evaporator - Receives the liquid refrigerant
- Condenser - Facilitates heat transfer
- Expansion valve - regulates refrigerant flow into the evaporator
- Compressor - A pump that pressurises refrigerant

The **cold side** of an air conditioner contains the evaporator and a fan that blows air over the chilled coils and into the room. The **hot side** contains the compressor, condenser, and another fan to vent hot air coming off the compressed refrigerant to the outdoors. In between the two sets of coils, there's an expansion valve. It regulates the amount of compressed liquid refrigerant moving into the evaporator. Once in the evaporator, the refrigerant experiences a pressure drop, expands and changes back into a gas. The compressor is actually a large electric pump that pressurises the refrigerant gas as part of the process of turning it back into a liquid. There are some additional sensors, timers and valves, but the evaporator, compressor, condenser and expansion valve are the main components of an air conditioner.

- **Man Safe Systems:**

Are about putting in place, systems that prevents the likelihood of falls and injury happening in the workplace or building. **Man Safe Systems** keep the operative safe when working at height by connecting he/she to the system using appropriate PPE. The system comprises cable, post and fixings that are tested to take the fall of the user. These usually take the form of a fall arrest system or a fall restraint system. A man safe system could be a single anchor fixing near a window or a line system installed on roof. The regulation requires the lifeline systems are inspected and certificated every 12 months by a competent height safety specialist company.

- **IT and Audio/Visual systems:**

Information Technology (IT) is all about storing, manipulating, distributing and processing information. IT helps business's to streamline operations, reduce costs, improve efficiency, maximise profit, and minimise waste. With the rapid innovations in technology, the workplace has been changing in many ways including how we communicate. Ensure the best IT and creative systems for your industry are installed. **Audio/ Visual systems** are an excellent aid in business communication that helps each to understand and retain information differently. Audio visual system communication technologies include sound, lighting, video, display and projection systems that are widely used in business today.

Audio Visual Solutions improve business communication via Interactive Displays, Touchscreen kiosks, Digital Signage, and Video Conferencing. They have a significant impact on your business if well implemented and will provide better service to your customers. The benefits these solutions provide are a great encouragement to the growth and thriving of your business in the marketplace. These benefits help improve your work experience and employee performance by:

1. Creating an atmosphere of interaction.
2. Helps people to comprehend the message delivered.
3. Allows engagement among the customers and employees.
4. Saves businesses a significant amount of money e.g. less travel.
5. Allows easy access from anywhere in the company.
6. Creates an impression of a growing company to the world of investors.
7. Extremely reliable to offer the services it has been programmed to give e.g. highly secured platforms.

- **Building fabric:**

Insulate your building. The condition of the building fabric can have a significant effect on energy use. A general maintenance programme should ensure that doors and windows are in good working order and draught stripping is fitted where appropriate. The building fabric is a critical component of any building since it both protects the building occupants and plays a major role in regulating the indoor environment. Consisting of the building's roof, floor slabs, walls, windows, and doors, the fabric controls the flow of energy between the interior and exterior of the building. Optimal design of the building fabric may provide significant reductions in heating and cooling loads which in turn can allow downsizing of mechanical equipment. When the right strategies are integrated through good design, the extra cost for a high-performance fabric may be paid for through savings achieved by installing smaller HVAC equipment. To optimise business or appeal to customers, ensure the fabric aspects of your building, including carpets, seating, curtains and blinds are aesthetically pleasing, fit for purpose and comfortable.

- **Drainage systems testing:**

Weather can change within minutes. Your building must be prepared for unexpected rainfall or drainage blockages. You must be aware of the dangers of **Standing Water.** As standing water increases on your roof, the weight your roof must bear, also increases. If the problem is not attended to, the standing water will likely exceed the designed weight threshold of the roof and potentially end in collapse leading to possible damage of electronic equipment, inventory, and worst of all, could take lives. Standing water weighs approximately 5 lbs. per inch per square foot of roof surface. An inch of standing water adds approximately 500 lbs. to a roof measuring just 10 feet * 10 feet. Another impact standing water can have on your roof is increased leaks.

Water is very seeking and will make its way into the tightest of spaces; a typical roof with an appropriate slope may have many pinhole imperfections in the roofing surface. These holes go completely through the roof surface, yet no leak results. Why are there holes but no leaks? Because the roof is sloped, and water is running off the roof. As the water runs, it isn't searching for tiny pinholes through which it can reach the ground, it simply keeps moving along the surface of the roof. However, once the water stops running and is standing on the roof, it now has enough weight and downward pressure to force it into those pinholes. Once these pinholes become active leaks, there is now water leaking into the building beneath the pinholes.

Statistics:

Statistics show that high rise buildings consume nearly 40% of the world's energy, and that approximately >30% of global greenhouse gas emissions are attributable to buildings.

Zero Net Energy (ZNE) buildings are ultra-efficient new construction and deep energy retrofit projects that consume only as much energy as they produce from clean, renewable resources. ZeroNet Energy buildings can be used autonomously from the energy grid supply – energy can be harvested on-site usually in combination with energy producing technologies like photovoltaic (PV) solar panels and Wind while reducing the overall use of energy with extremely efficient HVAC and Lighting technologies. Smart building technology has already been found to save companies up to 18% on utility bills.

Glass Selection: Windows connect building occupants to the outdoors via daylight, views, and natural ventilation while protecting them from the elements. Even so, in many older buildings, windows fail to weatherise occupied space and are prone to water leakage, condensation, or noise transmission. The consequences include occupant discomfort, water damage, and higher energy bills. Choosing the right energy efficient glazing for your building(s) will protect people and property whilst reducing operating costs by reducing heating and cooling bills.

Three properties are integral to evaluating glazing energy performance:

1. Insulating performance (U-factor)
2. Solar heat gain coefficient (SHGC)
3. Visible light transmittance (VLT or VT).

The ideal properties depend on the local climate, building type, and design. For instance, a low U-factor (less heat loss) is most important in a cold climate; a low SHGC (less solar heat gain) is a priority where overheating is a concern. Visible transmittance is important when daylight is incorporated into the project design.

The biggest challenge is while using new technologies and designs, how to make buildings safe, energy efficient and as environmentally friendly as possible without compromising human comfort. Companies are under pressure from rising energy costs as well as legislation and climate change. **Key Performance Indicators** (KPI's) must be to keep overhead costs under control and improve reliability, predictability, and profitability of its facilities.

Utility Services:

You must know where all the main services from your utility providers (e.g. Electricity, Natural Gas, Water) are located on your site. You must know their capabilities and what action must be taken either in normal operations or an emergency, when required.

Examples:

- MV/LV power supply
- Power Factor Correction
- Electrical power main distribution boards
- Standby generator
- Mains water
- Natural Gas
- Compressed Air
- Steam
- Smoke/fire detection systems
- Fire sprinkler water systems
- BMS systems

All these services may be in one **Central Utility Building** (CUB) or scattered across your site.

You must know exactly how to isolate/reset any utility system at any time. Do not wait until something goes wrong, or to carry out maintenance on a piece of equipment before you decide to familiarise yourself with these essential utility services. Remember, if an individual

fire water sprinkler head has been activated in error (e.g. accidentally hit by a ladder), water at high pressure will be spraying out. The water pressure in these types of systems can be > 10 Barg, at these pressures, very large volumes of water will be delivered into an area in minutes.

You must know how to stop the water flow as quickly as possible. **This could possibly involve draining the entire sprinkler system pipe network.** The water damage alone to buildings floor/wall/furniture areas and associated electrical equipment can be very substantial as well as the associated costs involved in the clean up.

Facilities operations and activities, if not managed correctly, can have significant safety and environmental impacts. It is easy to forget the things that rarely happen.

N.B. A lengthy period without severe incident can lead to a steady erosion of protection.

MV/LV Power Supply, Testing, and Servicing

- Medium Voltage (MV: Voltages from 600V- 69 kV)
- Low Voltage (LV: Voltages at 600V and below).

MV/LV Power Distribution Network Single Line Diagram example:

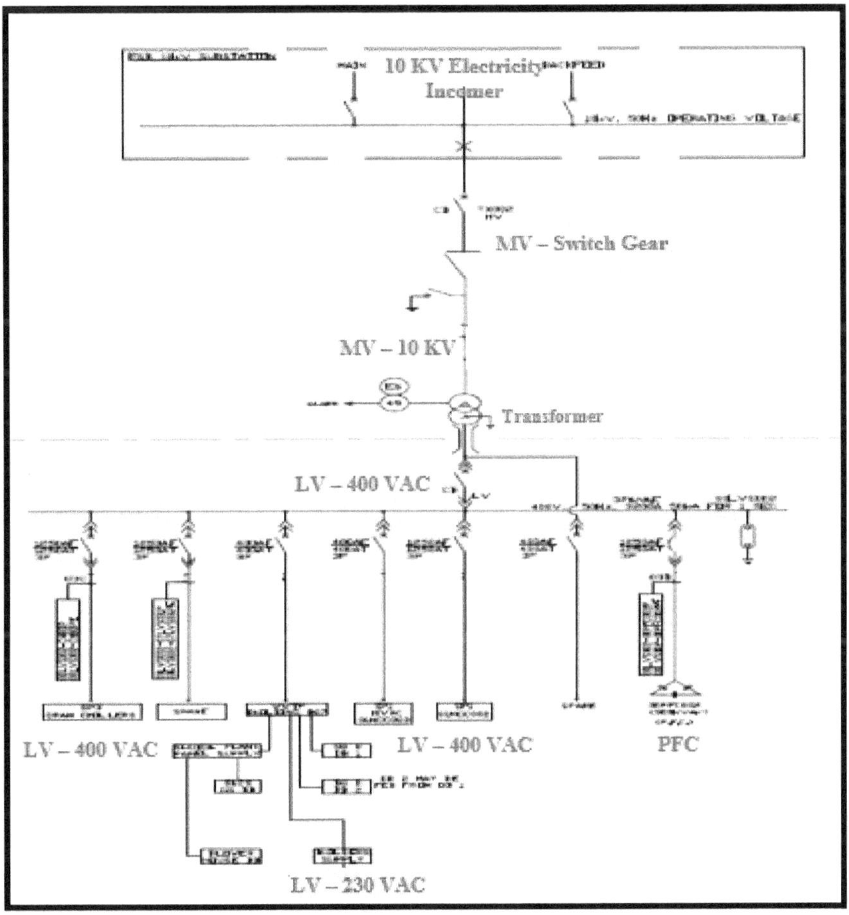

How MV is converted to LV via transformer and switchgear example:

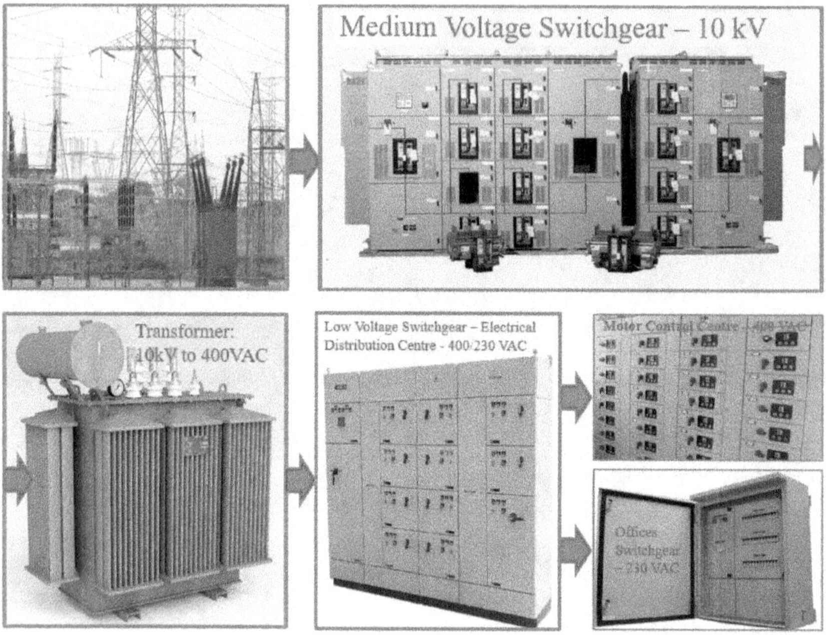

Typical Annual MV Checks:

Plant item	MV Maintenance Operation	Frequency
MV Switchboard - 00MVS001		
	Visual inspection and cleaning of exterior and gas vents.	Annual
	Visual inspection and cleaning of bus compartments incl busbar.	Annual
	Visual inspection and cleaning of cable compartments including fuses,CT's, VT's, cable terminations and earthing.	Annual
	Visual inspection and cleaning of LV compartments.	Annual
	Mechanical operation of circuit breaker, CB truck and earthing switch operating mechanisms and lubrication with proprietary lubricant.	Annual
	Record Circuit Breaker operating voltage and current.	Annual
	Mechanical operation of mechanical interlocks and lubrication with proprietary lubricant.	Annual
	Functional check on all indicating lamps and controls.	Annual
	Functional check on all electrical interlocks and auxiliary contacts.	Annual
	Partial discharge check (hand-held meter).	Annual
Ring Main Units - RMU001,RMU002 & RMU007		
	Visual inspection and cleaning including cable boxes,cable terminations and earthing.	Annual
	Mechanical operation of circuit breaker, and line earthing switch operating mechanisms and lubrication with proprietary lubricant.	Annual
	Record Circuit Breaker operating voltage and current.	Annual
	Functional check on all indicating lamps and controls.	Annual
	Functional check on all electrical interlocks and auxiliary contacts.	Annual

Plant item	MV Maintenance Operation	Frequency
Transformer's - TX No.1 & 2, TX007,TX008		
	Visual inspection and cleaning including cable boxes, busbars, bushings, fans, cable terminations, control cabinets, ancilliary devices and earthing. Visual inspection and replacement if necessary of silica gel.	Annual
	Recording of tap counter reading, traffo alarms standing, battery voltage and charge amps, oil level pressure, max running temp attained, mechanical damage and general condition.	Annual
	Oil sample including test for dielectric strength and moisture content.	Annual
	Visual inspection and cleaning of bund.	Annual
	Visual inspection and functional test of bucholz and gas sampling.	Annual
Control & Protection		
	Record protection relay settings and test with secondary injection including trip test on circuit breaker.	Annual
	Functional test on all control functions.	Annual
Report		
	Full report of all findings including, photographs, description of problems and remedial actions, method statements, test results, etc	Annual

Typical Annual LV Checks:

Plant item	LV Maintenance Operation	Frequency
LV Switchboard - LV Switchboard 05 -1,05 - 02, 00LVS001,00LVS002 ; Fesoterodine 01MCC001,01MCC002,01MCC003 ; Temporary Substation 00LVS007,00LVS008 ; Drum Store 26MCC001 & HTF MCC 47-MCC-001		
	Visual inspection of all busbars and cable compartments incld cleaning of Switchgear.	Annual
	Check Busbars for discolouration.	Annual
	Check the Switchboard to ensure all covers, flash guards and doors are in place and correctly fitted.	Annual
	Check the operating conditions and settings of all Circuit Breakers,Moulded case circuit breakers, coils,contactors, indicating lights, door locks, door hinges and fuses.	Annual
	Check to ensure that Labelling of all outgoing services are intact and correct.	Annual
	Check labelling and fuse charts are intact, up - to-date.	Annual
	Note: Operation Load per phase at the Main Switchboard Note: The Meter Readings	Annual
	Thermal Imaging of LV Switchboards	Annual
Power Factor Correction Units - 00PFC001,00PFC002		
	Visual inspection and cleaning including cable boxes,cable terminations and earthing.	Annual
Report		
	Full report of all findings including, photographs, description of problems and remedial actions, method statements, test results, etc	Annual
Note: This draft schedule is intended as a guide to the type of maintenance programme required.		

The importance of carrying out annual servicing and testing of your site's "Mains Power Supply" equipment cannot be stressed enough. Whilst at the same time checking that the UPS systems & Back-Up generators operate smoothly and are capable of supplying the site electrical load on loss of "Mains Power Supply" is critical. **Remember**, no matter how stable the power supply from the national grid is, your site's associated SWITCH GEAR and CIRCUIT BREAKERS are electromechanical and can FAIL to operate if not inspected, cleaned, tightened, lubricated, tested, and proven at 12 – 24 monthly intervals, or per manufacturer's instructions.

N.B. Every electrical conductor or circuit part must be considered energised until proven otherwise.

The Fundamental Formulas of Electrical Engineering:

I - Amps Current	=	$\frac{V}{R}$	or	$\frac{P}{V}$	or	$\sqrt{P/R}$
V - Volts Voltage	=	$R*I$	or	$\frac{P}{I}$	or	$\sqrt{P*R}$
R - Ohms Resistance	=	$\frac{V}{I}$	or	V^2/P	or	P/I^2
P - Watts Power	=	$V*I$	or	$R*I^2$	or	V^2/R

- **Current** – is the flow of electrons in a conductor.
- **Voltage** – is the electrical pressure causing the current to flow.
- **Resistance** – is the opposition to the flow of current in a conductor determined by its length, cross sectional area and temperature.
- **Power** – is the product of current and voltage, hence P = V x I.

Electrical Power Main Distribution Boards

Power Distribution is a system, consisting of a Main Distribution Board **(MDB),** Sub Main Distribution Boards/Final Distribution Boards, by which electrical power is transmitted via branches to reach the exact end user. A Sub Main Distribution board (also known as panel board, breaker panel, or electric panel) is a component of an electricity supply system that divides an electrical power feed into subsidiary circuits, while providing a protective fuse or circuit breaker for each circuit in a common enclosure. Normally, a main switch, residual-current devices (RCD) or residual current breakers with overcurrent protection (RCBO), are also incorporated.

The protection system of an electrical installation comprises of a hierarchy of circuit breakers that should be able to protect an electrical installation by switching off faulty circuits, while maintaining power supply to the healthy sections as far as possible.

N.B. Working on electrical equipment is a potentially dangerous operation and must only be undertaken by suitably qualified and experienced personnel (with the appropriate PPE e.g. arc flash clothing - protective clothing that meets all the requirements of ASTM F 1506) who are familiar with the operation of the equipment. The switchgear must be isolated, locked off, and earthed before beginning any maintenance work. However, UNDER NO ACCOUNT must it be assumed that isolation and earthing has made electrical equipment or conductors 'dead', a positive check must be made to ensure that both are not 'live' by using a voltmeter or other suitable test equipment. "Electricity can be very unforgiving."

Typical layout of a small to medium industrial electrical installation:

This layout shows a 400 VAC mains supply intake point in a small industrial installation (Fig.11). The main electrical supply is taken through an isolator to a bus-bar chamber. From this chamber, other sub main isolators control the supply to various other sections of the plant.

Fig.11:

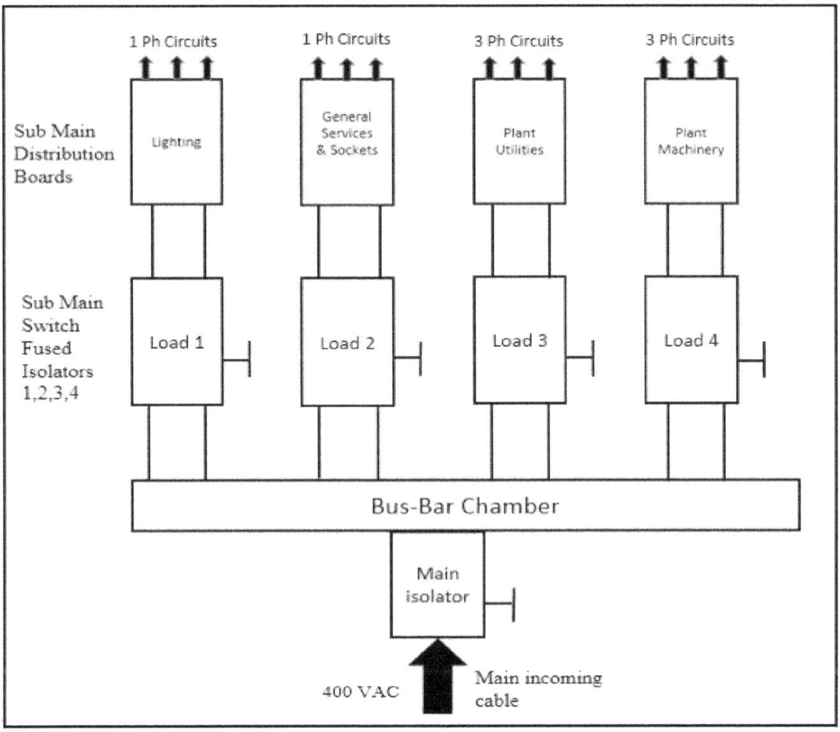

Division of Loads between Phases

When connecting single-phase loads to a three-phase supply, care should be taken to distribute the single-phase loads equally between the three-phases so that each phase carries approximately the same current and the neutral current is kept as low as possible. Distributing the single-phase loads equally across the three-phases is known as 'balancing' the load.

Power Factor Correction

Industrial customers are penalised by Utility Networks in the event of them having a BAD power factor of **<0.95 lagging** (Power factor: cos $\varphi = 1$. A power factor of one or "unity power factor" is the goal). The ratio of Active (Real) Power (kW) to Total Power (kVA) is called the Power Factor (PF = kW / kVA). It is a measure of the systems electrical efficiency in an alternating current circuit and is represented as a percentage or a decimal.

A bad PF can be overcome by the installation of an automatic power factor correction unit in your plant. Power factor correction is usually achieved by adding capacitive load to offset the inductive load present in the power system e.g. motors are inductive (approx. cos φ = 0.7 – 0.8) and require a magnetic field (Reactive Power: kVAR) to operate, the magnetic field is necessary but produces no work. The power factor of the power system is constantly changing due to variations in the size and number of the motors being used at one time.

The most inexpensive and widely used method of correcting PF is using a **variable capacitor bank** connected to your plant's LV switchgear bus. The advantage of the variable capacitor bank is that the PF controller monitors the system power factor and automatically regulates the amount of capacitive load (via electromagnetic contactors) connected to the system to offset the inductive load. Since the capacitive load is regulated, there would be no conflict with the Utility Networks. All major industrial and commercial customers should have these units installed.

Power Factor Correction Unit & Schematic Diagram Example

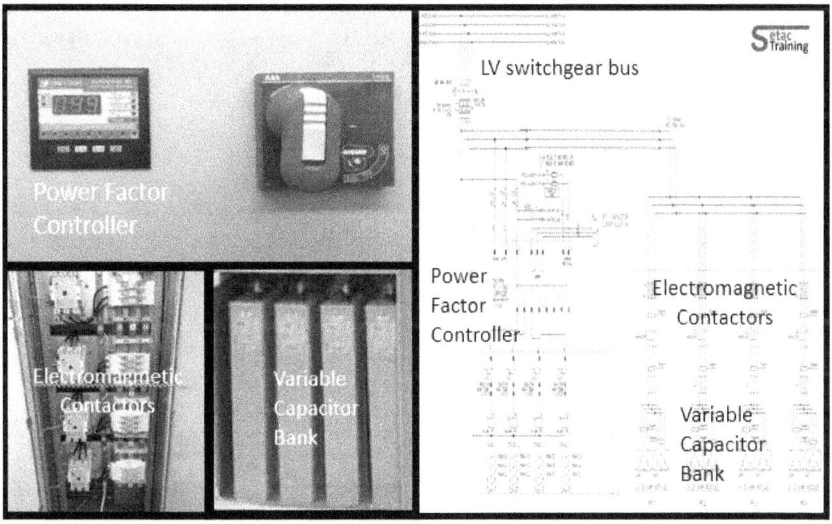

UPS (Uninterrupted Power Supply) systems:

Businesses rely on electrical power **e.g.** Data Centers are part of today's infrastructure providing technology support for the services that people and businesses rely on in their daily lives. A reliable source of electricity must be provided. Efficient, adequately sized UPS systems (Fig.12) with back-up electrical generators must be installed and maintained for your business, meaning that even during power cuts you have a safe supply of electricity to carry on as normal. An uninterruptible power supply is an electrical apparatus that provides emergency power to a load when the input power source or mains power fails.

A UPS differs from an auxiliary or emergency power system or standby generator in that it will provide instantaneous protection from input power interruptions by supplying energy stored in batteries. The 'on-battery runtime' of most uninterruptible power sources depends on the electrical load it is supplying but must be sufficient while another standby power source (Generator) comes online or to properly shut down the protected equipment. A UPS is typically used to protect hardware such as computers, data centres, telecommunication equipment or other electrical equipment where an unexpected power disruption could cause injuries, fatalities, serious business disruption or data loss. UPS units range in size from units designed to protect a single computer without a video monitor (around 200-volt-ampere rating) to large units powering entire data centres or buildings.

UPS System (Fig.12) - Double conversion single UPS in normal mode. *Applications: all sensitive loads that must be backed up without a break in the supply of power.*

Fig.12

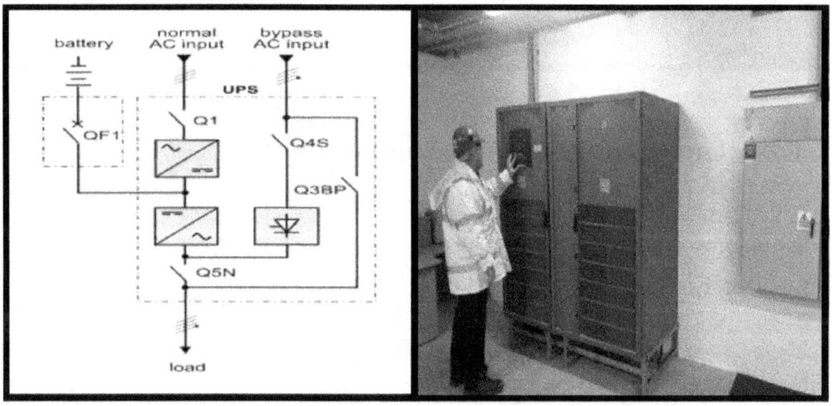

A single UPS is a modular unit made up of:

- two AC inputs (or one combined input)
- a rectifier/charger
- a battery
- an inverter
- a static switch (automatic bypass)
- a manual bypass (maintenance bypass)

In normal mode, the load is continuously supplied by the rectifier and inverter modules. If the normal AC source fails, the double conversion technology ensures no-break transfer to battery power (battery mode). The load is transferred to the bypass AC source only if the inverter fails, is overloaded, or the battery has reached the end of its backup time.

Standby Back-up Electrical Generator

A standby generator is a back-up electrical system that operates automatically within seconds of a utility power outage; an automatic transfer switch senses the power loss, commands the generator to start and then transfers the electrical load to the generator (Fig.13).

The standby generator begins supplying power to the circuits. After utility power returns, the automatic transfer switch transfers the electrical load back to the utility and signals the standby generator to shut off. It then returns to standby mode where it awaits the next outage.

To ensure a proper response to an outage, a **standby generator weekly test** should be carried out either automatically or by qualified personnel doing a 'manual test'. A full generator service must be carried out by the manufacturer annually, or per manufacturer's instructions.

N.B. The generator must always be left in 'standby mode' after a manual test is carried out. Most units run on diesel, natural gas, or liquid propane gas.

Automatic standby generator systems may be required by building codes for critical safety systems such as elevators in high-rise buildings, fire protection systems, standby lighting, or medical and life support equipment.

Fig.13 - Standby Back-up Electrical Generator – GA Drawing:

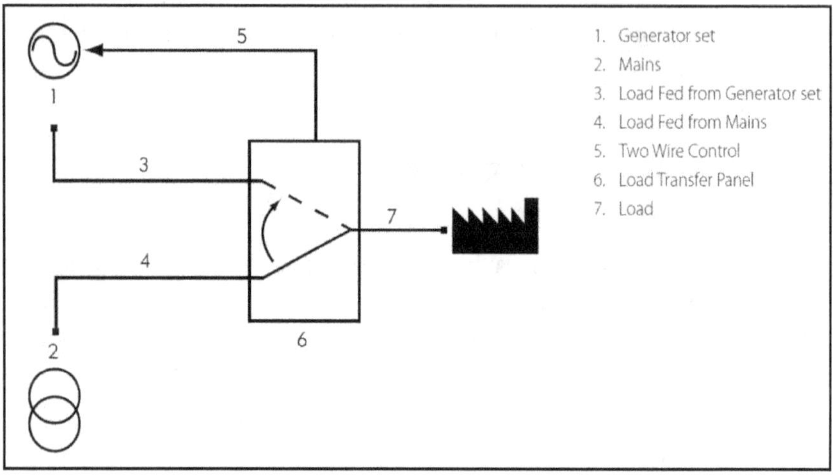

Standby Back-up Electrical Generator

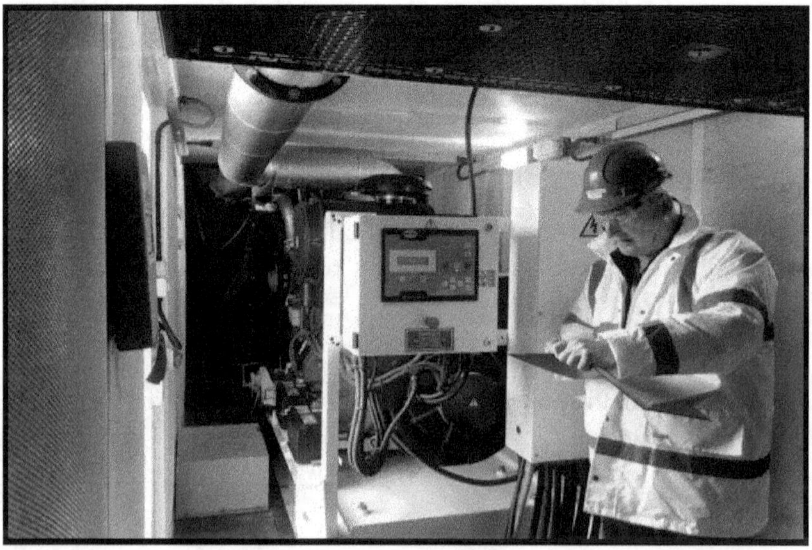

Load Transfer Panel

The **generator set** is required to automatically provide switching to standby power in the event of mains failure. The Load Transfer Panel senses when the mains have failed, it signals the generator set to start, switches the load from the **failed mains** to the generator set and then switches it back after the mains supply is re-established.

Summary

A reliable source of 'MAINS ELECTRICITY' is vital to any company. With increasing dependence on computers and automated processes, most modern facilities cannot afford failure-induced downtime. 'STANDBY ELECTRICITY GENERATOR' systems ensure a proper response in the event of a 'Mains Electricity' outage. UNINTERRUPTED POWER SUPPLY (UPS) systems exist entirely to protect other equipment, often on a large scale and running business-critical applications. In a 'Mains Electricity' outage, a UPS acts as a backup battery system protecting your equipment as well as your data. Although it won't provide power indefinitely, it'll help keep computers and essential machinery running long enough for your backup generators to kick in. All systems within the 'Power Protection Platform' must be maintained and serviced regularly. **MV/LV Power Distribution Network Single Line Diagram of a large industrial installation with multiple transformers, Back-Up Generators and UPS units example:**

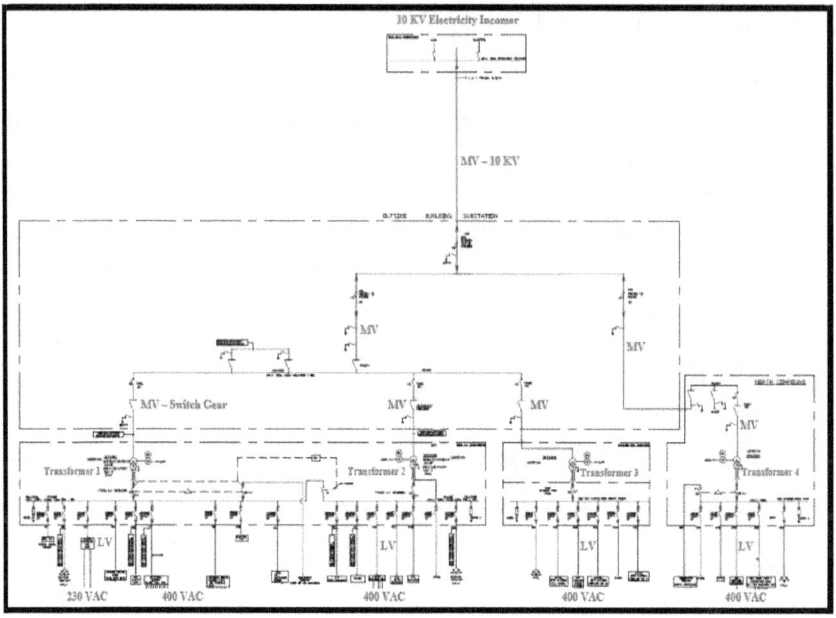
10 KV Electricity Incomer
MV – 10 KV
MV
MV
MV – Switch Gear
MV
MV
MV
Transformer 1
Transformer 2
Transformer 3
Transformer 4
LV
LV
LV
LV
230 VAC
400 VAC
400 VAC
400 VAC
400 VAC

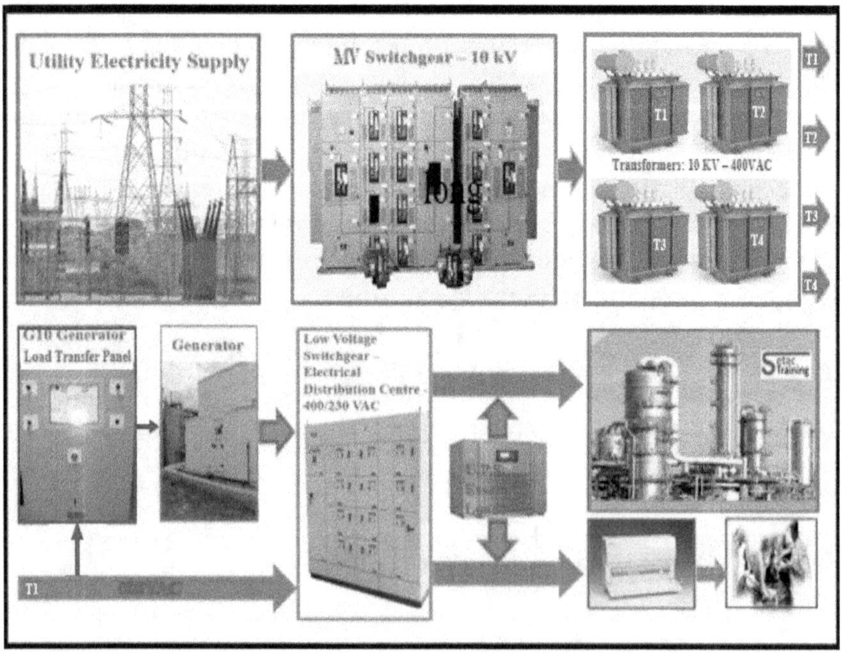
Utility Electricity Supply
MV Switchgear – 10 kV
T1
T2
T3
T4
Transformers: 10 KV – 400VAC
G10 Generator
Load Transfer Panel
Generator
Low Voltage
Switchgear –
Electrical
Distribution Centre –
400/230 VAC
T1

Tip: Condensation is the enemy of electrical equipment. Anti-Condensation Space Heaters are designed to maintain the temperature of an electrical enclosure. When condensation gathers on the inside surface of an enclosure, the risks of malfunction are high. It causes premature ageing, rusting, short circuits, and breakdowns in electric and electronic equipment that is housed in the enclosure. One of the most effective solutions in removing condensation combines a heater with a control device e.g. a thermostat. The thermostat turns on the heater when the ambient temperature drops below the set point. Outdoor equipment is particularly prone to condensation, especially when it is not thermally insulated, that is due to the change in temperature as daytime warmth turns into the chill of night.

Anti-Condensation Space Heaters example:

Water

Drinking Water, also known as **Potable Water** or improved drinking water, is water safe enough for drinking and food preparation. Methods and procedures for the maintenance of all water systems in your building(s) must be in place ensuring all that they are kept in a safe condition, and free from infection with legionella bacteria which could cause Legionnaires Disease.

Importance of Water Conservation - fresh, clean water is a limited resource. While most of the planet is covered in water, it is salt water that can only be consumed by humans and other species after undergoing desalination, which is an expensive process. People and businesses should do their best to conserve water.

Safe drinking-water is vital for good health. Water used for drinking, teeth cleaning, hand washing, bathing, showering, food preparation and cooking needs to be free from harmful germs and chemicals.

Typically, the water quality provided by a utility company to residential and industrial installations will be:

Conductivity: < 700 µS/cm at 20°C

pH: levels of > 6.5 < 9.5 pH units.

Companies must employ the latest practices and technologies being used in water treatment and hygiene e.g.:

1. Ongoing water filter replacement programme.
2. Drinking water fountains serviced at 6 monthly intervals.
3. Annual cold-water storage tanks inspection.
4. Daily safe operating procedures and temperature monitoring of both Hot and Cold-Water systems are performed.
5. Water Softening systems weekly inspection.
6. Chemical dosing systems weekly inspection.
7. Cooling tower water treatment and monthly detail monitoring by a water treatment specialist. Legionella testing at 3 monthly intervals.

8. 'Names of Site Responsible Person(s)' and their documented 'Allocation of Specific Responsibilities'.

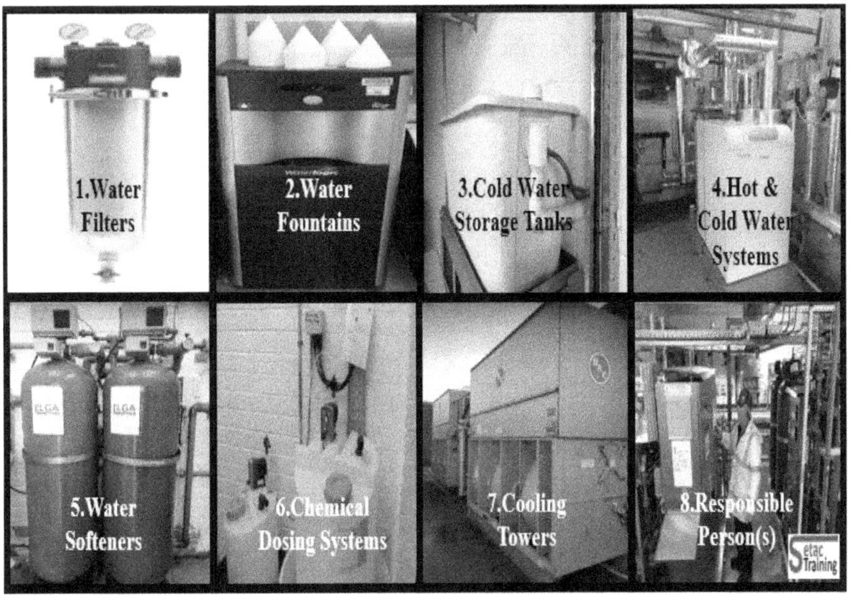

1. Water Filtration:

A service should be carried out at least **once a year** on all water filtration equipment. Changing of water filters can be based on **Differential Pressure.** Reading this pressure differential is important to monitor the condition of the filter element, as the filter becomes plugged with particulate matter and other various contaminants, the flow restriction through the element increases, resulting in reduced efficiency **or** via metered **water throughput based.** The water filtration units need to be sanitised by trained engineers or you run the risk of bacteria growth. An annual service can avoid parts getting damaged, which can be very expensive.

2. Drinking Water Fountains

Your water fountain should be kept cleaned and sanitised to ensure the safety of your employees, or the general public. It is recommended that you clean your drinking water fountain at least once a day with a disinfectant cleaning solution. Pay special attention to the mouthpiece and protective guard.

You should have a certified drinking fountain specialist perform a complete inspection of your unit every six months. This will increase the longevity of your water fountain, prevent downtime, and ensure that your unit is operating safely and free of defect.

3. Cold Water Storage Tanks

Cold Water Storage Tanks (Fig.14) must be located in areas that are 'readily accessible for cleaning'. Health & Safety Legislation requires regular Maintenance and Inspections.

Fig.14 - Typical Cold Water Storage Tank Installation example:

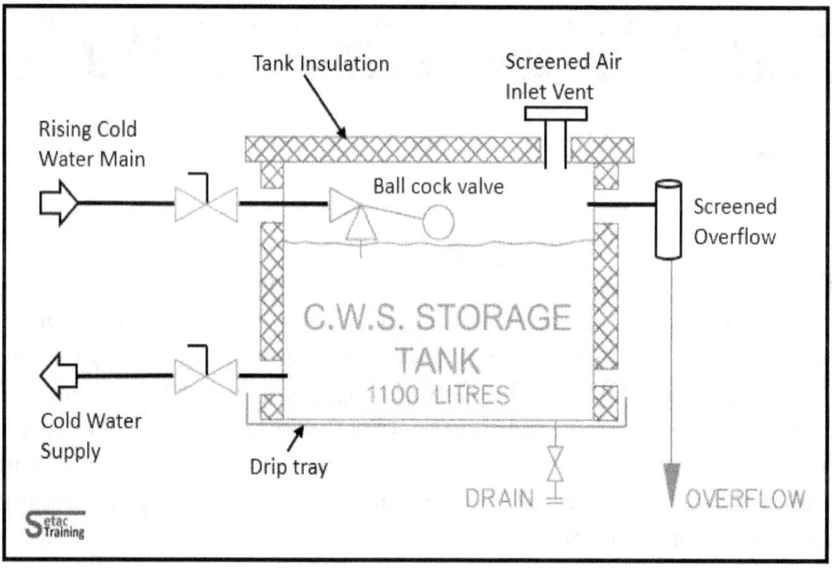

General

- Water tanks should be located to prevent water damage or consequential loss in the event of leakage howsoever occurring.
- All tanks which are located above water sensitive areas should have a bund wall around them with adequate evacuation ducts.
- If the tank is sited outdoors, check that the cover has not suffered structural damage and that it is securely bolted down.
- All other tanks should have Condensation or Drip trays to prevent nuisance damage and to keep floors dry. **N.B.** Appropriate overflows must be fitted as close as possible to the base of the tray. These must be fitted before the tank is filled.
- Install adequate insulation which will help prevent freezing and will also help keep the water as cool as practicable, ideally less than 20°C. **N.B.** Insulation only buys you time, it slows down but does not prevent heat loss or gain over protracted periods.

Cold Water Storage Tanks mostly **fail** due to age, lack of maintenance, failure of the structural supports underneath the tanks, contaminated water attacking the internal fittings and ball valve failure where overflows were not fitted.

Check List

1. All internal supports should be checked for corrosion, if corrosion is found, it should be rectified immediately.
2. If the tank has a cover and is sited outdoors, check that the cover has not suffered structural damage and that it is securely bolted down. Maintain integrity of tank insulation.
3. Check that all pipe work connected to the tank is suitably braced. Check generally for leaks or drips.
4. Check that the structural supports under the tank are in good condition. **N.B.** If it's a plastic cold-water tank, it must always be

mounted on a firm base supporting the complete underside on the tank; if this is not done, the weight of the water within the tank may cause the plastic to stretch, distort and eventually fail.

5. Check the **Screened Air Inlet Vent (Fig.15)** is open to the atmosphere to allow for the movement of air resulting from changes in the water level so that the water always remains at atmospheric pressure.
6. Check the **Screened Overflow Pipe (Fig.16)** that it hasn't become blocked. The primary purpose of rodent screens - as you might imagine - is to keep rodents and other unwanted pests from contaminating your water storage supplies.

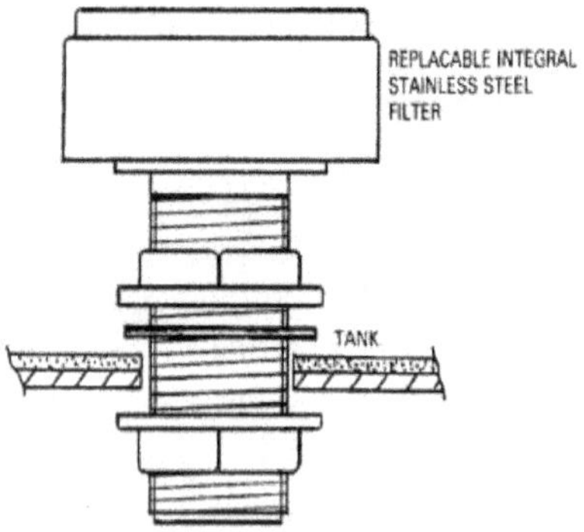

Screened Air Inlet Vent. (Fig.15)

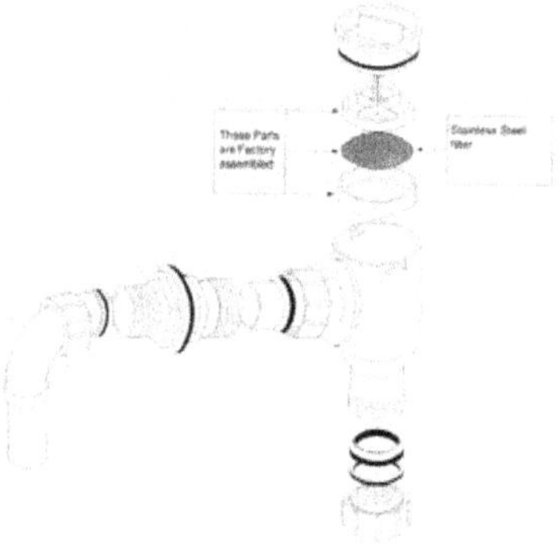

Screened Overflow (Fig.16)

Cold Storage Water Tank with Screened Air Inlet Vent & Screened Overflow Pipe example:

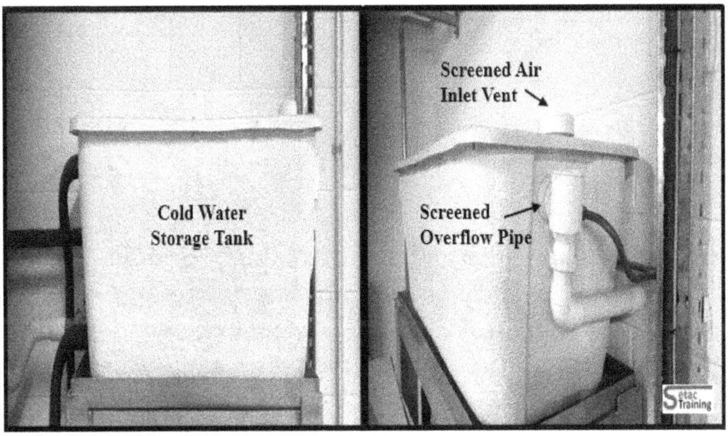

N.B. "Appearances can be Deceiving." A Cold Water Storage Tank might look 'Ok' on the outside, but there could be a **serious health issue** 'going on' inside.

4. Hot and Cold-Water Systems

A 'Responsible Person' must ensure that the correct and safe operational procedures of both Hot and Cold-Water systems are performed:

Daily Checks: Ensure all systems are running correctly.

Weekly Checks: Flush all little used outlets including all domestic and safety showers. All little used hot and cold-water outlets rarely used should be flushed for 2 minutes. Emergency wash stations must be kept contamination free 1. Run eyewashes until water is clear and cool. 2. Run safety showers until water is clear and cool. 3. Use some type of flow catcher to avoid potential slip hazards.

Monthly Checks: Monthly Temperature Monitoring of Sentinel Taps. In the case of low volume Point of Use Water Heaters (POUWH's) this would be the furthest tap from the unit.

- Hot taps should be a minimum of 50°C within 1 minute of turning on. (On POUWH's measure temperature as soon as the tap is turned on)
- Cold water taps should be less than 20°C within two minutes of turning on.
- Calorifiers when present should have a storage temperature of 60°C and more importantly a return temperature of 50°C minimum.

Three Monthly Checks: Shower Head/Spray Head maintenance/Safety Showers.

- All shower heads and spray heads should be dismantled, cleaned, and disinfected to an agreed method statement.
- Alternatively replace shower heads quarterly.

Annually: Tanks - Inspect, clean and disinfect all cold-water storage tanks to an agreed method statement.

Calorifiers: Purge any debris in the base of the calorifier to a suitable drain. Collect the initial flush from the base of hot water heaters to inspect clarity, quantity of debris, and temperature.

Typical schematic layout of an LPG Boiler Services, Hot Water Services, and Central Heating system in an office building:

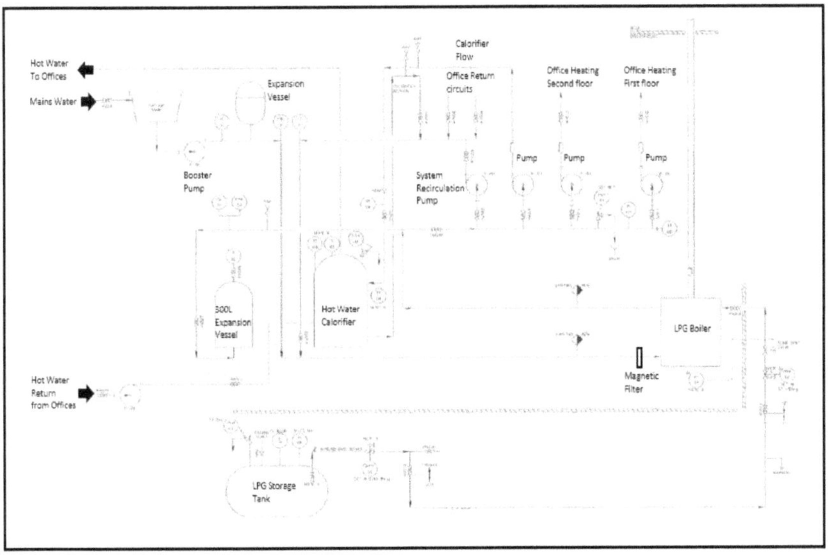

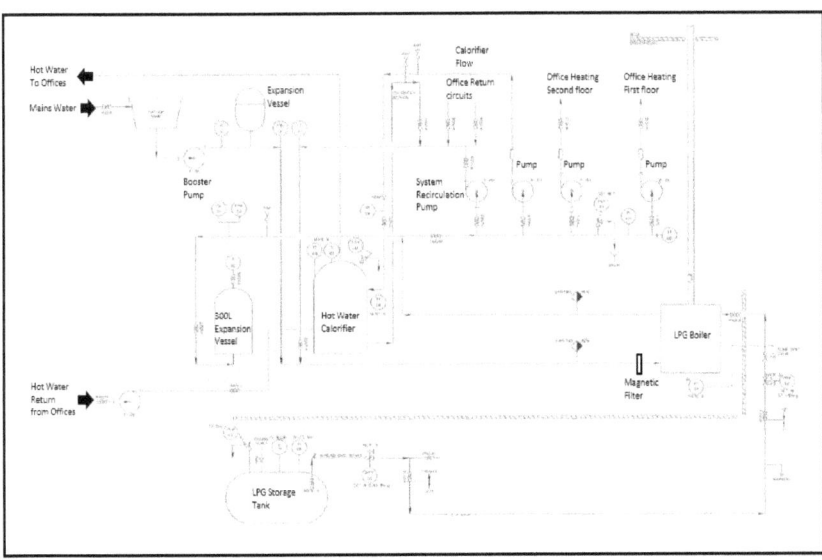

Typical layout of a Water Services Schematic and a Low-Pressure Hot Water (LPHW) Schematic in an office building:

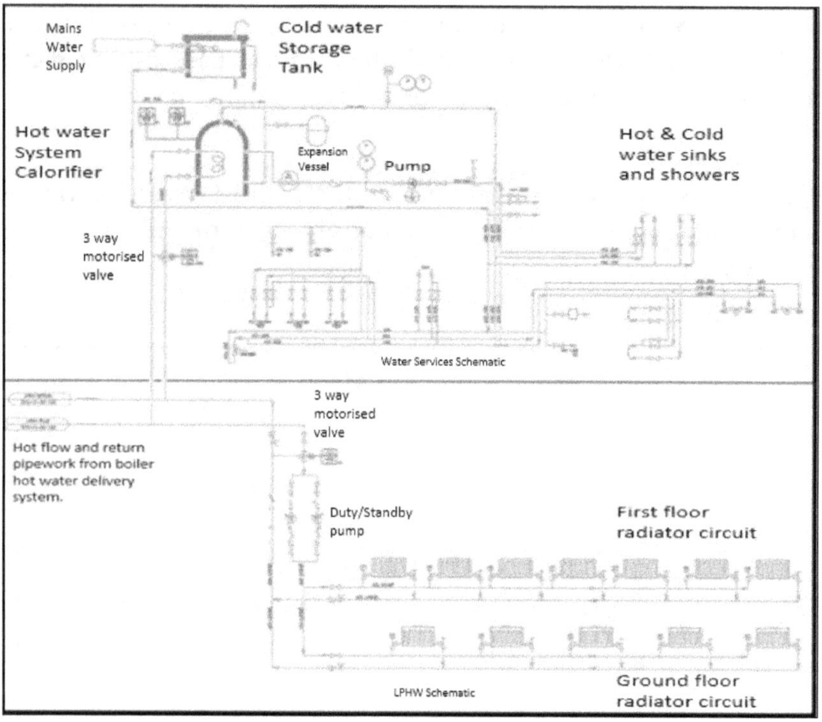

5. Water Softening

Hard water is water that has high mineral content (in contrast with soft water). Hard water is formed when water percolates through deposits of limestone and chalk which are largely made up of calcium and magnesium carbonates and makes its way into the groundwater supply.

Water with a high calcium and magnesium content are considered to be hard, and those waters with a low content are considered to be soft. The more Calcium and Magnesium the water contains, the harder it is.

Water hardness causes problems and leads to scale formation. Lime sticks to appliances and pipes creating a hard white or brownish scale, reducing its lifespan and in extreme cases, causing pipes to leak.

Limescale is a persistent problem in Boilers, Humidifiers, Dishwashers, Washing Machines, Reverse Osmosis systems, Thermostats and many more items. Limescale reduces the efficiency of Cleaning chemicals, affects the heating of water, and slows the operation of equipment such as RO plants.

Some of the effects of scale formation can be summarised as:

- Decreased plant efficiency
- Reduction in productivity
- Production schedule delays
- Increased downtime for maintenance
- Cost of equipment repair or replacement

Commercial / Industrial water softeners are used to remove hardness from water. An ion exchange hard water softener usually solves the problem.

6. Chemical Dosing systems

Chemical dosing equipment is used for applications such as boilers, cooling towers, and effluent water treatment systems. A water meter is often the ideal way to control the dosage of water treatment chemicals e.g. An effective cooling tower dosing system to control your cooling water chemical treatment programme is essential if problems with corrosion, scale formation, and Legionnaires' Disease are to be avoided.

7. Evaporative Condenser/Cooling Tower

The objective of a mechanical refrigeration system is to remove heat from a space or product, and to reject that heat to the environment in some acceptable manner. Evaporative condensers are frequently used to reject heat from mechanical refrigeration systems. The chilled water required is circulated through the tubes of the Chiller refrigerant evaporator where the temperature is reduced to its desired level (Fig. 17). The heat absorbed by the refrigerant in the evaporator is rejected by the cooling tower. The **Evaporative Condensing Heat Exchanger Coil** absorbs the heat from the refrigeration vapour thereby returning the 'high pressure, high temperature' vapour from compression to a 'high pressure, low temperature' liquid at the outlet of the tower. The **Cooling Tower** is the heat-rejection device, which discharges the warm air from the cooling tower to the atmosphere through the cooling of water. Water is pumped from the basin section and is distributed over the exterior of the condensing heat exchanger coil by a series of distribution sparge pipes and spray nozzles where the refrigerant vapour is condensed and heat is transferred to the water. The purpose of the cooling tower is to cool the warm water returning from the heat exchanger, so it can be reused. A fan system forces air through the falling water and over the coil surface. A small portion of the water is evaporated, removing heat from the refrigerant, and condensing it inside the coil. Therefore, like the cooling tower, all the heat rejection is by evaporation, thus saving about 95% of the water normally required by a 'once-through' system.

Fig. 17: GA Drawing:

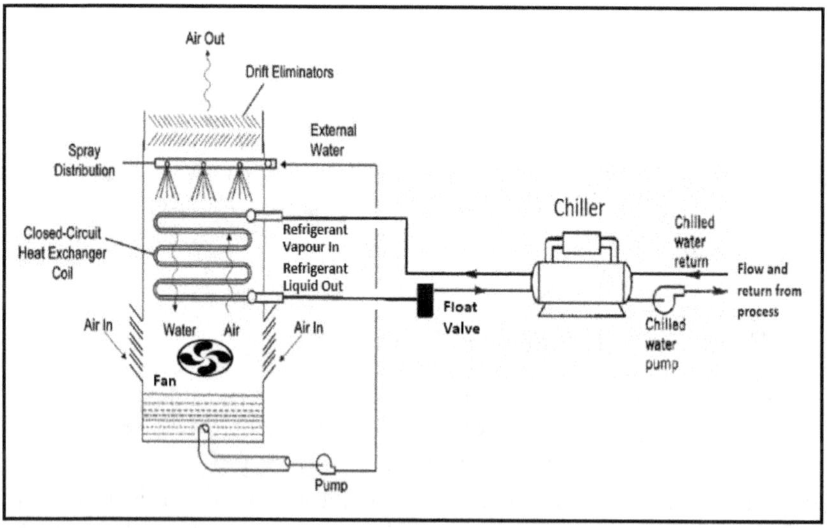

A **'Responsible Person'** must ensure the correct and safe operation of the system by having in place:

- A full and correct water treatment programme
- Weekly monitoring of key parameters
- Weekly inspection of water treatment equipment
- Monthly detail monitoring by a water treatment specialist
- Quarterly legionella analysis.
- Annual (sometimes bi-annually) clean and disinfection of cooling towers

Procedures - Cooling Towers

- **Daily Checks:** Visual check of cleanliness of the water in the system. Visual check that the systems are running correctly.
- **Weekly Checks:** Monitor the microbiological activity, pH, and conductivity of the cooling tower water. Check that the water treatment dosing equipment is operating correctly. Monitor water treatment chemical usage.
- **Monthly Checks:** Water treatment specialist should carry out a detailed analysis of the cooling tower and make-up water.
- **Quarterly Checks:** Cooling tower water should be checked for legionella bacteria by an approved laboratory. The physical condition of the cooling tower should be checked, this should include: Distribution system, drift eliminators, louvers, and screens.
- **Annual Checks:** A Cooling tower must have any scale removed. It must be cleaned and disinfected (sometimes bi-annually) to an agreed method statement. This work should be certified. It must also be monitored for any sign of corrosion.

Typical Maintenance & Monitoring programme for a Cooling Tower

Schedule

Type of Action	Action	Start-Up	Weekly	Monthly	Quarterly	Every Six Months	Annually	Shutdown
Checks and Adjustments	Cold Water Basin and Strainers	X			X			
	Operating level and make-up	X		X				
	Blow down	X		X				
	Sump heater package	X				X		
	Belt tension	X		X				
	Drive alignment	X					X	
	Locking collar	X				X		
	Rotation of fan(s) and pump(s)	X						
	Motor voltage and current	X					X	
	Unusual noise and/or vibration	X		X				
Inspections and Monitoring	General condition	X		X				
	Heat transfer section	X				X		
	Finned discharge coil (optional)	X					X	
	Drift eliminators	X				X		
	Water distribution	X				X		
	Fan Shaft	X			X			
	Fan Motor	X			X			
	Electric Water Level Control Package (optional)	X				X		
	TAB test (dip slides)	X	X					
	Circulating water quality	X		X				
	System overview	X					X	
	Record keeping	as per event						
Lubrication	Fan shaft bearings	X				X		
	Motor bearings *	X				X		
	Adjustable motor base	X				X		
Cleaning procedures	Mechanical cleaning	X					X	
	Disinfection **	(X)					(X)	(X)

8. Responsible Person(s)

Should any of the procedures or checks show that the water systems are not running within specified limits, the Responsible Person will decide what action is required to get the system back operating within specification. This should include a procedure should legionella bacteria be identified in any system.

Record System

The Responsible Person will ensure there is a Central Records System in place which will contain the most up to date Legionella

Control risk assessment.

The Central Records System should include:

- Names and positions of people responsible, and their deputies, for carrying out the various tasks under the written scheme
- A risk assessment and a written scheme of actions and control measures
- Schematic diagrams of the water systems
- Details of precautionary measures that have been applied/implemented including enough detail to show that they were applied/implemented correctly, and the dates on which they were carried out
- Remedial work required and carried out, and the date of completion
- A log detailing visits by contractors, consultants, and other personnel
- Cleaning and disinfection procedures and associated reports and certificates
- Results of the chemical analysis of the water
- Results of any biological monitoring
- Information on other hazards, e.g. treatment chemicals
- Training records of personnel
- Review meetings
- Audit of record system
- A nonconformance log detailing out of specifications results or readings and action taken reestablish control
- The name and position of the person or people who have responsibilities for implementing the written scheme, their respective responsibilities, and their lines of communication

- Records showing the current state of operation of the water system, e.g. when the system or plant is in use and, if not in use, whether it is drained down
- Either the signature of the person carrying out the work, or other form of authentication where appropriate.

Legionnaires' Disease

Legionella is a water borne bacteria that can grow in engineered water systems such as cooling towers, condensers, and hot and cold-water systems. Legionnaires' disease is a potentially **fatal** form of pneumonia carried by stagnant water which can affect anybody, but which principally affects those who are susceptible because of age, illness, immunosuppression and smoking. Legionella bacteria can also cause less serious illnesses, which are not fatal or permanently debilitating but which can affect all people.

According to the **World Health Organisation** paper on Legionella and the prevention of Legionellosis, temperature affects the survival of Legionella as follows:

Maintaining the temperature of hot and cold-water systems within buildings to prevent or minimise the growth of legionellae is an important control measure to prevent the risk of Legionella infection.

- Above 70°C (158°F): Legionella dies almost instantly
- At **60°C** (140°F) (Normal hot water storage temperature): 90% die in 2 minutes
- At 50°C (122°F): 90% die in 80–124 minutes, depending on strain
- At 48°C to 50°C (118°F to 122°F): Can survive but do not multiply
- 32°C to 42°C (90°F to 108°F): Ideal growth range.

N.B. To prevent Legionella infection, the recommended temperature for storage and distribution of **cold water** is below

25°C, and ideally below 20°C.

Water systems should:

- Avoid water temperatures between 25°C and 45°C, to prevent Legionella colonisation
- Ideally, maintain cold water below 20°C
- Ideally, maintain circulating hot water above 50°C.

N.B. Hot water is normally stored at 60°C with a minimum flow temperature of 60°C to be maintained in water leaving the heating unit, and above 50°C at the water tap (one minute after leaving the heating device). In domestic and public hot-water systems, control measures for reducing the proliferation of Legionella must not increase the risk of scalding, particularly for children, the elderly, or people with disabilities. To avoid this risk, install 'anti-scald thermostatic mixing valves' at each human point of use with a typical mixed water temperature setting of:

- Bidet at 38°C (100°F)
- Washbasin at 41°C (106°F)
- Shower at 41°C (106°F)
- Bath at 44°C (111°F)

Anti-Scald Thermostatic Mixing Valve (TMV) set at 41°C (106°F) washbasin example:

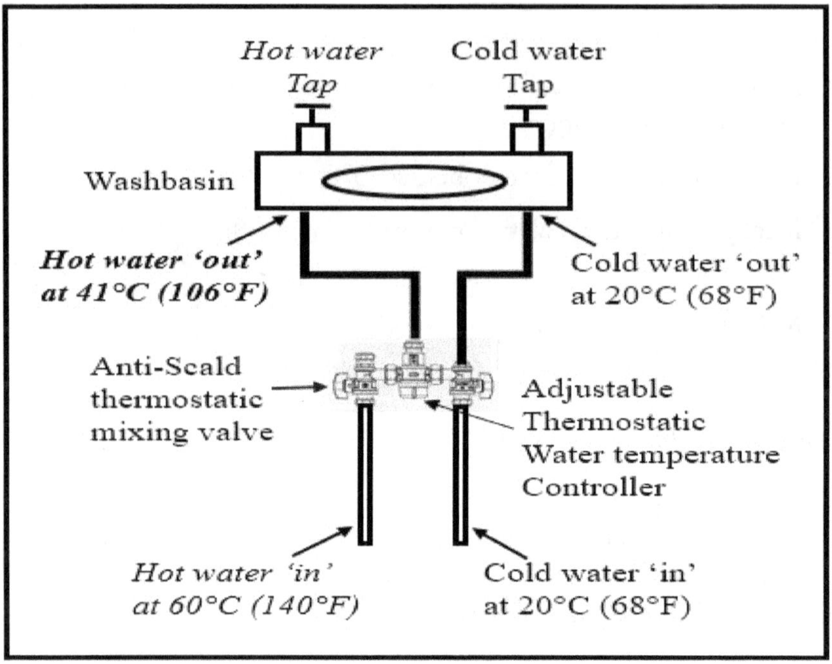

Factors that can lead to proliferation of or exposure to Legionella in piped water systems include:

- Poor water quality and water treatment failures
- Distribution system problems such as stagnation, water pipe dead legs (a redundant length of pipe, closed at one end) and low flow rate
- Construction materials that contribute to microbial growth and biofilm formation
- Inefficient or ineffective disinfection
- Water temperature of 25°C to 45°C
- Presence of biofilms
- Aerosol production.

To comply with their legal duties, employers, and those with responsibilities for the control of premises should:

1. Identify and assess sources of risk – this includes checking whether conditions are present which will encourage bacteria to multiply, **e.g.** Is the water temperature between 25°C to 45°C; is there a means of creating and disseminating breathable droplets by the aerosol created by a shower or cooling tower; and if there are susceptible people who may be exposed to the contaminated aerosols.
2. Prepare a written scheme of precautions for preventing or controlling the risk **e.g.** Water systems at risk from stagnation should be periodically flushed or disinfected, and temperatures that are optimal for growth of Legionella should be avoided. However, where flushing is used, the likely exposure of people to aerosols generated during flushing must be considered.
3. Implement, manage, and monitor precautions – if control measures are to remain effective, then regular monitoring of the systems and the control measures is essential. Monitoring of general bacteria numbers can indicate whether microbiological control is being achieved. Sampling for Legionella is another means of checking that a system is under control.
4. Keep records of the precautions; and appoint a person to be managerially responsible.

Purified Water

Water is an essential ingredient to pharmaceutical manufacturing. Purified Water must meet the requirements for ionic and organic chemical purity and must be protected from microbial contamination. Once purified, it must be stored and distributed in systems appropriately designed, installed, commissioned, and validated.

Proven technologies like Reverse Osmosis (RO), and Continuous Electrodeionisation (CEDI) provide your manufacturing processes with dependable and low maintenance solutions for water purification. System components and options are selected and configured based upon the product-water specifications (e.g. TOC & Conductivity) and specific properties of your feedwater. A service provider's Performance Qualification Maintenance Contract (PQMC) on your PW Generation system is normally carried out every four months or per manufacturer's instructions.

Purified Water Generation System GA Drawing Example

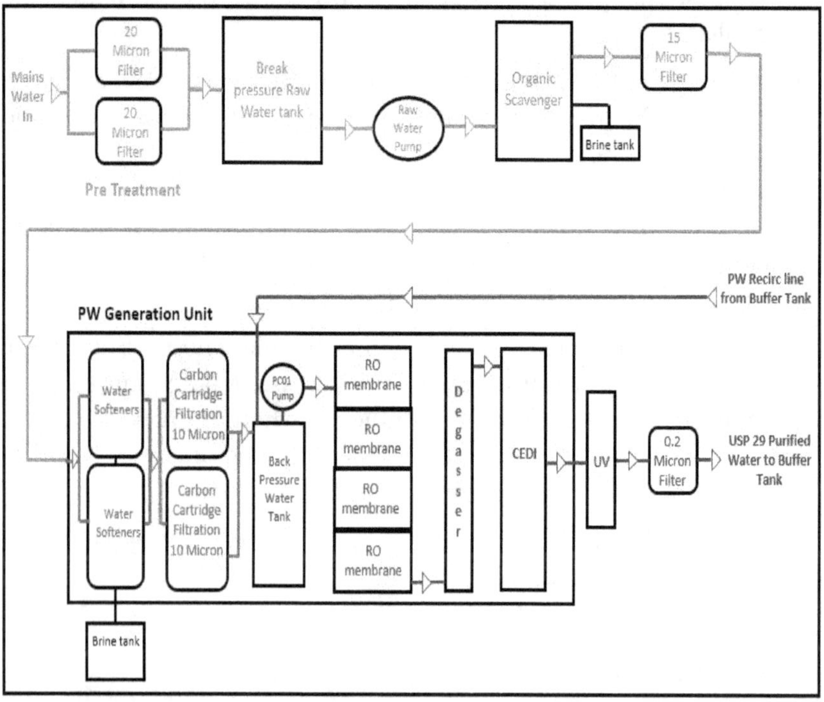

Foul and Storm Water drainage

The foul sewer is designed to carry contaminated wastewater to a sewage works for treatment, whereas the surface water sewer carries uncontaminated rainwater directly to a local river, stream, or soakaway.

What each sewer does:

Sewer system	Disposes of:	Transferred to:
Foul water	Toilet, bath, shower, kitchen sink, washing machine, dishwasher.	Local Sewage treatment works
Surface water	Rainwater from roof, driveway, patio, roads.	Local watercourse

N.B. Foul sewers can succumb to age related deterioration, distribution due to ground movement, and other forms of stress and may be in need of pipe renovation. A **Drain Integrity Study** must be carried out to ensure that your facility is compliant with any relevant quality standard for waters, trade effluent, and sewage effluent. You as the property owner are responsible for the repair and maintenance of the wastewater facilities and pipework within your site and up to the point where the pipework meets the public sewer.

Natural Gas

In its original state, Natural Gas (which is composed mostly of methane) is odourless and colourless. Distribution companies add an odorant, so that it is easily detectable in the event of leaks.

Under normal operating conditions, Natural Gas burns cleanly, producing heat, carbon dioxide, and water vapour. **N.B.** Natural Gas can cause death by suffocation if the gas displaces the air in a confined space. It is the cleanest burning fossil fuel available, but if natural gas isn't burning properly or the gas appliance has a mechanical problem, it could create a hazard.

How to Prevent Carbon Monoxide (CO) Poisoning:

Carbon monoxide (CO) is a deadly, colourless, odourless, poisonous gas. The incomplete burning of various fuels produces it, including coal, wood, charcoal, oil, kerosene, propane, and **Natural Gas**. Carbon monoxide can be generated by any combustion fuel, so it is important that all appliances are installed and maintained by certified competent engineers with proper knowledge, skills, and tools. Fumes you cannot smell like carbon monoxide are extremely dangerous. Besides furnaces, any gas-fuelled appliance produces it—ovens, grills, fireplaces.

The best way to prevent CO poisoning is to make sure appliances are installed and operated according to the **manufacturer's instructions and local building codes.** Have the heating systems professionally inspected and serviced annually to ensure proper operation. The inspector should also check chimneys and flues for blockages, corrosion, partial and complete disconnections, and loose connections.

Tip: Look out for problems that could signal improper appliance operation e.g. Burning or unfamiliar odours.

Audible carbon monoxide (CO) alarms are an essential back-up precaution, but they are not a substitute for the proper installation and maintenance of combustion heating appliances.

Approved carbon monoxide alarms must be strategically placed and installed where they will be most effective. Check the instructions from the manufacturer. Ensure that CO alarms are correctly located, are fitted at the right height, and close enough to the potential source of CO.

CO alarms are designed to detect and alarm before potentially life-threatening levels of CO are reached. These types of alarms will locally audibly alert personnel when a certain level of CO is in the air. They can also be linked to a BMS which triggers an automatic alarm/action and text in the BMS.

The most common problems that can occur with an 'Oil heating' system:

Fumes - If you smell oil, it generally means your system requires maintenance. The fumes can be dangerous and may signal a crack or misalignment in your oil burner. When an oil burner ignites, it pressurises the combustion chamber for a few seconds. The smoke from the **unburned oil** can move into the surrounding fresh air chamber (the heat exchanger) that then circulates into the general area. If there is a crack or a hole in the heat exchanger, you will smell oil fumes. A sooty wall around a heating vent is another sign of internal problems with your system that needs repair. With a cracked heat exchanger, vents can become blocked and inadequate air supply for combustion appliances can force contaminated air back into the area. You will smell an oil odour (and possibly see smoke and soot), which should prompt you to call a service engineer immediately before any carbon monoxide is released.

Many different CO alarms are available, but it is recommended that the CO alarm:

- Complies with European Standard EN 50291 (or equivalent) - This should be marked on the box.
- Carries the CE Mark.
- Has an **end of life** indicator - This indicator should not be confused with any fault indicator.
- Carries an independent certification mark – For example a kite mark, this indicates that the CO detector has been approved by an accredited testing and certification organisation.

Tip: Even though the dangers of carbon monoxide are much lower in a house with an oil heating system (as opposed to a gas furnace), a CO detector should be installed.

Compressed Air - A typical industrial plant Compressed Air Generation System P&ID:

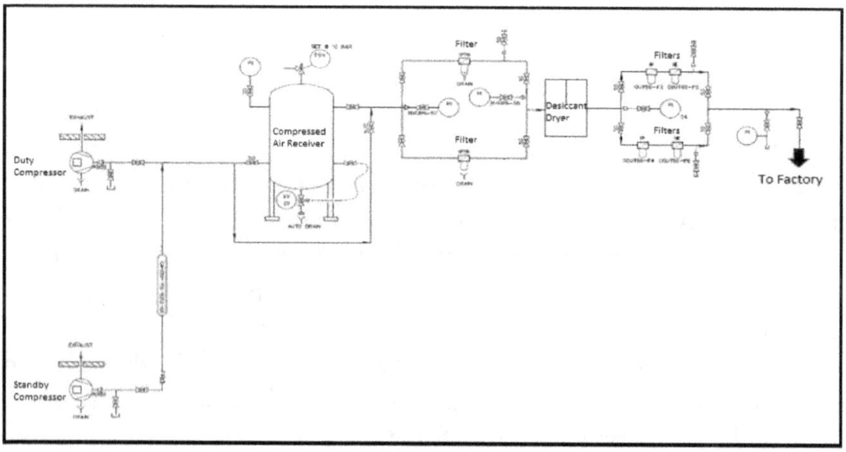

Compressed Air Generation System GA Drawing:

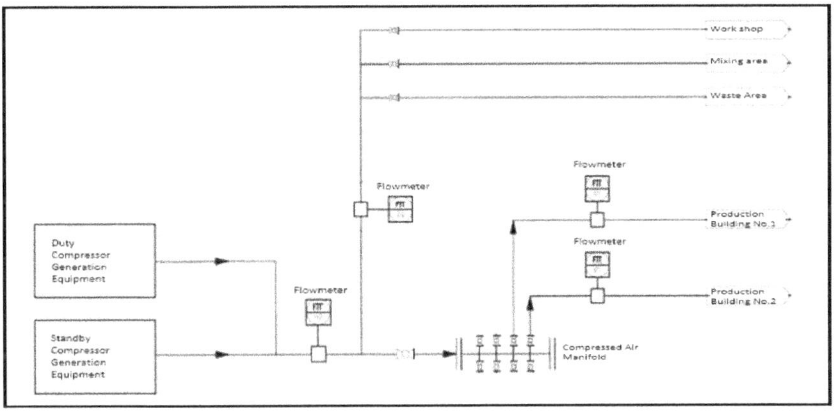

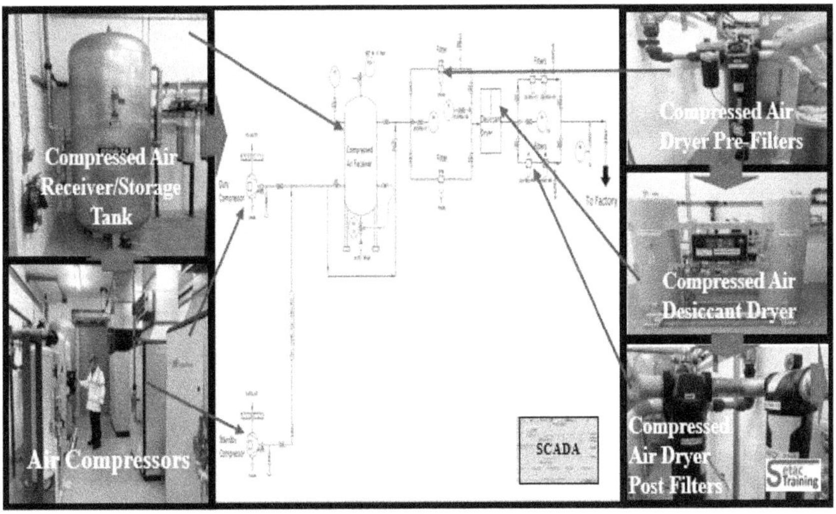

In industry, **Compressed Air** is so widely used that it is often regarded as the fourth utility, after Electricity, Natural Gas, and Water. A powerful characteristic of compressed air as an energy source is its ability to be generated and then safely stored for future distribution to point of use when required. However, compressed air is very expensive to produce and is more expensive than the other three utilities when evaluated on a per unit energy delivered basis.

Compressed air is used for many purposes, including:

- Pneumatic pistons.
- Air tools e.g. pneumatic drills
- Process control systems e.g. electro pneumatic valves, pumps.

Air Compressor: By installing a top quality oil-free air compressor and ancillary compressed air purification equipment to your system, you can ensure that your company will adhere to the following strict regulation: According to Section 820.70 (e) of Title 21 of the Code of Federal Regulations set by the FDA states that "each manufacturer shall establish and maintain procedures to prevent contamination of equipment or product by substances that could reasonably be expected to have an adverse effect on product quality." **Contaminated** air can be catastrophic for your production quality and your point of use equipment and machinery.

Compressed Air Storage: Optimum efficiency of a complete system demands that storage receivers are correctly sized to suit generation and demand. If no air receivers have been installed, the system's pressure profile and lack of storage control, limits the effectiveness of compressed air storage. The associated distribution pipework also forms a storage facility and as such the system must be designed and installed correctly to ensure system integrity.

Compresse Air facts:

- It takes 7 kWh of electrical energy to produce 1 kWh of compressed air. Electricity consumption being 60 to 70% of the total cost of compressed air.
- **Leaks** (the invisible energy thief) account for 40% of all losses but are simple to control. Losses through a 5mm diameter hole can be financially very costly, so a routine system of checking and repairing leaks should be established.

- More energy is needed to generate air at high pressure, so the generating pressure should be reduced to the minimum level acceptable. Reducing operating pressure by 1 bar can reduce operating costs by up to 5%.
- The intake air for a compressor should be cold and dry and taken ideally from outside the building through an inlet that prevents rain ingress. Every 5°C fall in air intake temperature reduces operating costs by about 2%.
- Capturing and utilising the **heat energy generated** when compressing air considerably improves the overall system efficiency. Heat is an inevitable by-product of air compression. Why let this thermal energy vanish into the atmosphere via the cooling system and radiation, when it could be reclaimed using an energy recovery system?

Dew point temperatures in compressed air, range from ambient down to -112 °F (–80°C), sometimes lower in special cases. Compressor systems without air drying capability tend to produce compressed air that is saturated at ambient temperature (Atmospheric air is usually moisture laden from ambient humidity). **N.B.** 'Moisture reduces efficiency'.

Systems with refrigerant dryers pass the compressed air through a chilled heat exchanger, causing water to condense out of the air stream. These systems typically produce air with a dew point no lower than 23°F (5°C).

Desiccant drying systems absorb water vapour from the air stream and can produce air with a dew point of **-40°F (–40°C)** and drier if required. **Compressed Air purification equipment** is essential to all modern production facilities. It must deliver uncompromising performance and reliability whilst providing the right balance of air quality with the lowest cost of operation.

Compressed Air Filters protect your equipment and delicate instruments from DIRT (dust, solid particles, rust particles, micro-organisms), WATER (water vapour, condensed liquid water, water aerosols, acidic condensates), and OIL (liquid oil, oil aerosol, hydrocarbon vapour), usually found in compressed air. The filters must be regularly serviced to prevent blockages. Filters are installed to provide contaminant removal to a specific air quality requirement, therefore the primary reason to change filter elements should always be to maintain air quality and they should therefore be **replaced every 12 months or when reaching a differential pressure (dp) of 5 psi/0.35 bar.** Consideration must also be given to system pressure losses (and therefore operating costs) as the cost of a replacement filter element is often significantly lower than the energy cost associated with operating with higher differential pressures. Maximum system pressure loss should be less than 0.7 bar/10 psi. Often a company's emphasis is on the cost of maintenance and replacement parts, when in reality, these costs are insignificant to those associated with product spoilage should a filter element fail. What seems like a cost saving in the short term can turn out to be a costly mistake.

Compressed air is energy. Leaking air lines, couplings, and hoses; simply put, are sources of energy waste. You're literally blowing money away. Most companies have a leakage rate of more than 20%. Apart from being completely unnecessary, it is expensive and contributes to your CO2 emissions.

Saving costs on compressed air is not simple. Why do Leakage Management?

- Eliminates waste / Direct savings!
- Environment – Significant CO2 reduction!
- Delay or avoid new investment
- Secure reliability of production equipment/tools
- Reduce noise level

What is the true cost of Compressed Air?

Electricity usage in kWh – (Compressors, Dryers)

+

Service & Maintenance - (Compressors, Dryers, Filters)

+

Investments – (Compressors, Dryers, Filters, Control System)

÷ *Consumption* = **€/m3**

Refrigeration

Regular Preventative Maintenance (3 monthly intervals, or per manufacturer's instructions) must be carried out on your site's "Refrigeration" systems e.g. Vapour Compression Refrigeration. To properly maintain a system, all the major components should be included in the maintenance schedule. These include 1. EVAPORATOR – (low pressure liquid refrigerant e.g. 'Anhydrous Ammonia' in evaporator absorbs heat from the Process Circulating Glycol Cooling liquid and the liquid refrigerant changes to a gas). 2. COMPRESSOR (the superheated vapour enters the compressor where its pressure is raised).

3. CONDENSER (the high pressure superheated gas is cooled in several stages in the condenser). 4. EXPANSION DEVICE (liquid passes through expansion device e.g. High-pressure float valve, which reduces its pressure and controls the flow into the evaporator).

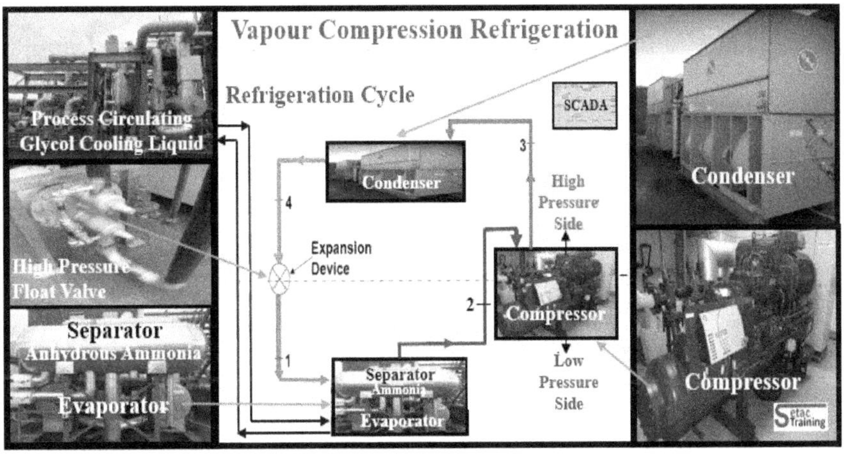

Condenser blockages, evaporator issues, heat transfer and valve expansion problems are some of the issues that can impact the performance, efficiency, and lifespan of your systems. Correct operation via a good automation strategy and regular routine maintenance is the best way to ensure your systems remain at peak efficiency, keeping your running costs to a minimum, and extending their working life e.g. Properly maintaining your refrigeration equipment can save between 5% and 10% on energy costs. Refrigeration systems can account for the largest energy usage, often up to 60% of your plant's total operating expenses. The older your refrigeration system, the more energy it consumes (consider replacing it, often the cost of a new system is quickly recoverable in energy savings).

Steam – Black Utilities - In industry, steam is used regularly for heating or as a driving force for mechanical power. Steam is one of the most common and effective heat transfer mediums used in industry. Steam-based heating processes use latent heat and transfer it to a given product. Steam is formed when water vaporises to form a gas. For the vaporisation process to occur, the water molecules must be given enough energy that the bonds between the molecules break. This energy given to convert a liquid into a gas is called 'latent heat'. Take the example of a kettle boiling water. Water temperature is raised using a heating element. As the water absorbs more and more heat from the element, its molecules become more agitated and it starts to boil. Once enough energy is absorbed, part of the water vaporises, which can represent an increase as much as 1600X in molecular volume.

Safe Operation of Boilers:

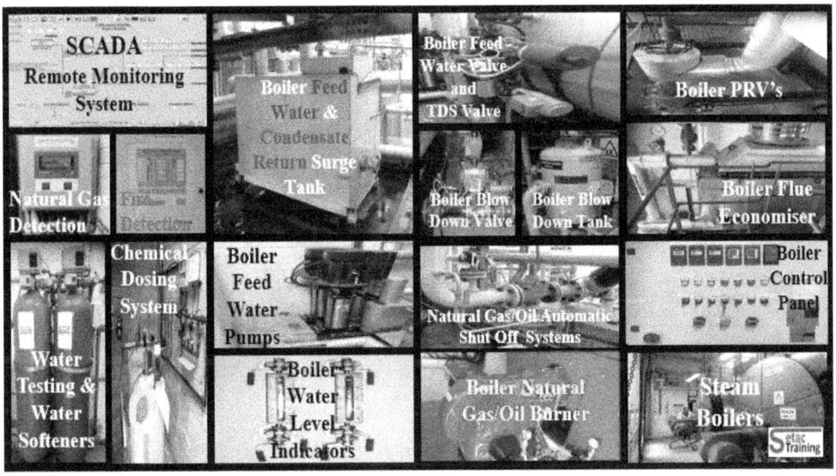

Boiler systems are required to comply with different legislation, including several health and safety regulations, which aim at ensuring that new and existing boiler systems are continually operated and maintained in a safe manner.

Who is a boiler operator?

A specifically trained worker who monitors and controls automatically operated boilers to generate steam that supplies heat or power for buildings or industrial processes. They observe many parameters such as water level, steam pressure, temperature, boiler alarms, feed water regulators, chemically treated water systems, Total Dissolved Solids (TDS), boiler 'blow down' and burner operation on the boiler's bespoke control panel to verify specified operation of the boiler's combustion control system.

A **Steam boiler** (Fig.20) or simply a boiler, is basically a closed vessel into which water is heated until the water is converted into steam at a required pressure. Fuel (e.g. Natural Gas or Diesel oil) is burnt, the hot flue gases that are produced, pass down the cylindrical internal furnace. The heat of these hot gases transfer to the water and consequently steam is produced in the boiler.

Fig.20 - Typical Boiler GA Arrangement:

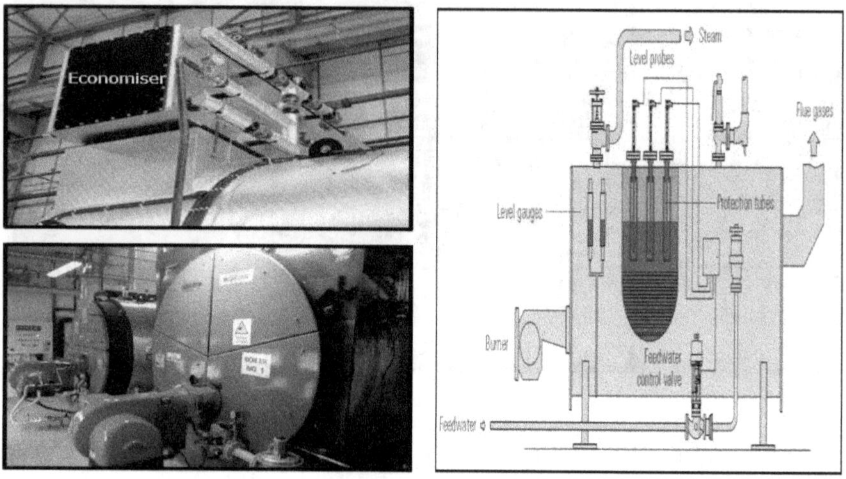

Tests and Examinations of Boilers

The results of all tests and examinations of boilers must be recorded and retained for a suitable period.

Examples of the type of records that must be kept and made available for insurance inspection include:

- Manufacturer's records and instructions
- Examination reports
- Record of periodic tests e.g. (Non-Destructive Testing (NDT), Hydraulic test)
- Written Scheme of Examination (WSE)
- Certificates of thorough examination
- Records of servicing & modifications

- Maintenance of controls
- Training records for boiler operators
- Audit reports for boiler operators
- Test log
- Water treatment records
- Risk assessment.

Maintenance personnel

Maintenance personnel must only carry out work for which they have been trained and are deemed competent. Suitable training courses and maintenance services can usually be provided or recommended by manufacturers of boilers, fittings, or control equipment.

N.B. Correct operation and regular routine maintenance is the best way to ensure that your boiler plant remains at peak efficiency, keeping your running costs to a minimum, and extending its working life.

Economisers

In the case of Boilers, flue gases or exhaust from the boiler outlet are at high temperature and water that needs to be preheated is at low temperature. **Economisers** are designed to recover heat from Boiler exhaust gases, and transfer this to the incoming Boiler feed water, thus increasing the overall Boiler Thermal Efficiency. The exhaust from a boiler is generally in the temperature range of 200°C to 250°C, so there is a huge amount of losses from the boiler if a **heat recovery device** is not installed after it. If the exhaust gases which are leaving the Boiler at such high temperature are made to pass through the Economiser in order to provide the required sensible heat to the water by increasing its temperature, it will reduce the heat load on the boiler to the greater extent. **Up to 6%+ fuel savings with an economiser can be realised.**

Boiler Water Treatment

Water treatment programmes must be properly designed and installed that maintain performance and integrity e.g. **Oxygen attack in boilers** - without proper mechanical and chemical deaeration, oxygen in the feed water enters the boiler. Much is flashed off with the steam; the remainder can attack boiler metal. Oxygen in water produces pitting that is very severe because of its localised nature. Water containing ammonia, particularly in the presence of oxygen, readily attacks copper, and copper bearing alloys. The resulting corrosion leads to deposits on boiler heat transfer surfaces and reduces efficiency and reliability. Effective water treatment includes a combination of filters, water softeners, chemical dosing techniques, and reverse osmosis.

Responsibilities

The owner/user of a boiler system (Fig.21) is ultimately responsible for ensuring the system complies with all the relevant Health & Safety legislation.

Typical Steam Distribution System: (Fig.21)

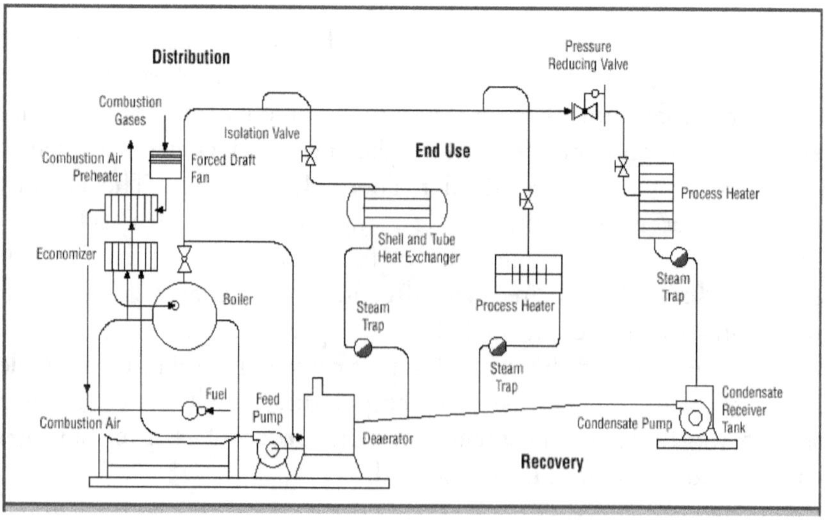

You must know the thermodynamics of the site's utility systems (Thermo meaning 'heat' and Dynamics meaning 'Power'). Thermodynamics studies the movement of energy, how energy instills movement and how that energy can be exchanged between physical systems as heat or work; it mainly involves changes in temperature, pressure, and volume.

E.g. A typical example of a **heat exchanger** would be an electric kettle, where a 2 kW electric heating element will bring 1.7 litres of water to a boiling point of 100°C in approximately 5 minutes. Electrical energy being converted to heat, thus boiling the water, other examples are immersion heaters and electric showers.

Now multiply the above kettle kW rating of 2 kW's by 3,000 or more where you have an industrial facility which may have many large **water converted to steam** boilers to superheat the water in the boiler to generate steam pressures of >8 Bar with temperatures of > +140 °C.

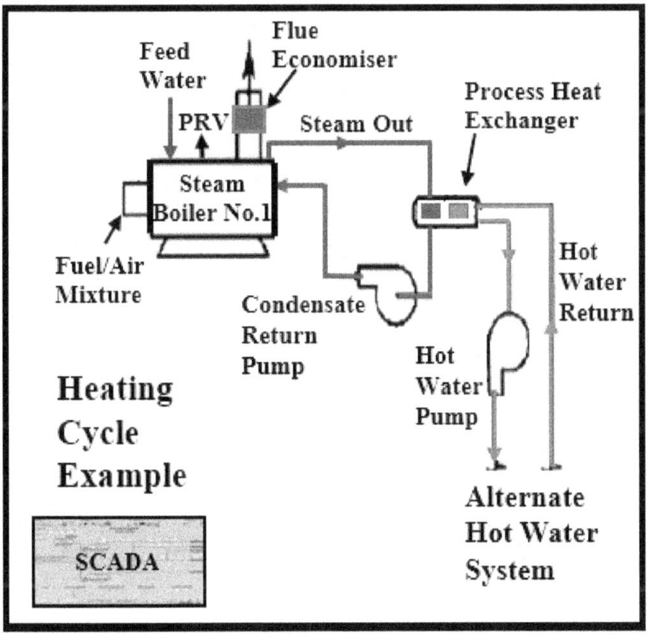

This **steam** is then used to heat a heat exchanger (Fig.22 & Fig. 23) which supplies hot water via pumping systems to the entire plant process equipment heat exchangers e.g. vessels, dryers.

Steam to Hot Water Generation and Distribution System: (Fig.22)

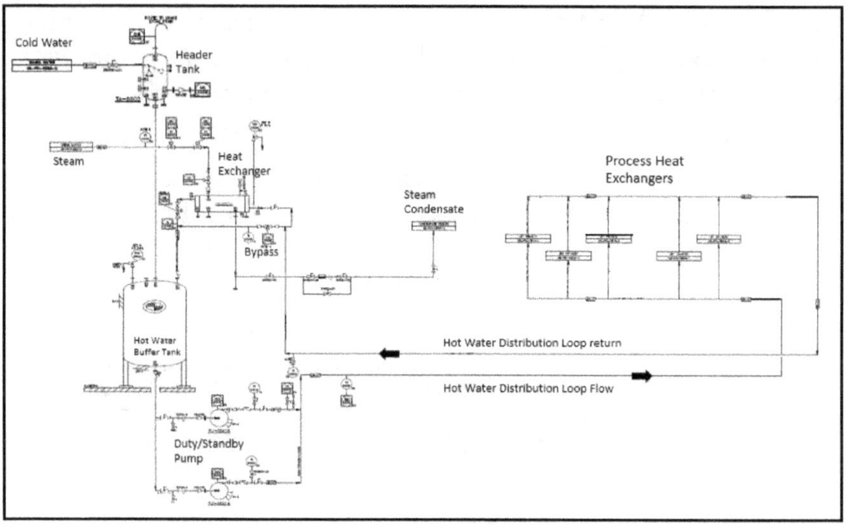

Steam heats the heat exchanger (see below example - expanded view) circulating liquid to a set temperature as dictated by the SCADA control system via TT's and automatic control valves. Steam gives up its latent heat in the heat exchanger and is converted back to water condensate through a steam condensate trap and back to the boiler for recycling.

Steam to Hot Water Generation and Distribution System (Example - Expanded View – Fig.23)

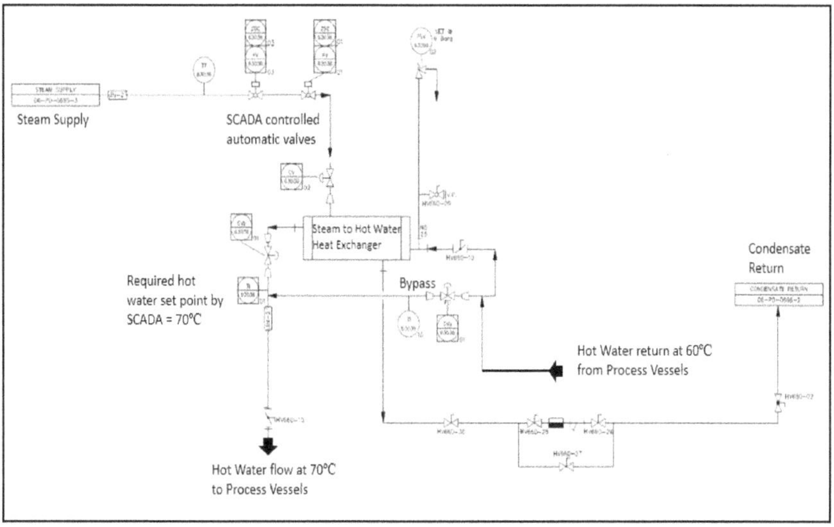

What is condensate? It is the liquid formed when steam passes from the vapour to the liquid state.

Condensate Tips:

Boiler feedwater - If condensate is not used as feedwater, the boiler must be continually topped up with cold water, which is costly in terms of both water and energy. Cold feedwater must be heated. In contrast, condensate is already hot, so not only does it reduce the need for (and cost of) fresh water and treatment chemicals, it also requires much less energy than cold make-up water does to be ready for use. Indeed, **reusing condensate in this way can reduce boiler fuel costs by 10–20%.**

Or, to look at things differently, **every 6°C boost in the temperature of the feedwater knocks 1% off a typical boiler's energy usage.**

Reducing boiler fuel demand has other benefits too. It can bring down emissions of carbon dioxide, nitrogen oxides and Sulphur oxides, which makes the entire process more environmentally friendly.

Safety Tip: Water hammer – Health & Safety Hazard

Water hammer must never be tolerated. Any system which is properly designed and operated will never experience this water hammer phenomenon, which is so often attributed to **Steam Systems**.

If water (condensate) collects in the bottom of a badly aligned pipe (Fig.24), the steam velocity (steam flow is approximately 35 m/s) will cause ripples on the surface of the water. The steam will blow these ripples into waves until a wave is high enough to fill a pipe (Fig.25). There is then an incompressible liquid (water is virtually incompressible) piston travelling along the pipe at steam speed. This whole pocket of water is picked up by the steam and carried forward in a solid column. It is taken to some point down the line where there may be a change of direction or an obstruction such as a valve (Fig.26). The water is brought to a sudden halt at this point and the energy it contains, by virtue of its movement (kinetic energy), is suddenly converted into pressure energy. This sudden pressure causes large banging/rattling sounds and vibrations throughout the system that may cause considerable damage.

If the water hammer is due to the condensing of steam during distribution in steam pipes, it can be eliminated. **Proper alignment** of the piping system ensures a continuous fall in the direction of flow and this must be combined with an adequate number of good drain points. It is particularly important to drain any low points in a system and to make sure that **steam traps** are working properly.

Poor alignment of a steam piping system:

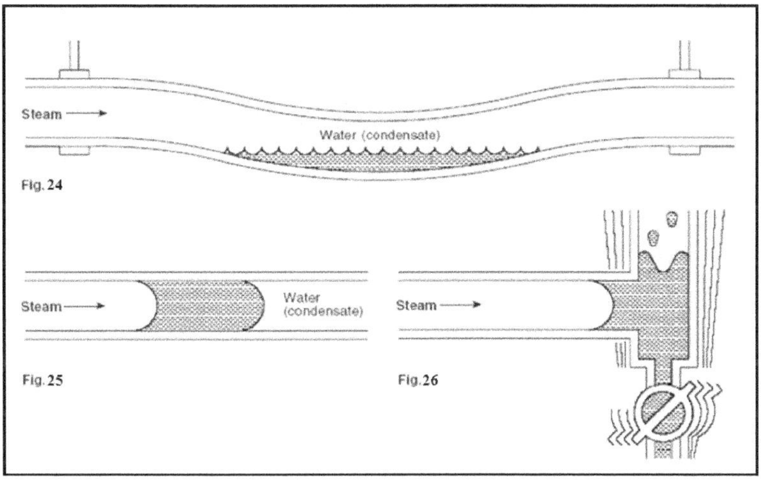

N.B. Water hammer can damage steam traps, valves, steam meters, pressure reducing valves, make joints leak, fracture pipes and **can cause serious injury to personnel.**

Steam Trap: Steam traps are a type of automatic valve that filters out condensate (condensed steam) and non-condensable gases such as air without letting steam escape. Steam traps are used in such applications to ensure that steam is not wasted. When the work is done (steam has given up its latent heat), steam condenses, and becomes condensate.

GA Drawing - Steam Trap:

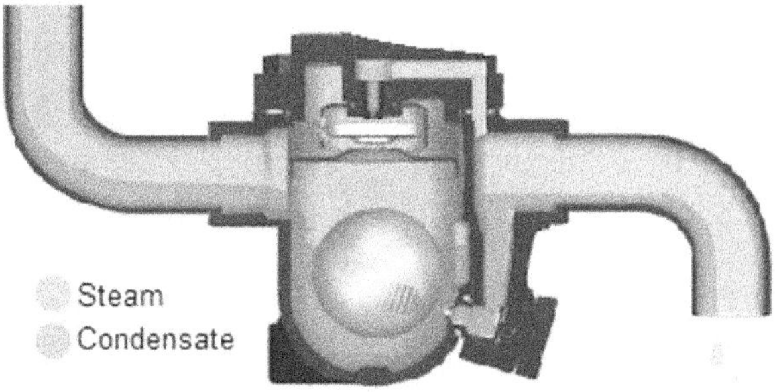

In other words, condensate (as important as it is to boiler feedwater) does not have the ability to do the work that steam does. Heating efficiency will therefore suffer if condensate is not removed as rapidly as possible.

Effectively working steam traps are very important. They can help lower energy consumption, maintain steam and product quality, and increase productivity and efficiency.

Regular surveys carried out by steam trap specialists are important. A **steam trap survey** involves a detailed inspection of your onsite steam traps, assessing your system's current efficiency. Once completed, they'll provide you with a comprehensive report, identifying an inventory of your steam trap populations, total costs of steam losses, a programme for replacing faulty or incorrect steam traps and a calculation of potential savings.

Steam pressure is very powerful and aggressive, when opening a wheel valve to pressurise a system, **do it very slowly**. Turn the wheel valve a quarter turn every 60 seconds or so or maybe longer, let pressure build up slowly.

N.B. The key to all the above is to ensure that the steam system's entire network and associated equipment throughout the plant is properly installed, maintained, and serviced. You must know the plant's heating systems capabilities: flows; temperatures; pressures; volumes; pressure reducing valves; steam traps and ensure it can maintain the heating needs on site.

Liquid Thermal Expansion – Hazards and Safeguards

Thermal expansion of liquids is a phenomenon with which we are all exposed due to its presence in everyday life. For instance, the mercury thermometer's operation is based on this principle.

Thermal expansion of trapped liquids in piping systems creates hazards when the pipe or vessel is filled completely with liquid and the liquid is blocked in.

Hazards include:

- Large pressure increases with only modest temperature changes (for example morning to noon temperature change).
- Leaks of flammable, toxic, corrosive, or environmentally damaging liquids.
- In extreme circumstances, catastrophic failure of piping or process equipment.

This diagram shows that a 10°C increase in temperature of trapped water can increase the pressure in the pipe by over 100 bar or 1450 lbs per square inch!

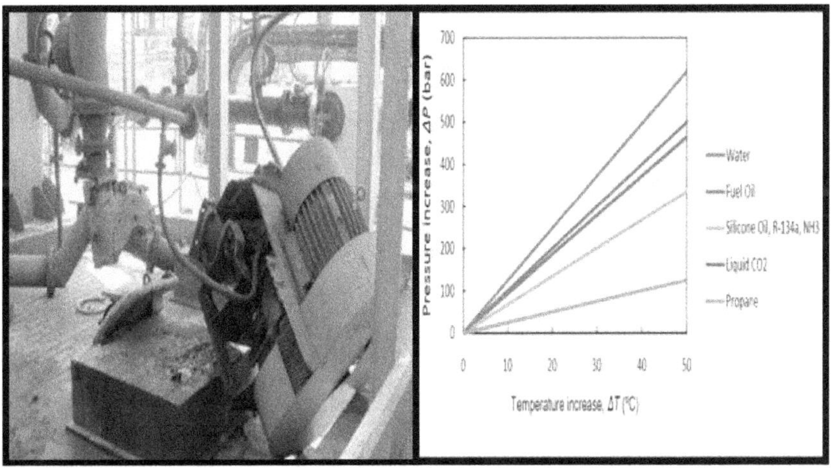

Safeguards:

- Do not allow the liquid to become trapped in.
- If the liquid can become 'trapped in' then provide pressure relief. This could be an expansion tank, a pressure relief valve, or space for the liquid to expand into.

- When performing maintenance or project work take special care to ensure that the work does not present the opportunity to trap liquid in the flow circuit.
- Guard against blockages in the flow path (valves, defective non-return valves, blocked filters, or strainers).

N.B. If isolating safety critical devices even for a short while, ensure that you understand the purpose of the device and that alternative precautions are in place while it is isolated.

Fire Safety

Most fires are preventable. Those responsible for workplaces and other buildings to which the public have access can avoid them by taking responsibility for and adopting the right behaviours and procedures. Employers (and/or building owners or occupiers) must carry out a **Fire Safety Risk Assessment** and keep it up to date. This shares the same approach as health and safety risk assessments and can be carried out either as part of an overall risk assessment or as a separate exercise. Fire Safety refers to precautions that are taken to prevent or reduce the likelihood of a fire that may result in death, injury, or property damage. Fire Safety measures include those that are planned during the construction of a building or implemented in structures that are already standing.

Fire Safety in buildings is determined by a number of factors:

- The provisions of means of escape in case of fire
- The ability for a building to resists the effects of fire
- To minimise the spread of fire and smoke
- The provision of means of access to enable fire fighters to effect rescue and fight fire.

General Fire Safety precautions include provision of:

- Adequate and appropriate means of detection and giving warning in case of fire
- Suitable means of fighting fire
- Specifying the action to be taken in the event of fire
- Appropriate and adequate training of staff in company fire safety procedures.

The Law: The Dangerous Substances and Explosive Atmospheres Regulations 2002 (DSEAR) require employers to assess the risk of fires and explosions arising from work activities involving dangerous substances, and to eliminate or reduce these risks.

Many factors influence the risks from a fire involving dangerous substances. **Employers and property owners should consider:**

- Whether a fire could lead to an explosion
- How fast a fire might grow
- What other materials might be rapidly evolved
- Any dangers from smoke and toxic gases given off
- Whether those in the vicinity would be able to escape.

The main elements of a Fire Safety Programme are:

- Fire Safety Register and Regular Fire Safety inspections
- Emergency procedures and evacuation drills
- Maintenance and Servicing of Fire Safety Systems
- Staff instruction, training, and emergency planning.

An effective, reliable **automatic fire detection system** linked to a fire warning system is essential in any apartment block building, commercial office area, or industrial workplace. The threat from fire carries one of the highest risks to loss of life, and the potential to damage or shut down a building or business. The FM department must have in place; regular training and maintenance, inspection and testing for all the fire safety equipment and systems, keeping records and certificates of compliance.

Example: Extract from I.S. 3218 – 2013: FIRE DETECTION AND ALARM SYSTEMS FOR BUILDINGS - **Fire alarm system test frequency:** Test all devices once per annum. Test central equipment four times per annum; **Interval:** Min 2 Month - Max 4 Month. Total visual inspection on phased basis per annum.

Fire Alarm Panel example:

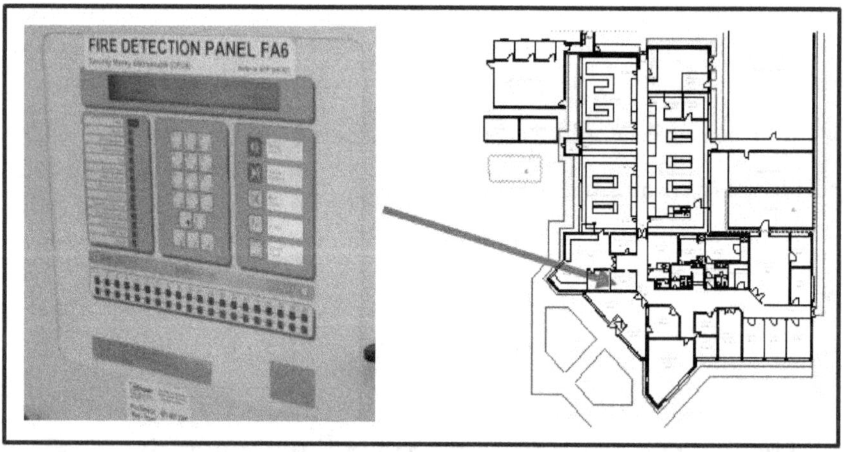

The fire alarm system will go into **alarm**, for example, following the activation of a 'Break-Glass Unit' (also called a Manual Call Point or MCP), a 'Smoke Sensor', a 'Heat Sensor', a 'Flame Detector Sensor', or a 'Fire Sprinkler Flow Switch'.

Trained personnel must know the exact location of the Fire alarm panel, the layout and operation of the building fire detection system, and be familiar with its operation e.g. Locate, Interpret, Act on, Silence, and Reset alarms.

Smoke Ventilation Systems

Smoke ventilation systems are Life Safety Systems designed to facilitate the safe escape of occupants in the event of a fire and also enable the fire to be fought in its early stages. There is well defined and strict legislation and design guidance that defines and governs the installation, performance and maintenance of Smoke Ventilation Systems that must be considered and adhered to when specifying any system.

Smoke Control Systems save lives and help protect property by:

- Keeping escape and access routes free from smoke
- Facilitating fire-fighting operations
- Reducing the risk of the fire developing further
- Protecting the contents of the building
- Reducing the risk of damage to the building

Whether the building application is residential, commercial, healthcare or educational, a smoke ventilation system must be included as part of the overall fire strategy for the building e.g. **Automatic Opening Vent (AOV)** smoke vents are designed to automatically extract smoke in the event of a fire. The AOV opens to 140° in less than 60 seconds complying to the requirements of the Smoke Vent Standard BS EN12101-2. AOV's can be activated by dedicated smoke detectors/ Smoke Vent Call Points or through integration with existing fire alarm systems, or an existing Building Management System. **AOV example:**

Automatic Opening Vent (AOV) Example:

Fire Sprinkler Systems

A fire sprinkler system is an active fire protection method consisting of a water supply system providing adequate pressure and flow rate to a water distribution piping system onto which fire sprinklers are connected.

Regular maintenance, inspection and testing of your Fire Sprinkler Systems must be in place e.g. Weekly operational test of Diesel engine fire pumps. The NFPA recommends that you hire experienced fire protection experts to come and inspect your sprinkler system at least four times every year. Regular inspection will assure you that the system will work optimally whenever there is a fire emergency and help to minimise loss of property and prevent loss of life. The results of all tests and examinations must be recorded and retained for a suitable period.

FM department personnel must have in place; regular maintenance, inspection and operational testing for all the fire sprinkler equipment e.g. Water booster pressure pumps, Diesel engine fire pumps, foam cannons, and keeping records and certificates of compliance.

Diesel engine fire pump example:

Foam cannon example:

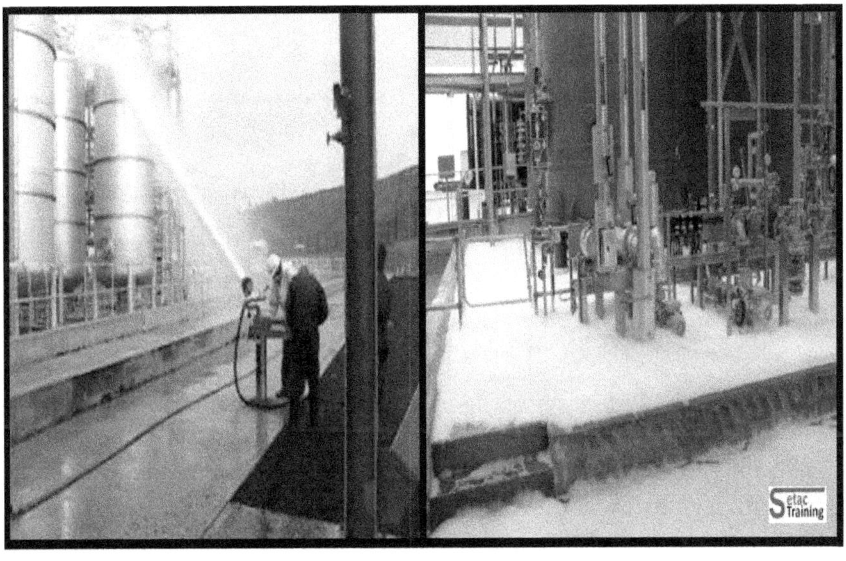

Fire Sprinkler or Sprinkler Head

A **fire sprinkler or sprinkler head** is the component of a fire sprinkler system that discharges water when the effects of a fire have been detected, such as when a predetermined temperature has been exceeded.

Sprinkler Head Temperatures

- **68°C or 155°F:** Bulb Colour - **Red**
- **79°C or 175°F:** Bulb Colour - **Yellow**
- **93°C or 200°F:** Bulb Colour - **Green**
- **141°C or 286°F:** Bulb Colour – **Blue**

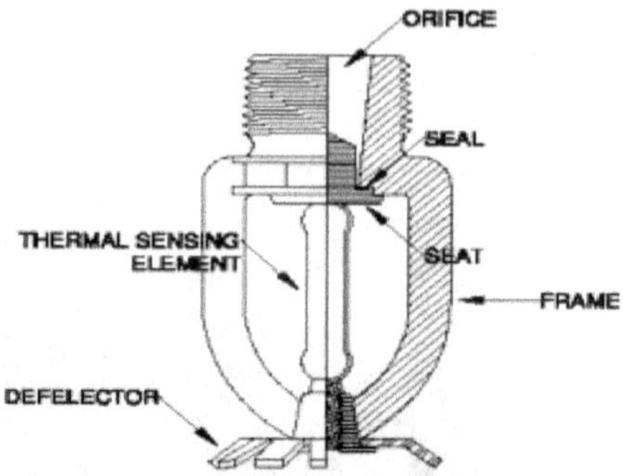

TYPICAL SPRINKLER

An inline flow switch on the associated pipework feeding water to the sprinkler head will detect a flow and signal to the fire alarm panel, activating the general 'fire alarm' sounders.

Wet Pipe Sprinkler Systems

Wet Pipe systems are the most common and reliable sprinkler systems. Installed in premises where temperatures are permanently maintained **above 4°C (39°F).** Automatic sprinklers are attached to a piping system which contains water and is connected to a water supply. ***Only*** those sprinklers activated by the fire, will discharge water.

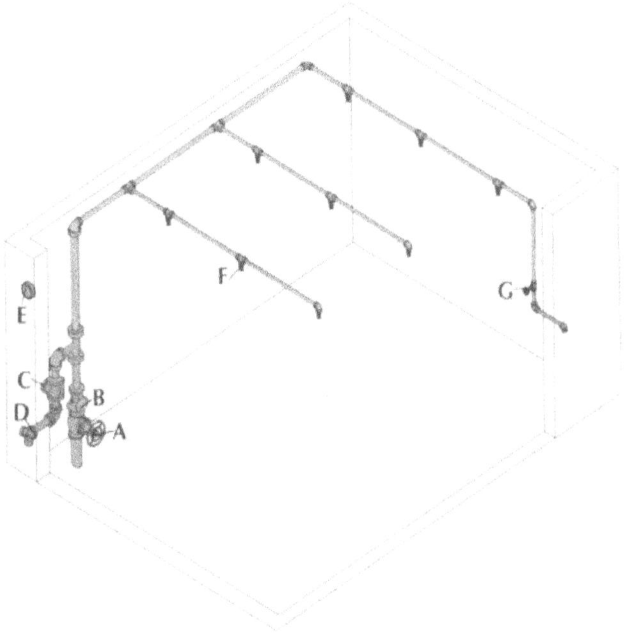

A Indicating valve

B Alarm check valve (Sprinkler Valve)

C Fire department check valve

D Fire department connection (Optional)

E Water motor alarm

F Automatic sprinkler

G Inspector's test valve (End of Line Valve)

Dry Pipe Sprinkler System

Dry Pipe systems are installed where sprinkler piping is **subject to freezing**. These systems use automatic sprinklers attached to piping that contain pressurised air. When the sprinkler opens, it releases the pressurising agent. Without the mechanical mechanism on the dry pipe valve being held shut by the pressurising agent, water forces it open, fills the system and discharges ***only*** through the sprinklers activated by fire.

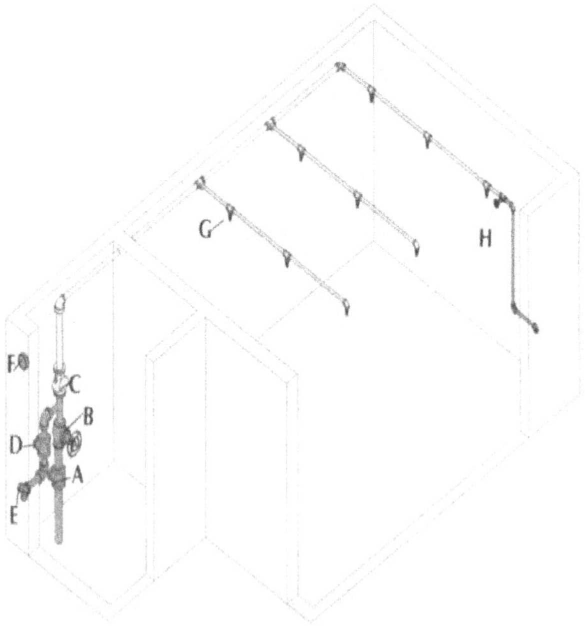

A Supply check valve

B Indicating valve

C Dry pipe valve (Sprinkler Valve)

D Fire department check valve

E Fire department connection (Optional)

F Water motor alarm

G Automatic sprinkler

H Inspector's test valve (End of Live Valve)

Deluge Sprinkler System

Deluge systems are used for protection in areas where it is necessary for the complete area or equipment to be drenched simultaneously. Open sprinkler heads or nozzles are attached to a piping system connected to a water supply through a valve that is opened by the operation of a detector system installed in the same area's as the open sprinklers. When this valve opens water flows into the piping system and discharges from all open sprinklers or spray nozzles attached to the piping system.

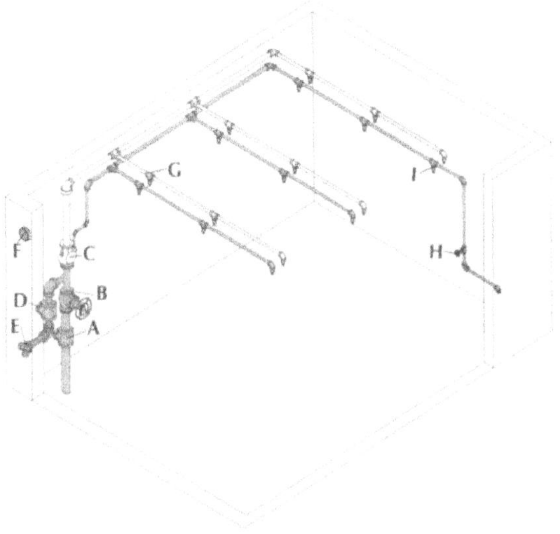

A Supply check valve

B Indicating valve

C Deluge valve

D Fire department check valve

E Fire department connection

F Water motor alarm

G Deluge sprinkler (open)

H Inspector's test valve

I Pilot sprinklers (automatic)

Fire Extinguishers

There are 5 main fire extinguisher types: Water, Foam, Dry Powder, CO2, Wet Chemical.

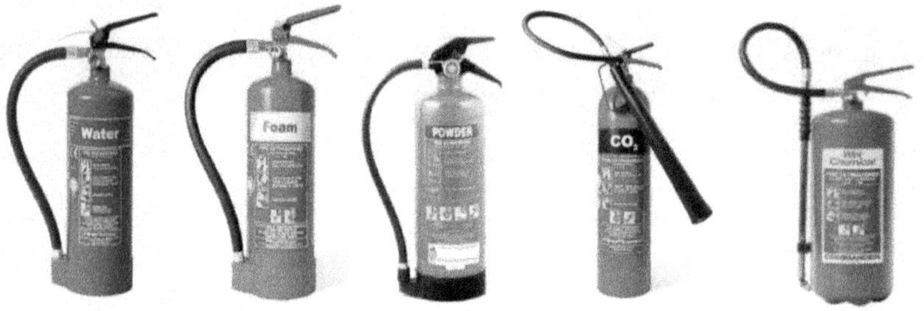

Companies must have the right types: classes of fire/colour coded; size; and weight of fire extinguishers (e.g. Powder Fire Extinguisher/Blue can be used on Class A, B, and C fires) on their premises for their particular line of business, or they may not meet current regulations.

Fire Extinguisher Inspection and Maintenance

Like any mechanical device, fire extinguishers must be maintained on a regular basis to ensure their proper operation. Fire extinguishers must be inspected or given a "quick check" every 30 days. For most extinguishers, this is a job that you can easily do by locating the extinguishers in your workplace and answering the three questions below.

- Is the extinguisher in the correct location?
- Is it visible and accessible?
- Does the gauge or pressure indicator show the correct pressure?

In addition, fire extinguishers must be maintained annually in accordance with local, state, and national codes and regulations (e.g. I.S. 291-2015. Standards). This is a thorough examination of the fire extinguisher's mechanical parts, fire extinguishing agent, and the expellant gas. Your **fire equipment professional** is the ideal person to perform the annual maintenance because they have the appropriate servicing manuals, tools, recharge materials, parts, lubricants, and the necessary training and experience. **N.B.** A fire extinguisher must always be visible, located in its designated area, and never used for propping open a fire door, or as a coat hanger.

Fire Hose Reels:

Fixed and hinged fire hose reels in manual and automatic versions are strategically located to provide an accessible and effective supply of water to combat a fire risk. Fire hose reels require connection to a pressurised source of water either from the mains supply or a storage tank. Fire hose reels are supplied for both open surface and recess installation, they come complete with a jet spray nozzle enabling the operator to control the direction and the flow of water to the fire.

The construction and performance of a fire hose reel is kite marked BS EN 671 making it ideal for fighting all types of Class A fires. A fire equipment professional will provide the hose reels, and additionally, a full service package which includes:

- Hose Reel Flow Testing
- Hose Reel Pressure Testing

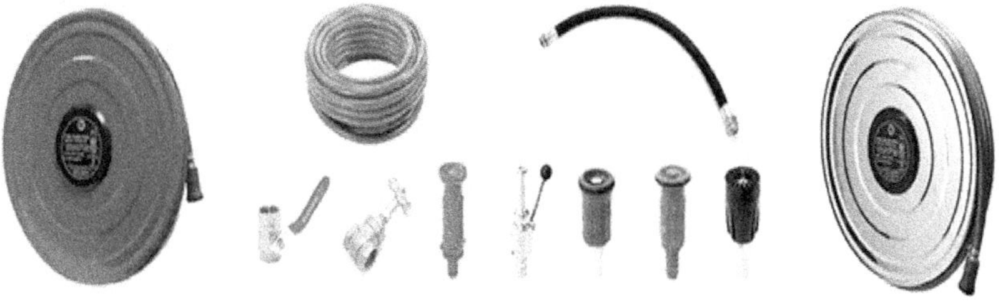

Their technicians will be fully certified to EN 671-3 2000 to test and maintain all types and sizes of semi-ridged hose reels. They will also provide hose reel maintenance service and can provide any replacements required for existing malfunctioning equipment. Replacements will also be installed by their technicians.

Fire Doors: are a crucial part of any public, commercial or multiple-occupancy buildings protection strategy. They resist fire and smoke, protect escape routes and essentially save lives, but only if they work properly. Legislation dictates that Fire Doors should be checked every 6 months – If you have been designated the "Responsible Person" for a building, it is your duty to ensure your Fire Doors are working effectively and to protect occupants in your building from the risk of smoke and fire.

Emergency Lighting:

This is the lighting provided for use when the supply to the normal 'mains' lighting installation fails. Emergency lighting luminaires must comply with all standards regarding non-flammability (resistance to flame and ignition) conditions. It can be a self-contained emergency light or form part of a **Central Battery System** in which batteries for several emergency luminaires are housed in one location.

All emergency lighting systems should be tested **monthly**. It is important to note that the entire system doesn't have to be tested at the same time. The system can be tested in sections, over a testing schedule, so long as each luminaire is tested each calendar month.

Annual Emergency Lighting Tests

A test for the full rated duration of the emergency lights (e.g. 3 hours) must be carried out. The emergency lights must still be working at the end of this test. The result must be recorded and, if failures are detected, these must be remedied as soon as possible.

Emergency Lighting is critical in **'Escape Route Lighting'** via exit signs and exit direction signs to ensure that the means of escape to the final exit point can be effectively identified and safely used.

Emergency Plan

The purpose of an Emergency Plan is to describe the procedures to be followed and the roles and responsibilities of key personnel in the event of an incident on your site. The Emergency Plan provides a blueprint for the actions to be taken during an emergency. Your company must provide evacuation/relocation plan information and routinely schedule and hold drills at least once a year.

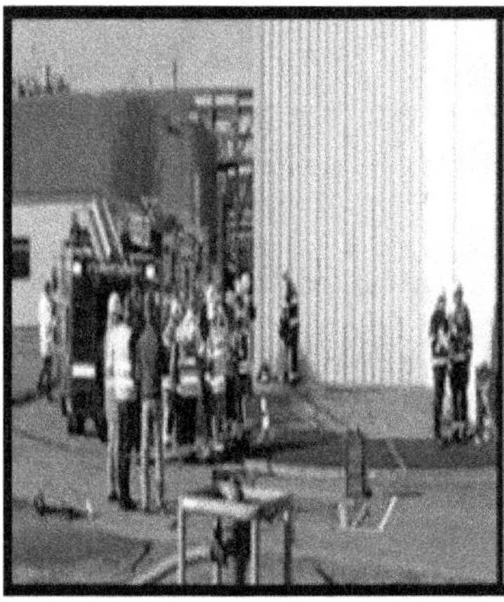

The 3 key elements of Emergency Preparedness:

1. **Early warning** (typically through an alarm or voice communication system).

2. **Adequate means of egress (exit routes)** and be familiar with them.

3. Personnel familiarity with the **Emergency Plan** through knowledge and practice.

Avoid '**Shock Response**': inhibits your ability to think clearly, and act decisively. **Be Prepared !!!**

Computer-Aided Facility Management:

Computer-Aided Facility Management (**CAFM**) is the support of facility management by information technology using proactive asset management tools. A CAFM software platform streamlines facilities management and maintenance, it spans space management, real estate planning, project management, building operations, preventive maintenance, and more.

These systems leverage facilities data into performance metrics and planning tools to optimise the process of managing facilities to keep a plant and its associated buildings running optimally.

Monitoring applications help avoid incidents and minimise risk. Cost can be prohibitive installing this type of technology, but, the benefits will far out way the initial installation cost in plant uptime, reliability, efficiency, and operating costs.

Energy Supply Contracts:

A large **Energy Consuming** company should consider employing an outside consultancy firm to negotiate and manage its **Energy supply contracts**. Negotiating an energy supply contract is a complex business. Securing the best deal possible in a market with traded futures is about much more than driving a hard bargain. It is essential to understand the dynamics operating within that market to identify when market conditions are favourable to securing a contract that brings competitive advantage. **Outsourcing energy purchasing** allows a company to choose exactly how involved they wish to be in the process.

At one end of the spectrum **the consultancy firm** could fully manage the entire process. Alternatively, there is the flexibility to be more directly involved in negotiations and timing decisions with the option to choose online delivery if preferred. They will tailor the level of support a company needs to assist in-house management to secure the best value for its business. The consultants' knowledge of the electricity and gas markets and the contract flexibility available from the various suppliers in these markets is normally excellent.

The consultancy firm will use its market knowledge, activity and influence to continually track what is happening in the market and to predict what is likely to happen in the future by bringing together a full complement of industry experts from across all disciplines of energy procurement, utilisation, research, and analysis. They are ideally positioned to provide more than information alone, identifying future trends, and adding value to market intelligence. Any specific information or reporting requirements that a company may have, particularly with reference to gas and electricity markets, or oil markets can be facilitated.

In today's complex and volatile energy markets, companies are adopting a more sophisticated approach to their purchasing strategy in an effort to ensure that cost competitiveness is maintained in a controlled and measured manner.

Tip: It's no longer enough to strike a good deal on energy supplies or achieve high levels of productivity: now you are expected to do both, and at the same time comply with a complex web of regulation. Striking the right balance between these, of course, can make that crucial and tangible contribution to your company's revenue flow.

Waste Management System

Waste Management is the control of materials and equipment that have become redundant and therefore must be disposed of in an environmentally, safe and cost-efficient manner. Waste management of consumables and maintenance components is necessary. This can include anything from waste oil to a complete demolition and disposal of an entire machine and its associated components. Waste issues are becoming increasingly significant to organisations.

Three key areas forcing organisations to address waste management include:

1. Economic Motivation
2. Environmental Impact
3. Compliance Requirement.

Economic Motivation - The cost of waste disposal has significantly increased in recent times. Factors responsible for the increasing cost of waste disposal include:

- lack of landfill capacity
- increasing costs of waste disposal due to the higher environmental standards now being applied to landfills sites.

It is unrealistic to expect waste disposal costs for your organisation to decline if current practices with regard to waste management continue. Thus, there is a strong incentive for all who produce waste to minimise it by **Reducing, Reusing, and Recycling.**

Environmental Impact

Waste should be managed responsibly because:

- materials that are limited in supply should be treated with care to ensure that valuable resources are not exhausted
- landfilling is unsustainable - we cannot continue to bury the problem.

All waste management options (from recycling to landfill), impact on the environment.

There is increased public concern about environmental and health impacts of landfills and incinerators. The European Landfill Directive limits or bans the disposal to landfill of certain waste and that of some organic waste, yet the volume of waste we generate in our homes and places of work is continuing to increase.

Minimising the production of waste in the first place is the only way of ensuring there is no environmental impact.

Compliance Requirement

Waste Management Legislation places responsibilities on all parties to deal with their waste in a responsible manner. Leaving aside cost issues, fines, and other enforcement procedures – for lack of compliance – means that organisations must take waste management seriously.

The process includes sorting, recycling, clearance, collection, transportation, and disposal of waste materials according to waste disposal regulations. Waste management includes substances that are in a solid, liquid, or gaseous state, and their management techniques differ for each state.

Waste Management is complex but very important and if not properly managed, can be very expensive due to the multiple varieties of waste produced by industry. Different types of waste require special management techniques. There are licensed waste management vendors who will provide a total waste management (TWM) package and will deal with handling, transportation, and documentation of both hazardous and non-hazardous waste.

Key Objectives:

- Work towards ZERO landfill target
- Reduced volumes will reduce land fill & associated transport costs
- Encourage and promote Recycling by employee's onsite
- The greater the understanding of where the waste is ultimately going, the greater the understanding of how and why certain wastes must be packaged for transport and for easy treatment at a waste disposal facility.

Bunds

A **bund** is a containment around an area where hazardous liquids are handled, processed, or stored. **Maintaining your site bunds is just as important as maintaining the equipment within them.** The type of bund most often seen consists of four walls and a base surrounding tank. If the tank leaks, the leak will be contained by the bund. For bunds to be effective, they must be of sufficient size to contain the volume of any likely leak and be impervious to the liquid. Bunds are usually constructed of reinforced mass concrete; other types include prefabricated steel or plastic bunds.

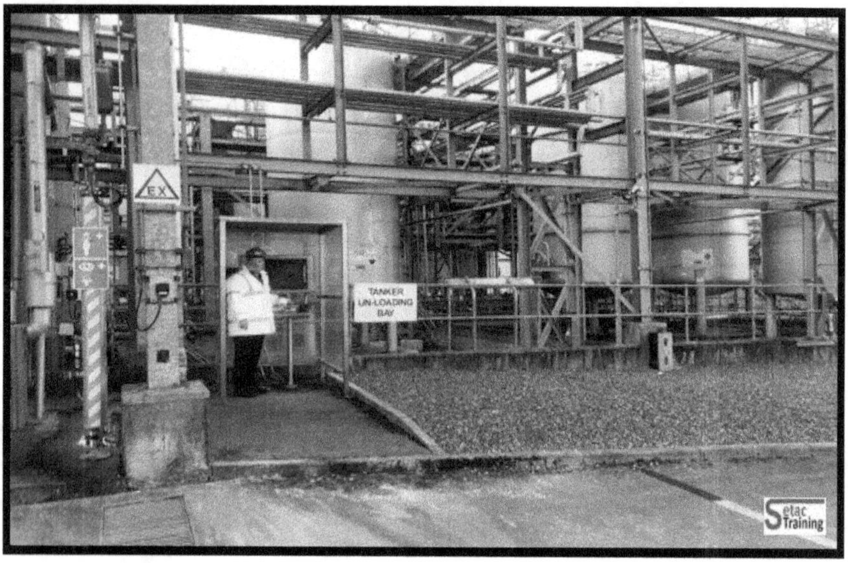

The general sizing rules for bunds requires that they hold 110% of the tank volume where there is just one tank in the bund. Where there are two or more tanks in a bund, the general rule is that the bund must be greater than 110% of the largest tank, or 25% of the total tank volume, whichever is the greater.

In order to comply with current legislation, all bunds must pass an Integrity Test at least every 3 years. A tank that contains hazardous liquid **e.g.** Diesel Oil or a solvent such as Acetone could potentially contain the highest risk material in your facility. It is critical that staff, visitors, and the environment are protected from a potentially harmful chemical leak.

Tip: Tank protection and venting are key safety elements on any industrial site. **Bursting disk** selection is a critical part of this.

Any leak into a bund from a tank containing a hazardous liquid where the bunds' integrity is compromised could pollute the ground and groundwater. Not only does this put drinking water supplies at risk, but it has serious cost implications as the site clean-up costs and soil remediation can be significant. The EPA **will** carry out site inspections on your facility, a poorly maintained bund may result in corrective measures being enforced.

Spills & Spill Kits

Spills happen!! People sometimes knock over five-gallon drums of chemicals, run forklifts into sides of one thousand litre IBC totes, break hoses when they are fixing machines and drums get overfilled when they get distracted. **N.B. Just be ready when the inevitable happens.** Employees who respond to emergency spills should be trained in accordance with OSHA's Hazardous Waste Operations and Emergency Response (HAZWOPER) standard. Knowing what liquids are stored onsite and in what quantities are two of the first steps in choosing the right equipment for the job. Spill response supplies are available for just about every spill scenario, they range from simple easy-to-use solutions like absorbents to elaborate vacuum systems and other specialised equipment that are specifically designed for responding to corrosive, explosive, and other harmful chemicals. Employees should know how to use spill kits and be comfortable using them. Determining 'how big' spill kits should be; what to put in them, and where to store them will be key.

The three keys to successful spill response efforts:

1. Choose the correct spill kits and response supplies.
2. Have them available in 'spill prone' areas.
3. Ensure that personnel know how to use them properly.

WASTEWATER TREATMENT PLANT (WWTP)

A Waste Water Treatment Plant exists to treat **industrial** process effluent in compliance with the Industrial Emissions (IE) licence. The EPA set Emission Limit Values (ELVs) e.g. Toxicity, COD (Chemical Oxygen Demand), Suspended Solids.

Sewage is basically the flow of used wastewater from a **community**. Effective sewage treatment combines physical, chemical, and biological processes to accomplish several goals:

- to reduce "aesthetic pollution"- unsightly or smelly organic matter
- to kill pathogenic microorganisms and remove toxic wastes
- to reduce organic material or BOD (Biochemical Oxygen Demand)
- to remove inorganic nutrients (nitrogen and phosphorous) that can cause eutrophication.

Regular Preventative Maintenance (PM) must be carried out on the treatment plant's bespoke systems e.g. Biological tanks, Tertiary Settlement tanks, Clarifier Settlement tanks, Sludge Treatment System, chemical dosing with associated equipment: agitators; mixers; pumps; blowers; instruments (pH, temperature, Dissolved Oxygen (DO)); automated valves.

An industrial Waste Water Treatment Plant (WWTP) example:

REGENERATIVE THERMAL OXIDISER (RTO)

Regenerative Thermal Oxidisers are used to destroy air toxins, odours, VOCs, and hazardous air pollutants (HAPs) that are discharged in industrial process exhausts. Emissions from the RTO to atmosphere are monitored via a Continuous Emissions Monitoring System (CEMS) and are licenced under the terms of the Industrial Emissions Licence.

Correct operation and regular Condition based Maintenance (CbM) monthly routines must be carried out on its bespoke systems including Fresh Air Fans, Rich Fume fans, Combustion fans, RTO burners and Scrubber and associated instrumentation.

A full 5 Day annual shut down including servicing, inspection, and testing completed by the equipment vendor is the best way to ensure your RTO systems remain safe, run at peak efficiency, thus keeping your running costs to a minimum, and extending its working life.

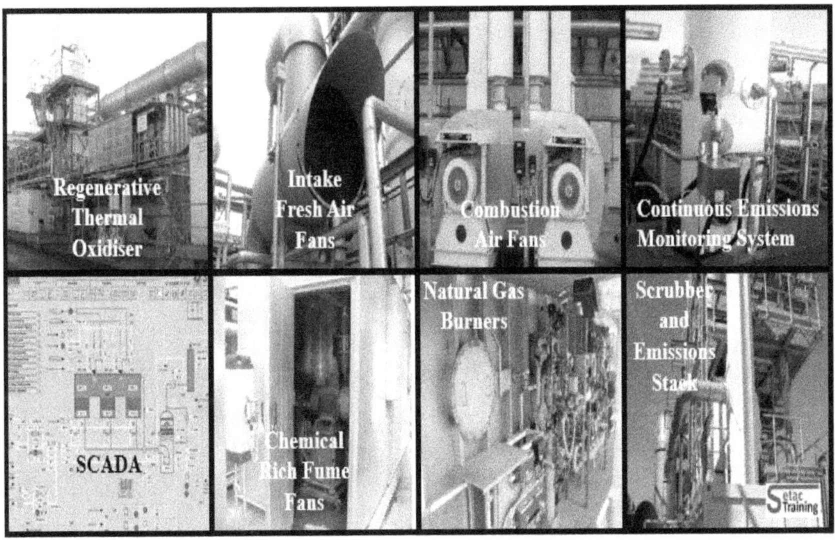

Facilities Department Environmental Goals

- Prevent or reduce pollution to air, land, and water through programmes that reduce environmental impacts and conserve natural resources using Continuous Environmental Management Systems (CEMS) e.g. ground water sampling points, air emissions points, liquid effluent discharge points.
- Review environmental objectives and targets and set guidelines for reduction of environmental impacts.
- Continuously reduce environmental impacts and conserve natural resources.

The fundamental questions you must ask yourself:

- If any one of your critical FM systems fails, can the **failed system** recover quickly (with very little disruption) and get the plant and its processes back on line to ensure business continuity?
- Are all the necessary procedures, duty/standby systems, onsite personnel and **on call** contractors/vendors in place to ensure the **day to day** routine monitoring and servicing works are carried out in a safe and environmentally compliant manner?
- Have the contractors/vendors coming onsite been properly vetted and approved to ensure they are suitably qualified and insured before carrying out any work?
- Is your company doing all it can to prevent or reduce pollution to air, land, and water through programmes that reduce environmental impacts and conserve natural resources.

Typical Industrial Facilities Systems:

Plant Potable Water distribution system
Natural Gas distribution systems
Plant Boilers and Steam generation system
Hot Water systems
Chilled Ammonia systems
Waste Water Treatment Plant systems
Sludge treatment plant systems
Purified Water systems
Chilled Water systems
Compressed Air Systems
Nitrogen inert gas distribution systems
HVAC - Heating, Ventilation, and Air-Conditioning systems
Smoke Ventilation Systems
Waste Management
Foul and storm water drainage
Bund Management
Energy Management
Process drainage systems
Fire Water Retention Tank systems
Fire Sprinkler Systems including (Diesel Generator & Pump)
Fire Detection Alarm systems
Foam and Chemical Fire Suppression Systems
Fire Extinguishers & Fire Hoses
Plant small power LV distribution systems
Plant Electrical distribution systems up to MV.
Plant Diesel engine powered Electricity Generators
Wind and Solar Powered Electricity Generators

10. BREAKDOWN/ ESCALATION PROCEDURE

Outside normal office hours when staff numbers on site are reduced, there is a requirement to ensure that identified critical alarms and critical utilities are monitored and that personnel are available to respond to, and deal with such alarms in a timely manner thereby mitigating any negative consequences which could arise.

A site escalation procedure must be in place should a problem arise that onsite personnel cannot resolve **e.g.** the provision of off-site management technical support and the provision of a timely response in the event of a site alarm activation/emergency. Management must make sure all essential relevant vendors names, phone numbers and critical site information documents are updated regularly, easily accessible (kept in at least two known safe locations: main security office & main archive room) and available to relevant personnel. This will limit downtime dramatically as opposed to trying to find them when an issue arises. Having this phone list accessible to all will help others resolve problems quickly as well.

When shift personnel cannot resolve an issue, they will now have to start making phone calls 'out of office hours' to managers or vendors. When shift personnel decide to make these calls, they normally are looking for technical advice or clarification on the issue that must be addressed immediately. The 'Responsible Person On Call' being contacted must be technically qualified, has access to relevant information (e.g. Remote PC) so they can make an informed decision and are trained to give the shift personnel clear direction, clarification, or guidance in a major breakdown or incident.

Shift personnel must ask themselves these questions before they make a call offsite:

1. Is the call warranted?
2. Why couldn't they resolve the breakdown issue themselves – Specific questions: What RCA approach was used? Have they been properly trained? Do they need more training?
3. Is the proper information (FDS, P&ID's, electrical, mechanical drawings, O&M manuals) readily available, easily accessible and has it been used in trying to locate the fault?
4. Have I all the known facts of the issue and its current status?

Shift personnel must submit a written report to their manager, outlining the reasons why they had to make the phone call, referencing that all above points had been questioned and clarified before the call was made.

Criteria that must be met by onsite shift personnel who are responsible for the operation of the plant 'Out of Office Hours':

Personnel:

- Must be competent and confident in their own abilities as they are accountable for their actions.
- Must keep their discipline, no matter how much they would like to press a button or open a valve. *They must keep their hands to themselves.* Ensure any action taken is based on competent process/engineering fact.
- Must try not to suffer from **Analysis Paralysis** either **(thinking gets in the way of doing)**.
- Cannot be expected to trouble shoot or deal with a problem on a piece of equipment if they have not been trained how to deal with the problem in the first place or if inadequate alarm systems warning of impending shutdown or danger are not installed.

- Must try to understand and establish exactly what they are doing and the consequences of their decisions before any actions are taken.
- Must always exercise caution. Use logical and deductive reasoning at all times.
- Must **think on their feet**, adapt to a situation and deal with problems as they arise. Stay focused, calm, and remember their training. Give precise answers to personnel, both on or off site, based only on accurate process/engineering fact.

Personnel must use their senses:

- to look out for anything unusual **e.g.** oil stains, leaks, burn marks
- to listen out for anything that doesn't sound right
- to use their nose to smell anything out of the ordinary **e.g.** burning oil, rubber
- to get to know the noises the piece of equipment generates, check for any excessive vibrations by actual touch, this could be first sign of wear and tear of a bearing
- to watch out for hazards, identify, report and rectify them.

Criteria that must be met by the off-site 'Responsible Person On Call':

N.B. The **Responsible Person On Call** must be competent enough to be able to make effective decisions about what to change and how to change it, during a situation. He/She must be able to convey their knowledge and experience to onsite personnel when a call is made.

Responsible Person On Call:

- Should be able to verbalise, organise and plan with regard to personnel and equipment in order to manage a given situation. They must show leadership, confidence, and control in making the best use of their available resources.
- Must be accountable, give proper clear direction, and work through each issue until it is resolved.
- Discuss the issue with the onsite shift personnel and try to come to a mutual consensus on the best way to resolve it.
- Must realise, a possible wrong course of action taken by site personnel, **as advised by them**, could possibly exacerbate a problem, this must be prevented at all times. If the 'Responsible Person On Call' cannot resolve the issue over the phone, they must report to site and take the necessary steps to resolve the issue.

Severe Weather Preparation to ensure Business Continuity

Severe Weather examples: High winds, hail, excessive precipitation, and wildfires are forms and effects of severe weather, as are thunderstorms, downbursts, tornadoes, waterspouts, tropical cyclones, and extratropical cyclones. Regional and seasonal severe weather phenomena include blizzards (snowstorms), ice storms, and dust storms.

Knowing your business has 'emergency' plans in place to help cope with a natural disaster like a severe storm will make it easier for your business to minimise losses, maintain business continuity and recover quickly.

A company's severe weather preparation plan should include:

- Assigning responsibility for planning and preparedness to a senior executive and deputy e.g. they may make the decision to scale back or suspend operations during weather-related disruptions
- monitoring the latest and predicted future weather forecasts

- ensuring effective crisis communications with employees, vendors, suppliers and others to ensure their safety, quickly react to events and aid efforts to return to normal operations
- Minimising potential damage to physical property. Identify essential structures, products, equipment, inventory and utility services and install prevention and safety systems to protect them
- Making sure the severe weather preparation plan in place has been tested and is prepared for all potential 'worst case' scenarios and specific essential personnel know their roles and responsibilities.

11. VENDORS

When a vendor supplies an **equipment package**, be it large or small (e.g. a water purification package), they normally cooperate with a company's commissioning and qualification inspection team. The vendor is responsible for the provision of documentation detailing the design (that it meets the requirements of the user as detailed in the provided specifications and in sufficient detail to avoid any ambiguities), fabrication, testing and pre-commissioning of the equipment before the equipment arrives on site. The vendor is also responsible for preparation and execution of the SAT (Site Acceptance Test) and commissioning protocols in conjunction with the company. They must provide full 'As-built' GA drawings, P&ID's, Electrical & Mechanical drawings, O&M manuals, training presentations/videos, commissioning spares, and also provide a detailed list and quotation for 2 years' critical operational spares.

The fundamental question a company must ask itself is: "Will the piece of equipment integrate with existing onsite automation infrastructure?". If the answer is 'No' and it is a stand-alone system without any consolidation of data, then this means data **cannot** be checked or added in real-time to any computerised management system. As a result, the equipment is cheaper but there is little or no production data to intelligently manage the manufacturing process.

Tip: Whether you are a vendor or a mechanical/electrical contractor delivering a service to a factory, the quality of the work you provide must always be of a very high standard. Just as important is the follow up associated **service history paper work** provided (hard copy or electronically or both). This is important in maintaining business continuity. Ensure the paperwork looks professional, it is printed on headed paper, fully detailed with all works done, costs involved, parts used, is unambiguous and open to any scrutiny. If providing reports on the status of equipment at regular service intervals, ensure all writing is relevant and easily readable with proper spelling. These reports are normally scanned and sent to the customer. All service history paperwork must be stored safely and indexed properly for easy retrieval by the customer.

N.B. Customers will very much appreciate these types of vendor reports as will the auditors who view them.

A company must set up service contracts with vendors and insist on a good service package being provided. 24-hour cover should be considered if the piece of equipment is deemed critical and downtime must be kept to a minimum. This type of 24-hour cover does not come cheap. A decision to implement such a service contract must be weighed up against the cost or length of downtime that can be tolerated by plant operations.

N.B. A good maintenance contract programme on any piece of equipment is the key to:

- Ensuring that the safety features are in correct working order
- Adding longevity to its parts
- Optimising its performance
- Reducing breakdowns
- Ensuring a long service life.

Purpose of a Vendor's Maintenance and Servicing Instructions:

The maintenance and servicing instructions are intended to convey information considered as necessary by the vendor for the implementation of maintenance and servicing jobs or for the compilation of in-company maintenance instructions to specialised personnel. The maintenance and upkeep instructions are a summary of the most important technical data which are required for maintaining and servicing the equipment supplied by the vendor. The information supplied in the maintenance and servicing instructions should be considered as reference values which may need to be adapted and possibly corrected in line with operating conditions.

Commissioning & Qualification (C&Q).

The maintenance and servicing instructions make it possible for the equipment technician to implement preventive maintenance. Information is provided relating to periodical replacement of wearing parts in order to avoid damage to the piece of equipment. This **flags** to production planners to make arrangements for strategically controlled production stoppages for this essential work to be carried out whilst at the same time reducing process downtime.

Special operating instructions provided by the vendor will contain maintenance and repair instructions. These instructions must be given priority. The specialist personnel responsible for operation and maintenance must be completely and comprehensively informed of the contents of these maintenance and servicing instructions. A copy of same must be kept directly in the vicinity of the equipment and in the engineering office.

Vendors will normally offer you 3 types of service agreements:

1. Inspection.
2. Maintenance.
3. Full service.

1. Inspection: equipment inspections carried out by the equipments' vendor service technicians provide users who perform their own maintenance work with the peace of mind in knowing that their system is operating correctly. The vendor service technician checks all main components and safety related systems.

2. Maintenance: this type of service agreement will ensure that equipment provides maximum reliability, availability, and long-term value retention. Maintenance components are changed in accordance with checklists by the vendors' service technicians. Main components and safety related systems are checked and if necessary, adjusted or replaced after consultation with the customer.

3. Full Service: A vendors' full-service contract ensures that complex systems deliver optimum performance throughout their entire service life and retain maximum value. All maintenance, servicing, and inspection appointments, as well as commissioning work (e.g. all electrical apparatus is correctly installed, tested, maintained, and used in accordance with its specified characteristics) are carried out by the vendors' service technicians according to a company's specific needs. Main components and safety related systems are checked and if necessary, adjusted, or replaced as required. Covered under this **full-service contract** also would be:

- Consumables and maintenance components including environmentally responsible disposal. (N.B. Hazardous materials must not be allowed to discharge into natural watercourses or drainage systems. All hazardous material waste must be kept separate from normal waste and be disposed of in a specialist disposal facility and in accordance with any statutory provisions that may apply).
- Scheduled replacement of parts (e.g. service kits, drive belts, motor bearings).
- Spare parts essential for system operation.

Benefits of a good Maintenance Contract is to maximise the product life cycle while consuming less, save money on repairs or replacements, & eliminate unscheduled interruptions to your operations. In all 3 types of service agreements, comprehensive service photo documentation provided by the vendor after the inspection (including current condition updates, recommendations for repairs and upgrades, highlighting of any issues found which will include potential cost savings, energy and usage improvements) provides further reassurance that all is in order and carried out from a health and safety regulation perspective also. The associated labour, journey and overnight costs are calculated in agreement with the customer.

Tip: Routine recommended servicing of machines is expensive, if questioned by the company's finance department as to the high servicing costs of the machinery that has only been purchased recently, compare it to that of a taxi company; it buys a complete new fleet of cars and they have them **serviced every six weeks (**approximately every 12,000 miles or 19,000 kilometers - as a rule of thumb, all vehicles **should** be **serviced** at least once a year and/or every 12,000 miles). **Why?** Unlike a privately owned car, their taxis are on the road almost 24 hours a day, seven days a week, if they are not properly serviced as per manufacturer's instructions, they will break down, which will result in **loss of revenue, loss of customer repeat business, and loss of reputation.** The same applies to your machinery, if it is not reliable and breaks down repeatedly. You cannot honour your commitment to deliver quality product 'on time' to your customers. You will **lose repeat business** and lose your reputation as being a producer and supplier of top-quality products.

Always give advice to vendors supplying new machinery to your site. Inform them of the **site standard** regarding equipment and what control gear must be installed e.g. VSD's, MCB's, contactors, digital timers, interposing relays, current monitoring devices, overload protection devices.

A normal O&M manual on a piece of equipment may be up to 25mm thick, full of parameters and listings. There may be only 3 pages dedicated to troubleshooting, quoting a number on an LCD screen with text. Personnel may spend valuable time trying to decipher what is the actual problem. Insist that a vendor supplies **easy to digest** training material: power point presentations; training videos; software-based learning programmes; computer-based training (CBT); or web-based e- learning training with their equipment. This type of training material is essential for training purposes and can also be referenced in the future. Remind the vendor of the importance of the quality of this training material and its subject matter. Use it as a negotiating leverage tool when awarding a contract.

All original signed documentation during the contract vendor's regular service intervals must be kept on site for company records and for auditing purposes. The photocopies of the documentation can go with the contract vendor. These documents will outline the machines operation and the replacing and servicing of all associated machine parts. This vendor documentation must be read thoroughly by the engineering manager and any noted observations must be followed up **e.g.** site compressed air being supplied to the vendor's equipment could be contaminated with oil or water moisture which could be damaging parts of the equipment and causing premature failure of internal parts.

Initially, at installation and set up, vendor support should be very intensive, but gradually, once your technical personnel's knowledge grows and develops, less support should be required.

Many technical issues can be resolved with telephone support. Most vendors provide advice and assistance without charge. It is given in good faith, but normally without responsibility, this maybe enough to satisfy a company's production needs.

You must ensure **business continuity** of your plant and be cognisant of technological discontinuities. Ensure that spare parts are still available for older critical pieces of process equipment. As newer models come on line, the manufacturer may no longer keep spares for the older model. Be prepared to upgrade as recommended by the manufacturer. Don't wait until the piece of equipment stops and then find the faulty part is no longer manufactured.

There are important factors that must be considered when choosing a service provider and some of the most important factors are listed below:

- Do they manufacture the equipment or just import it?
- **Dependability of Supply -** if they import equipment, how secure and long standing is their source of supply for future spare parts?
- Are spare parts kept on the shelf in your country or do they have to come from abroad?
- Length of time in business, bearing in mind some vendors may have previously been bankrupt
- How many service engineers are directly employed by the vendor? remember, any warranty is only as good as the vendor

- Do they provide accurate and timely technical information when required by the customer?
- Are they a member of an accredited agency?
- Ask the vendor to provide references from their existing **happy** customers. Has the vendor a good record of accomplishment in addressing customer-specific requirements?
- What is their level of **repeat order** from existing customers?

Buying directly from a reputable service provider (preferably they manufacture the equipment) means work is performed by factory-trained technicians who know the systems you have installed onsite better than anyone. You will have access to the engineers who actually designed your systems and are acquainted with all the latest firmware/software upgrades. That translates to a faster **mean time to repair**, the assurance of factory-certified spares, faster delivery via the supply chain, guaranteed continuity of service, and peace of mind that work is performed right the first time. Field service engineers use proven operational best practices and state-of-the-art software tools to design, implement, maintain, and optimise your environment quickly, accurately, and cost-effectively. Go with the service provider who has the experience, expertise, processes, and personnel to assist you in keeping the plant equipment online.

A proven, time tested, easy to apply engineering method in capturing information is to ensure all new equipment being supplied comes with the following software documentation format:

- URS_FDS_Design Specifications
- P&ID's_3D models
- O&M manuals and Digital Photos
- Instrumentation & Calibration
- Electrical drawings in PDF
- GA & Mechanical Drawings in PDF
- SOP's_PFD's_PM's
- IQ_OQ_PQ Documents

- Training PPP's & Videos
- Spare Parts Inventory_Vendor info & Certs
- Operational issues.

The OEM must populate the sub folders with the bespoke information and supply it to you on disk. This can then be uploaded onto your site hard drives where it can then be accessed (read only) by site personnel via the company intranet. As machine upgrades occur, old information can be archived by the engineering department who can then update the sub folder with the latest information. If this type of engineering format can be created for all site equipment, a **Knowledge Management** system will now exist on your site.

Tip: Try to set up an engineering network with other companies who have the same equipment/machinery, so that information can be shared regarding a machine's operation, serviceability, and reoccurring engineering issues.

Follow equipment manufacturers recommended maintenance schedules and log them. This will help maximise equipment life. Detailed maintenance records will also be useful to get better pricing if the equipment can be sold at the end of its useful life. Try to get manufacturer information on expected failure rates and repair costs. These can be useful benchmarks to measure your performance with the equipment.

Remember, a guarantee is only as good as the vendor who supplies it and this can vary widely from vendor to vendor. Also, ensure any warranty includes parts i.e. **Return to factory** - Vendor site visits are chargeable. Make sure the vendor has service engineers on the road, which means that should there ever be a need to call upon a warranty, a company can be assured of a **fast** response, typically same or next day. This is a key requirement when considering a machine supplier because a company will have bought a machine to do a specific job and so if it does break down for what ever reason, they do not want their production processes stopped for an extended period of time while waiting for either a spare part or an engineer.

Backup and support is important because a company must have absolute security that if an issue arises of any description that it will be resolved quickly. **Downtime costs money and possibly lost customers** so a company must have the peace of mind of on-site backup if required as well as access to locally held spare parts. Think about the company's future needs, don't limit focus to today's business or today's problems. Purchasing a piece of equipment is often a 5 to 10 year (or longer) decision, so the vendor the equipment was bought from is often as important as the equipment itself.

All service provider companies must supply the level to which their employees are trained in order to prove competency before they come on site to complete any designated specific task. The vendor will normally provide a good level of detail and put a lot of effort into compiling a training record for each of their employee's and should have no problem supplying this type of information. Any reputable vendor will see training and the upkeeping of the respective training records as being very important to maintain compliance and win new business. This vendor information must be held and updated as required by a company's HR department and if needed, be presented to an auditor to prove that the vendor is fully trained and competent to carry out the specific work e.g. Purified Water or HVAC systems in the pharmaceutical industry.

Tip: There are vendors who will provide **Predictive Maintenance (PdM)**, they will go online or via IoT, download data (possibly from the vendors headquarters in another country) from onsite equipment via a telephone modem or a mobile device on a regular basis and analyse it. The vendor will then give their expert advice and recommend when a service is needed or when parts are starting to wear out, thus avoiding unnecessary downtime. This information can also help the vendor improve their design if required. A vendor may also provide a full customer package which represents the greatest value for asset management by transferring operational risk to them. They will be responsible for scheduled maintenance as well as use predictive analytics and analytical tools to help prevent unexpected interruptions in your operations.

Main Points:

Ensure vendors come to site with the equipment needed to complete their tasks. The vendor must make sure all relevant spare parts are confirmed and accounted for on site before their arrival or they bring the spare parts with them.

- Ensure to receive a vendor's report or any other relevant documentation of confirmation, on works done or to be done before they leave site.
- Make sure to understand the vendors' written report (if anything is not clear or is illegible, ask the vendor to explain or rewrite). Ensure any recommended follow up actions are carried out as soon as possible.
- This report must be filed in the piece of equipment's maintenance history folder for future reference.

Tips to remind Vendors when Tendering

A 'Vendor' must:

- Only tender for work they can do
- Read the tender document carefully and make sure they understand it. Seek clarification if necessary
- Understand the specification, identify the mandatory (key) requirements. Check that they can meet the mandatory requirements which may be highlighted by words like - Shall, will, must. Can they address them?
- **Consider what the client wants:** evidence that they can meet the needs of the client by demonstrating capability, capacity, relevant experience. The tender will meet the tender specification, will deliver what's being asked for, and provide the requested information in the format specified. The tender will agree to the contract Terms and Conditions
- Ensure their tender is received on time and at the correct address of the client
- **Maintain a Tender content library to include:** Company organisation chart, Financial Accounts/Statements, Insurances, Accreditation Documents, References, Key Staff Biographies.

12. KNOWLEDGE MANAGEMENT

Knowledge Management is the efficient collecting, handling, leveraging and distributing of information and resources within a commercial organisation using technologies that support continuous improvement.

Information Infrastructure

Does your business have the security to succeed? DATA is GOLD and must be protected e.g. 43% of cyber-attacks target SME's. Business is easier with great data especially when it is managed and analysed effectively. Data, of itself, is of very little value, you must convert it into information and understanding! Managing **Big Data** will be critical to a company and having the ability to interpret vast amounts of data, and being able to bring data to life, visually. The valuable insights gained from data must be clearly communicated to everyone in the company ecosystem.

Knowledge Management will play a key role within your company in achieving, maintaining, and demonstrating full proof of control of all your engineering information. This minimises the risk of non-compliance, saves time and costs when dealing with regulators. It allows a company to quickly adapt to new and changing regulatory requirements. The information infrastructure that's in place will determine how easy or difficult any intended work to be carried out is going to be. **Information must flow seamlessly throughout the organisation.** There is no point in a company having all the relevant information if it is very difficult to find. E.g. Studies have shown that up to 70% of the time required to complete a maintenance job is spent on searching for information – only 30% is spent on actually doing it. Companies that value knowledge want to 'know how' and 'where to' access it.

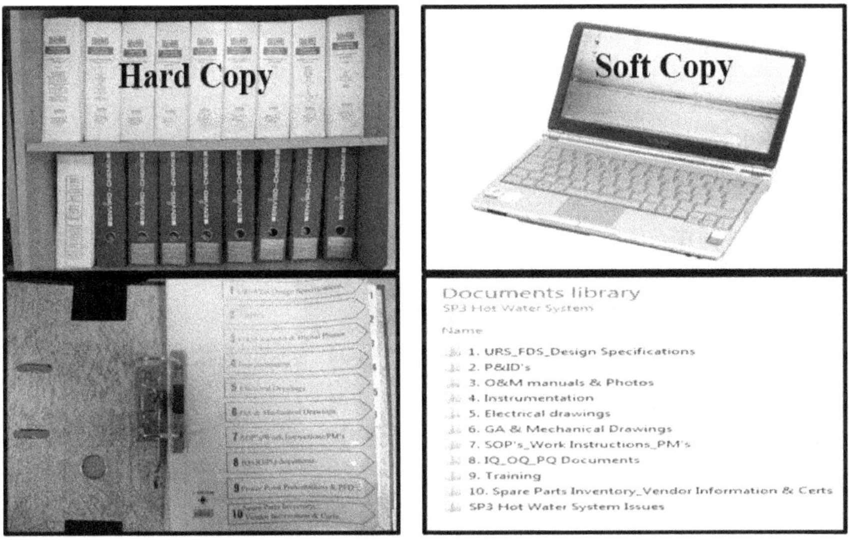

N.B. Erosion of Expertise - Critical Site Operational Knowledge must **not** reside in one person, a knowledge vacuum must never be allowed to exist.

Company management and its employees must continually seek more knowledge through training, education, and career development. A company must innovate or fail. Their ability to learn, adapt, and change becomes a core competency for survival.

It is important to understand how knowledge is formed, and how people and organisations learn to use it wisely.

Understanding knowledge is the first step to managing it effectively.

Knowledge Management is different from **Information Management.**

- **Knowledge Management** targets collecting and distributing knowledge - both explicit (that which is set out in tangible form), and **tacit** (that which is within people) throughout the organisation.
- **Information Management** deals mainly with documented **explicit** knowledge – or information - only.

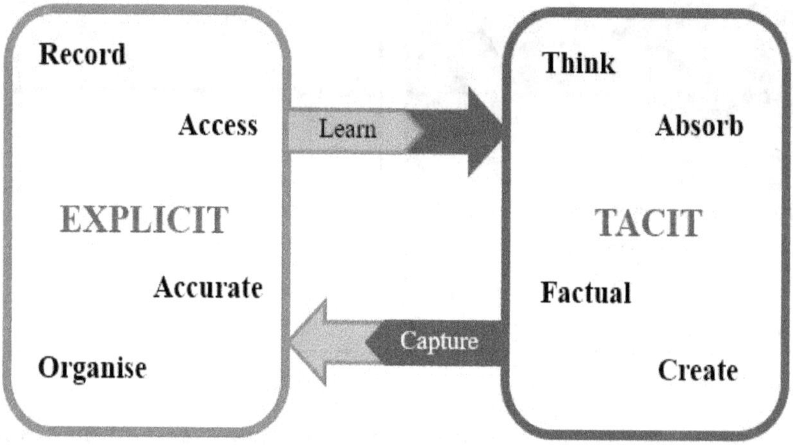

Knowledge Management comprises a range of strategies and practices used in an organisation to identify, create, represent, distribute, and enable adoption of insights and experiences. Such insights and experiences comprise knowledge, either embodied in individuals or embedded in organisations as processes or practices.

KM efforts typically focus on organisational objectives such as improved performance, competitive advantage, innovation, the sharing of lessons learned, integration and continuous improvement of the organisation.

KM efforts overlap with organisational learning and may be distinguished from that by a greater focus on the management of knowledge as a **strategic asset** and a focus on encouraging the sharing of knowledge.

Every business is a knowledge business; every worker is a knowledge worker. **Knowledge is perishable, Experience is not.** The shelf life of expertise is limited because new technologies, products, and services continually pour into the marketplace, but remember, the experience you've gained over many years has no shelf life. What you've learned and experienced from one system you'll bring to another. No one can hoard knowledge. People and companies must constantly renew, replenish, expand, and create more knowledge.

In a supportive environment, knowledge will take care of itself, for it to become a reality, you must create a climate of trust in your organisation.

Knowledge is embodied in people. It's impossible to talk about knowledge without addressing the way people work together, learn together, and grow in knowledge individually and collectively. Companies that are serious about knowledge foster an environment and culture that supports continuous learning.

Knowledge Protection - An investment in knowledge protection measures will pay off now and in the future. A loss of knowledge today can be painful, but it can threaten the very existence of your company in the future. Through knowledge protection precautions you reduce the risk of knowledge outflow and ensure your competitiveness and innovative strength.

One strategy to Knowledge Management involves actively managing and storing critical 'knowledge'. In such an instance, individuals strive to explicitly encode their knowledge into a secure shared knowledge repository **IT database**, where they can also retrieve knowledge they need that other individuals have provided.

Codification (the act of codifying; arranging in a systematic order) approach to KM relies upon the explicit knowledge and is more focused on the sharing of knowledge mainly through reutilisation of existing knowledge. "Codification could be described as a 'people-to-document' approach or as 'capturing knowledge' from many for reuse by many others."

Knowledge-based organisations seek guides, maps, and pathways for building knowledge across multiple performance levels. They understand the processes that support the creation, acquisition, sharing, and renewal of knowledge. The knowledge and experience you learn about how to maintain your equipment can become quite valuable. In today's global manufacturing world, even greater value can be extracted if you have a global knowledge capture and distribution system such that this knowledge of machinery maintenance can be effectively shared across your organisation – letting you reap even greater benefits on a much wider scale.

One of the hottest technologies for knowledge sharing is a corporate intranet. A company's intranet significantly improves productivity, process efficiency, and workflow; enhanced knowledge capital; strengthened teamwork across boundaries; and increased employee satisfaction.

There must be vessels or vehicles to support knowledge exploration.

- Technology support (information systems, databases, communication technologies, Web technologies, and email).
- Equipment (groupware, whiteboards, video conferencing equipment and flexible manufacturing systems).
- Tools (Job aids, knowledge maps, and computer-based performance support).
- Physical structures (learning centres, libraries, meeting rooms, and executive strategy rooms).

The learning environment must be suited to the needs of learners and to the expected learning outcomes.

13. PROJECT MANAGEMENT

What is a Project?

A project is a sequence of unique, complex, and connected activities having one goal or purpose. For a large-scale project to be successful, you need a clear plan, expertise, good communication & collaboration, and sufficient resources. **The failure to design a system properly is the same as designing for failure**. Remember, no amount of maintenance can overcome poor design.

Insist on the installation of key equipment and extol the benefits of them in the long run. Always have the justification facts to hand when questioned by stake holders and finance managers.

The time given to writing a good URS and FDS, be it a small project involving 2 weeks' work or a new installation which may take 12 - 18 months to complete cannot be underestimated. Try to get it right first time. Dedicate as many meetings as it takes with the relevant personnel who have crucial experience, information, or abilities; all are important criteria for a successful project outcome.

It is important to have a team of people who between them understand the entire operation of the business. Get the relevant departments input on the engineering, operations, environmental, health and safety, financing and quality. This in-depth knowledge of the business is critical to ensuring the processes established are going to be practical for everyday use. Having diversity in the group helps as different viewpoints about how the goal is attained will surface and various approaches should be encouraged. This removes the 'we should have thought of that' scenario after the project has been completed.

What is Project Management?

Project Management is an approach and a set of techniques based on accepted principles of management used for planning, estimating, and controlling work activities to reach a desired end result on time, within budget, and according to specification.

It is the application of knowledge, skills, tools, and techniques to **project activities** in order to meet or exceed stakeholder needs and expectations.

Project Management can mean different things to different people. Quite often, people misunderstand the concept because they have ongoing projects within their company and feel they are using project management to control these activities.

Project Management is about delivering Project Management best practice. It aims to provide:

- A more routine and planned approach to project work, rather than fire-fighting
- More clearly defined deliverables and success measures
- Make the job easier: you are managing the project, rather than the project managing you.

Tip: Proper planning in the beginning pays off in the later stage of the project.

Project Manager:

The **Project Manager** and the **key stakeholders** are responsible for defining the required project milestones. While the project follows the process groups and the project life cycle, the **Project Manager** must also be competent in various dimensions or knowledge areas in order to manage the project effectively. A project manager must be knowledgeable in a range of areas and have specific skills in order to deliver projects efficiently, effectively and successfully.

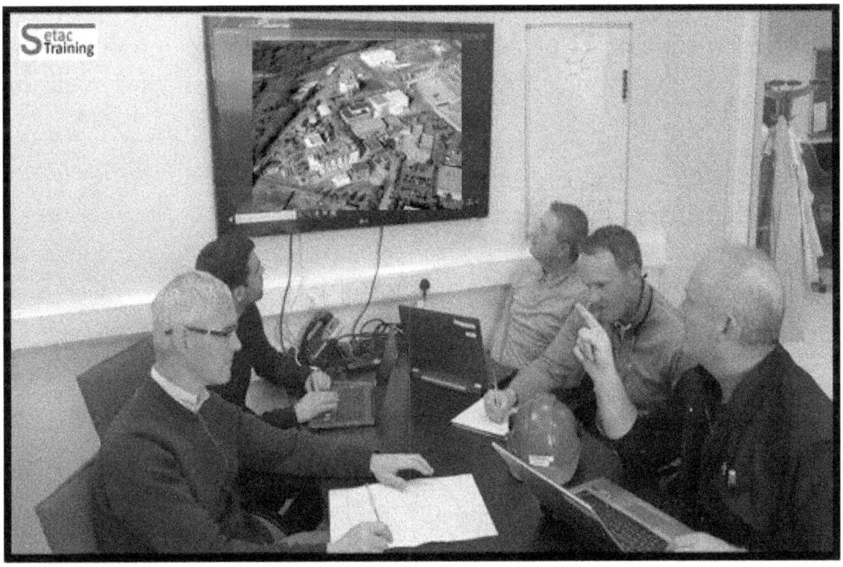

Specifically, there are 10 areas where the **Project Manager** must be skilled:

1. Project Integration Management
2. Project Scope Management
3. Project Time Management
4. Project Cost Management
5. Project Risk Management (Risk reduction/avoidance)
6. Project Communication Management
7. Project Human Resource Management
8. Project Quality Management
9. Project Procurement Management
10. Project Knowledge Management.

RACI (Responsible, Accountable, Consulted, Informed): describes the participation by various roles in completing tasks or deliverables for a project or business process. It is especially useful in clarifying roles and responsibilities in cross-functional/departmental projects and processes.

The Process of Time Management is about agreeing the overall project deadlines and identifying how and when they will be delivered.

The 5 process steps are:

1. Define activities
2. Sequence activities
3. Estimate activity resources
4. Estimate activity durations
5. Develop the Schedule and Control the Schedule.

Resource Management deals with the human, financial, distribution and demands of project resources.

Project Management Methodologies (Best Practice)

A methodology is a set of standards and procedures to follow for a given action. The purpose of the Project Management Programme is to create repeatable processes as well as provide consistency. It must be easily understood and allow flexibility. An agreed common discipline of organising, planning and managing resources to achieve the predefined project goals and objectives leads to a faster and proper execution of a project.

The growing demand for project management has encouraged the development of several standard methodologies. Although they differ in detail, these share many common fundamentals (e.g. good scope definition).

Advantages of using a Project Management Methodology

Individual projects are all unique and because projects can vary widely, a systematic standard approach to managing them is necessary. Every step of the project is planned, checked and acted upon to achieve the predefined project goals and objectives.

The advantages of using a project management methodology include:

- The outputs that each project is to deliver are clearly defined
- The objectives of each project are also clearly defined and linked to the business objectives
- The responsibilities for different parts of each project are understood
- A structured approach to managing the project
- A consistent means of monitoring and control
- A standard format & set of processes for reporting progress to senior management.

Project Management Tools

Critical Path Analysis (CPA)

The Critical Path is the term given to the sequence of **tasks** that are critical to the overall duration of the Project. If a task on the Critical Path is extended, it will delay the overall Project completion. The Critical Path is the longest path through the tasks and the least flexible.

Example - CPA **tank installation phase** of a project:

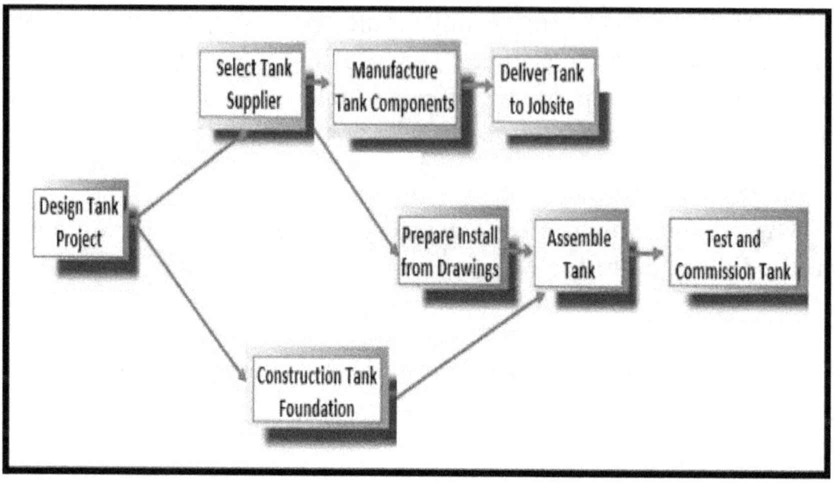

These require consideration of the earliest and latest possible finish date for each task in the Project. Where these dates are the same, there is no room for movement and the task is deemed critical.

The best analysis of Critical Path will always rely on computer software to undertake all of the iterative calculations required.

MS Project is often the most preferred software for project schedules.

Gantt Chart

What is a schedule?

A Gantt chart is a type of bar chart, developed by Henry Gantt that illustrates a **Project Schedule.**

A Project Schedule, whether prepared by computer or manually, should have the following features:

- Timescale shown from left to right along the top in whatever units are appropriate to the level of planning required, typically weekly or monthly, but not daily
- Individual Projects and their component tasks listed down the side and shown as separate bars, with start and end dates and duration
- Each individual task bar identified with a reference
- Milestones identified.

Schedules are the ideal medium for presenting both the project structure and timeframe.

Gantt Chart Example

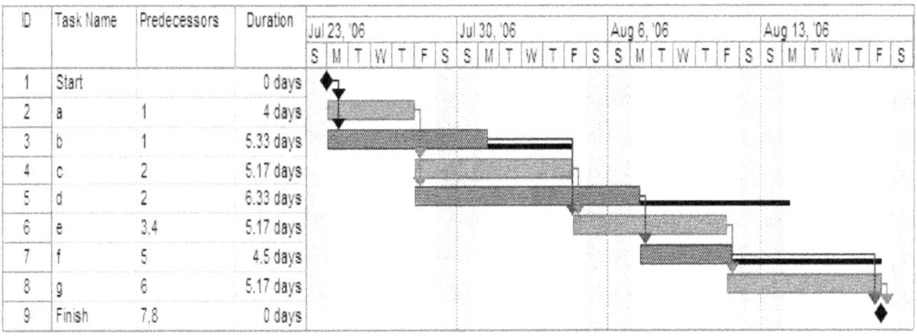

An underlying concept for the interaction among these processes is the **Plan-Do-Check-Act** Cycle e.g. PDCA Execution stage:

- **Plan** - The fundamental of project planning is built by the Work Breakdown Structure (WBS). The complete project scope is broken down into manageable work packages with clear allocation of responsibility.
- **Do** - Carrying out the project work according to the plan.
- **Check** - Identify deviations between plan and actual.
- **Act** - Introduce appropriate corrective actions.

Phase closures/gates are the phase-end reviews of key deliverables that allow the organisation to evaluate the project's performance and to take immediate action to address any issues or problems. Moreover, projects are typically reviewed at the end of each phase before the project manager can proceed to the next phase of the project. This cycle is linked by **results** - the results from one part of the cycle become the input to another.

Plan-Do-Check-Act Cycle example:

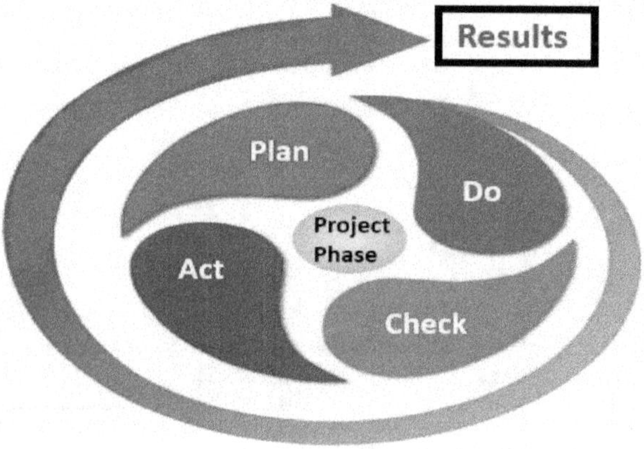

The Triple Constraint:

When considering a project success criterion, the 'Triple Constraint' (or often referred to as the project management triangle) is introduced. The triple constraint is the combination of the three most significant restrictions on any project:

1. Scope
2. Time
3. Cost

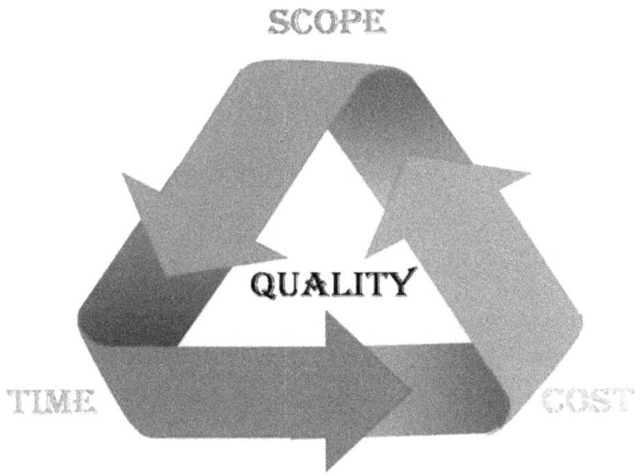

1. Scope: The scope of a project is about the end product, and the tasks to be done to create it. Defining the scope of a project involves capturing the client requirements and then using techniques such as Decomposition and Work Breakdown Structures (WBS - is a key project deliverable that organises the team's work into manageable sections) to document those requirements, clearly and unambiguously. If requirements are not included in the agreed scope definition, they must be considered 'extras', and subject to Change Control.
2. Time: The project Schedule and Resource Plan is developed from the work defined in the scope definition. If the scope increases once the project is underway, it may be necessary to delay the completion, or consider other methods to increase resource. For most projects, time is second to scope in priority, but being **on time** is always a major success. A realistic Project Schedule will always assist in achieving this success.
3. Cost: The cost of projects is estimated based directly on the scope. Sometimes the constraint is the scope required rather than the budget available, and therefore scope often takes a higher priority than cost.

The three most common measures of project success are:

1. Delivers the scope of the project as per the initial project charter.
2. Completed on time.
3. Within budget.

N.B. Poor or ineffective communications is one of the top reasons for project failure.

Unfortunately, when a project installation budget is tight, and money is running low, the first cutbacks will always be made to the non-essential equipment - 'the nice to haves'.

Learn from mistakes – Review What, Why and How AND Contributing factors.

After the project is completed, set up an **After Action Review,** call another **lessons learned** meeting and document the successes, and if any, the shortfalls that may have occurred:

Successes:

1. URS & FDS were excellent - good design day one - which made the PM relatively easy.
2. Project was completed before time which left extra time for testing.
3. Project was completed within budget.
4. Bonuses were given to personnel for excellent work and early delivery of project within budget.

Short Falls:

1. **Bad Project Management** - poor design day one – wasn't designed or planned properly, insufficient resources, not enough **know how** or time was given to this phase of the project.
2. **Project not completed on time** - unrealistic time frames.
3. **Project ran over budget** – unrealistic budget figure, not enough or no contingency allowed for, rising labour and material costs not considered.
4. **Personnel issues** – accidents, sickness, strikes, key personnel leaving for other jobs, loss of motivation.
5. **Weather** – poor weather conditions, personnel unable to work.

Document project files with as much information as you can:

1. Why did we do it?
2. What was our thinking?
3. On what data did we base our calculations?
4. What were the success criteria and by how much did we fail?

All of those things may have been clear then, but if they are not documented at the time, then it can be very hard to evaluate any of it from memory a year or more later. And what if you have moved on and it is some other person trying to work it out? Such **historical documentation** is very useful in being able to look into the thought patterns of an earlier failed project to see whether or not you could do it better makes sense. Keeping good records will make things easier for you in the future and you will also be helping others too.

• **N.B.** Learn from Successes (success always leaves clues), but, just as important …. Learn from Shortfalls (put shortfalls down to experience and try to ensure they are not repeated).

14. ENERGY MANAGEMENT

For plant managers, two things in life are certain: most buildings and manufacturing processes use a lot of energy; and that energy costs money. By managing their assets effectively, they can not only meet legislative requirements but also reduce operating costs i.e. as costs go up, so too does the expectation to keep them down, which means greater efficiency, less waste, and more productivity. Individual organisations cannot control energy prices, government prices or the global economy, but they can improve the way they manage energy in the here and now.

The Four Fuels:

1. Hydrocarbons (Petroleum/Natural Gas)
2. Coal
3. Nuclear
4. Renewable Energy (Wind/Solar)

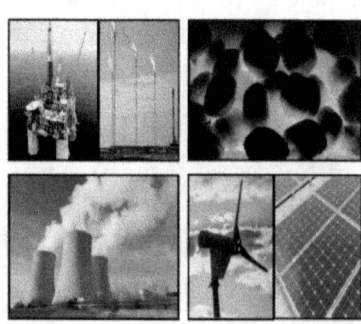

What is the Fifth Fuel?

Energy Efficiency (Invisible Fuel) - The 'Fifth Fuel' as energy efficiency is sometimes called, is the cheapest of all. The cheapest and cleanest energy choice of all is not to waste it. The largest single chunk of final energy consumption, >31%, is in buildings, chiefly heating, and cooling. Much of that is wasted, not least because in the past, architects had paid little attention to details such as the design of pipework (long, narrow pipes with lots of right angles are far more wasteful than short, fat and straight ones). 'Energy Efficiency' was nobody's priority: it took time and money that some architects, builders, and property owners would rather spend on other things. This mindset is now changing.

Energy efficiency is an ecological and economic obligation, it conserves valuable resources, reduces the environmental impact, reduces carbon emissions – as well as the cost-related implications of carbon taxes. It also minimises operating costs for each and every business. Increasingly, companies are trying to develop a 'Green' approach to reduce the environmental impact of their manufacturing processes in reducing energy and waste. Your company must be committed to the objective of employing the best available conservation practices to curtail the excessive consumption of resources in supplying utility services to its facilities and in turn, reducing greenhouse gases.

ISO 50001:2011 Energy Management Systems (EnMS) – Requirements with guidance for use is a specification created by the International Organisation for Standardisation (ISO) for an Energy Management System. The standard specifies the requirements for establishing, implementing, maintaining, and improving an energy management system, whose purpose is to enable an organisation to follow a systematic approach in achieving continual improvement of energy performance, including energy efficiency, energy security, energy use, and consumption. The 'Standard' aims to help organisations continually reduce their energy use, and therefore their energy costs and their greenhouse gas emissions.

The Plan, Do, Check, Act (PDCA) cycle:

Plan: a new Energy policy must be put in place. Top Management commitment is extremely important when it comes to establishing a new EnMS. An EnMS champion must be chosen along with his/her team to direct activities throughout the whole organisation.

Do: implement the energy management action plans by conducting an energy review and establish the baseline energy performance indicators (EnPI's), objectives, targets and action plans necessary to deliver results in accordance with opportunities to improve energy performance and the organisation's energy policy.

Check: monitor and measure processes and the key characteristics of its operations that determine energy performance against the energy policy and objectives and report the results. Action only makes sense if it leads to the desired result.

Act: the constant measurements are broken down in reports. Reporting of results and management review is crucial. Take actions to continually improve energy performance and the Energy Management System.

With a little investment, many **Energy Using Systems** have the scope to achieve higher efficiency through improved control, more regular maintenance, and closer monitoring. What manufacturing companies are looking for is tighter control on how to increase plant capacity and improve product quality. Both improvements coupled with effective energy management can have a big impact on a company's running costs and its bottom line.

Building Energy Performance must be a top priority. The performance gap between the predicted and actual energy performance of buildings will not reduce unless clients and end users see the issue as a priority. If you want a quick win, it will be by revisiting existing complex buildings, retuning them using system intelligence via a digital platform, and training and educating people who use them.

Energy Management isn't just about turning off lights and machinery when not in use. An integrated approach must be undertaken from initial design during the **Front End Study** (FES) to the actual servicing and optimal running and monitoring of the site machinery itself via Conditioned Based Monitoring (CbM). It involves ensuring you understand metering, your utility bills, and the tariffs you're currently paying for, using tools such as Balanced Score Cards and KPIs. Most important of all is to ensure the people whose job it is to ensure proper **Energy Management** is employed, have awareness, are environmentally conscience, are well trained, and are very knowledgeable on how their bespoke systems operate.

The **Site Energy Management team** must have energy consumption visibility and transparency and the ability to invest in energy savings measures. Understanding and monitoring energy consumption enables the team to implement actions throughout the consumption phases.

"If you don't measure it, you can't manage it."

"If you don't measure it 'Accurately', you can't manage it 'Accurately'."

E.g. If you have an Annual Energy Cost of €3,000,000:

- If 10% Accuracy is accepted, then Uncertainty of €300,000 per year is accepted.
- If 5% Accuracy is accepted, then Uncertainty of €150,000 per year is accepted.
- If 2% Accuracy is accepted, then Uncertainty of €60,000 per year is accepted.

Accuracy pays off!

Energy Management fundamentals:

A comprehensive, proactive, and sustainable energy conservation strategy must be in place, it will help save the environment, conform to laws and regulations, cut down on the company's running costs, and improve the corporate image. A company must try to educate all its employees into thinking about energy management the way they think about it at home. Everyone on site can play their part by being more energy conscious and switching off pieces of equipment that are not in use. Energy reduction must become a shared and embedded responsibility. Every employee is an **Energy Manager**, all staff consume energy in some form **e.g**. electrical power for their PC's, washing hands, canteen facilities, water for drinking …. and it's only by everyone taking personal responsibility that substantive savings can really be made. What is your personal energy footprint? Think how you can play your part in reducing your daily energy consumption to benefit both the company and the environment?

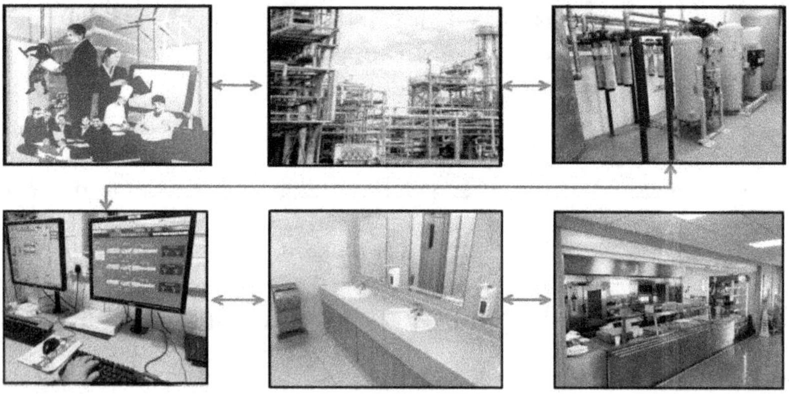

Understanding Power: Power is the rate at which work is done. It is the rate at which energy is consumed and hence is measured in units that represent 'energy per unit time'. The unit of Power is the **Watt**. The utility company charges you for the power you use based on the monthly readings of your electric meter that **measures the current (I)** passing through it due to your onsite operational equipment electrical energy usage.

So, if P (Watts) = V (Volts)* I (Current) then I = P/V:

P - Watts Power	=	V*I	or	$R*I^2$	or	V^2/R

Electricity is measured in Watts, or more specifically, kilowatt hours (1 kilowatt hour = 1,000 watt-hours). A Watt is the unit of electrical power equal to one ampere under the pressure of one volt (1 Watt = 1 Volt x 1 Amp). But this formula represents merely the measure of the electrical potential. Electricity meters operate by continuously measuring the instantaneous voltage (volts) and current (amperes) to give energy used in kilowatt-hours. To measure actual energy usage, you have to add an element of time. Therefore, electrical usage is a measurement of watts used over a period of time. Kilowatts describe how much energy an appliance is using at any given moment, while a Kilowatt-hour describes how much energy it has used over a period of time. **E.g.** the rate at which a light bulb converts electrical energy into light is measured in Watts—the more wattage, the more power, or equivalently the more electrical energy is used per unit time. When a light bulb with a power rating of 100W is turned on for one hour, the energy used is 100 watt hours (W·h), 0.1 kilowatt hour, Other examples, 1 kWh is the amount of energy used by a 1kW (1000 watt) electric heater for 1 hour or ten 100-watt light bulbs used for 1 hour.

A simple way of starting an 'Energy Management Programme' is by regularly reviewing energy invoices, graphing data, and identifying anomalies such as high non-production processing energy usage. Corrective actions should then be taken, and their impact evaluated when the next set of invoices arrives.

Where possible, feed the information into a PC and set up an excel **Run Chart** where monthly peaks and troughs will be clearly visible in energy usage (**kWh**) and costs. A basic monitoring system has now been created where 'Statistical Thinking' can now be employed.

Run Chart example:

SETAC ELECTRICITY USAGE 2016															
	Jan	Feb	Mar	April	May	June	Jul	Aug	Sept	Oct	Nov	Dec	Average	Total	Daily Average
2016 (Euros)	€4,046	€2,440	€2,628	€1,663	€1,630	€1,642	€1,577	€1,641	€1,564	€2,364	€2,724	€2,356	€2,190	€26,275	€72
2016 (Kwh's)	33,291	17,193	20,704	8,971	8,815	8,567	7,320	7,926	7,253	13,129	16,659	12,275	13,509	162,103	444
Days per Month	31	28	31	30	31	30	31	31	30	31	30	31			
Daily Avg (Euro)	€131	€87	€85	€55	€53	€56	€51	€53	€52	€76	€91	€76			
Daily Avg (Kwhrs)	1,074	614	668	299	284	286	236	256	242	424	555	396			

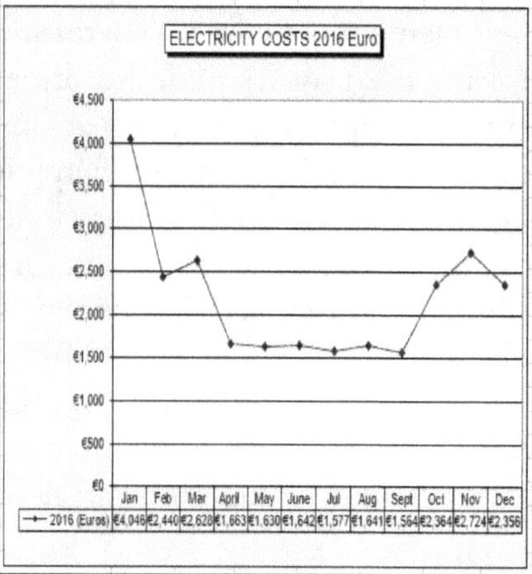

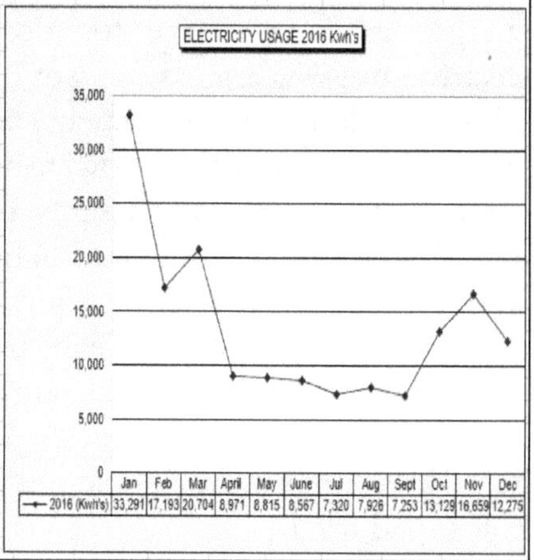

Energy Budget Development - Anticipate your spend with greater accuracy:

1. By closely managing your energy budget, you ensure that you maintain a process of continual improvement.
2. Maintain best-practice energy management in your day-to-day operations by reducing your energy costs **e.g.** daily monitoring of energy usage to maintain budgets.
3. Identify and prioritise energy efficiency projects that can generate capital. This improves your performance and productivity thus improves your company's bottom line.
4. Ensure that senior managers **commit to energy efficiency** within their own departments and that all staff play a role in the process.
5. Standardise processes so that improvements are sustained over time.

6. Helps you to comply with your energy-efficiency and emission-reduction obligations **e.g**. reduce carbon taxes.

The first step in improving **Energy Efficiency** is to take precise and detailed measurements of energy consumption.

BMS (Building Management System) – Energy Management

Tip: Regardless of their focus, healthcare facilities are among the largest energy consumers in the country. As sophisticated needs and extended uptime lead to high-energy consumption of most hospitals and clinics, the healthcare sector offers an unparalleled opportunity for facility managers to optimise and increase their energy savings. Whether you are targeting upgrades in an existing system or working to deliver a system-wide solution, a BMS system offers the services needed to meet both the HVAC and energy saving needs of today's healthcare facilities.

Technology is of great assistance in proper energy management. It can read, alarm and trend 'real time' energy consumption via a BMS PC screen to view and act on. It is costly to implement proper energy consumption monitoring systems e.g. installing power, oil, gas, and water meters. Then having to return all the real time power meter and flow meter readings from the instruments via 'hard wiring' to a PLC for deciphering and trending on a PC for viewing and dissemination purposes. Ensure to have the installation costs in implementing such a system versus payback complete before looking for money from the management team. The cost of installing such a system will always be the driving factor to implementation. **Wireless sensors** and networks can now be used in a wide range of process measurements, usually at dramatically lower costs as compared to 'hard wired' alternatives, with faster installation time and minimal disruption.

If this technology does not exist on your site, use the energy meters your utility providers have installed instead. The site energy management team can start reading and logging the kWh consumption daily. This will give a company real time (measure the amount of kWh's expended in 1 hour and relate to plant activities) data instead of waiting until the monthly bills arrive. This recorded data can be reported to other key departmental personnel in the company **e.g.** Operations.

Ask your **utility provider** about their online power monitoring service and can you gain access to it. (**Tip**: they normally take your site's energy consumption readings every 15 minutes remotely, so if you want more granularity on the energy bill, ask them to provide it. You can then compare these granular readings to when and at what time your largest energy usage occurs and relate it to your site activities).

Full energy efficiency cannot be introduced overnight. It will be the result of a determined effort over time. To design and implement an efficient energy management system via control strategies designed to provide a more intelligent and sustainable building isn't easy. Small energy management activities can be done almost immediately **e.g**. switching off lights, heating appliances, air conditioning unit's in areas during periods of unoccupancy. Putting up energy saving posters on notice boards will also remind all staff that energy saving is everyone's responsibility.

N.B. The real focus must be on the 'out of sight' large power using equipment **e.g.** chillers, air compressors, roof top air handling unit fans and circulating pumps: 20 kW; 50 kW; 250 kW motors that run 24/7.

Duty/Standby 250 kW Chiller motors example:

These costs can be astronomical; a simple ***rule of thumb*** *is*:

For every 1 kWh at a cost of 11.42 cents per kW (industrial rate) of power a motor uses running 24 hours a day, 365 days of the year will cost approximately **€1,000 per year**. Now multiplying that figure by 20 kW's (20*.1142 * 24 *365) = €20,000 to run one 20 kW motor at full load capacity for one year. Energy efficiency of different pieces of equipment in an industrial plant is not always easy to detect without in-depth study and analysis.

Implement an Energy Management plan – the plan will only be as good as the people who are implementing it, the time dedicated and the systems installed to monitor it. The gathering of data, knowing what to do with it and what changes will be needed to adapt to this new-found information will determine the successful outcome of the plan.

People like to see value for the money they invest. They like to see results, facts, and graphs. Don't bombard them with figures. Have graphs, bar or run charts ready to show what energy is being expended now versus what was previously being expended before the energy monitoring systems were put in place. Ensure also to have all the data available for those who may want to scrutinise it further.

E.g. the amount of energy expended versus the amount of product made

In one year, a company expended 1,000,000 kWh's of energy for every 10,000 units of product produced.

The following year, after successful energy management implementation, the company expended only 900,000 kWh's of energy for every 10,000 units of product produced.

A reduction of 100,000 kWh's of energy to produce the same amount of product as the previous year with a cost saving of over €11,420 (100,000 kWh's*.1142), which can be further invested in more energy savings projects.

The Environmental Protection Agency (EPA) will be very interested in this type of information and a company is also showing their commitment to energy reduction and a cleaner environment.

Tip: There are Government grants available in some countries for the installation of energy management systems in industry, why not make some enquiries?

Review and evaluate – if there are multiple stand-alone production plants, laboratory, maintenance, and administration buildings on a company site, an evaluation and review must be done on the amount of energy each area is expending. Individual energy meters should be fitted to the utility services for each area thus allowing energy costs for running these areas to be charged to their respective cost centres.

A great way to get company management concerned about energy usage for their areas is to make them **accountable for energy expended** and charge their yearly allocated department budgets directly for it.

The bills received monthly from an energy provider, should tally with a company's internally recorded energy figures. If this type of control and measurement exists on site, a company is well on their way to successful energy management implementation.

Knowledge transfer and operational understanding of a company's bespoke energy consuming equipment to the **site energy management team** via training is the effective approach required to ensure energy management procedures are embedded within an organisation and deliver tangible bottom line results on an ongoing basis.

The 3 A's of Energy Management

1. **Audit:** Energy audit - monitoring energy input, measuring energy consumption of individual units/systems, idle running period of energy intensive machines. 'If you can see where you're using your energy, you can see where you're wasting it'. What can be easily forgotten when it comes to audits is their effectiveness in providing a framework for the productivity and profitability of a business and its employees.
2. **Analysis:** If you don't understand how your plant's bespoke industrial processes work or know what the energy consuming usage baseline figures are from date of installation, it will be difficult to accurately measure, analyse, and manage the energy consumption usage of the individual supplied energy utilities, which leads to ineffective energy efficiency management and ultimately higher running costs. Data Analytics is crucial.
3. **Action:** Evaluation and decision making. Planning, and controlling energy consumption, and identifying areas of control.

Energy Champion Audits:

An Energy Champion must have a **Heightened Energy Awareness** of Energy Efficiency Measures (EEMs).

- Comparison of energy usage
- Energy use index and energy cost index
- SCADA, PDA's or paper logbooks for recording information
- Collecting and analysing data, and use for Condition based Maintenance (CbM) strategies
- Walk-through inspections. Instrumentation/data logging systems, meters, sensors.
- Business model, finance, cost saving opportunities assessing energy and looking out for stealth operational changes in real time.

Tip: The installation of an Energy Management System can provide continuous measurements in your electrical network and detailed reporting on the **site power load**. An Electrical Network Monitoring System can also be installed for continuous monitoring of the **site power supply** quality.

Monitoring & Targeting (M&T)

M&T is another useful technique for managing energy consumption. It usually involves predicting levels of energy performance against which future energy performance can be measured, thus allowing the impact of energy-saving actions to be evaluated.

An **Energy Performance Indicator** (EPI) is a point of reference for making comparisons on energy consumption. In general, an EPI may be based on consumption, cost, or environmental measures. EPI's can be used to develop relative measures of energy performance, track changes over time, and identify best practice in energy management.

Effective **Operations & Maintenance** (O&M) is one of the most cost-effective methods for ensuring reliability, safety, and energy efficiency. Inadequate maintenance is a major cause of energy waste in both the industrial and private sectors. Energy losses from steam, water, and compressed air leaks alone plus, uninsulated lines, maladjusted or inoperable control valves and other losses from poor maintenance are often considerable.

Valves (e.g. Steam, Water, Compressed Air)

N.B. Valves are not a 'fit and forget' piece of kit. **Poor** maintenance can create the risk of injury. Like any product, a steam valve for example should be regularly serviced to maintain optimum performance. Appropriate safety checks must be part of the maintenance programme. There are obvious cost benefits associated with refurbishment. Valves are a key part of any system. Properly used and maintained, they can improve process efficiency and lower costs, so by refurbishing on a regular basis, you can be sure they'll be kept in good condition. Not only will that help you benefit from improved efficiency and operating costs, it'll also mean the chances of system downtime are reduced.

Tip: "All Valves can fail." Manual isolation valves (Butterfly, ball valves) are essential parts of any piping system used to control the flow and pressure of contents, whether that is oil, gas, liquid, or vapours, they should be serviced regularly and tested to ensure that they open/close fully and are not seized. Manual valves are normally controlled with a handle attached to the stem. If the handle is turned ninety degrees between operating positions, the valve is called a **quarter-turn valve.** Butterfly, ball valves, and plug valves are often quarter-turn valves. If the handle is circular with the stem as the axis of rotation in the centre of the circle, then the handle is called a **hand wheel.**

Valve Operation: Never use excessive force to open or close a valve. The use of wheel spanners should be discouraged as valves are provided with hand wheels large enough to give a turning force sufficient to open and close the valve. If a valve is so tight that it cannot be fully operated by the hand wheel alone, it is defective, and the application of more force may make matters worse.

Operations and Maintenance (O&M) activities relate to the performance of scheduled and unscheduled actions aimed at preventing equipment failure or decline, with the goal of increasing efficiency, reliability, and safety.

Good maintenance practices can generate substantial energy savings and should be considered as a resource. Striving constantly for cost efficiency driven by reduced maintenance will play a key role in any company's long-term business strategy. Any piece of equipment that is running at low efficiency will incur year-round penalties in the form of increased power consumption and higher operating costs.

If energy management strategies are employed on all five critical utility services - **Water, Compressed Air, Gas/Oil, Electricity, Steam** (W.A.G.E.S) to any site, cost competitiveness can be realised resulting in more profits.

Energy / Utilities – W.A.G.E.S.

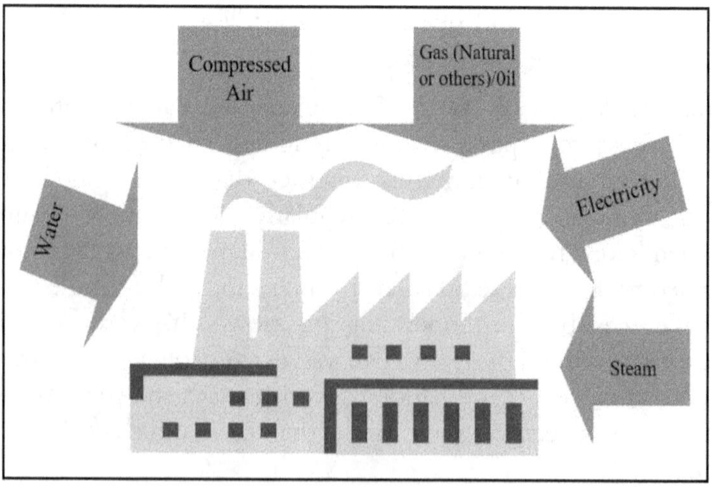

Moreover, improvements to facility maintenance programmes can often be accomplished immediately and at a relatively low cost.

Down time = Money: The sooner a faulty piece of apparatus can be brought back on line safely, and fully operational, the better. Maintenance, downtime, and spare parts all cost money and can add substantial costs to the daily operating budget of the facility. Ultimately the profits of the company will suffer. Personnel must run a plant as cost effectively and efficiently as possible but not to the detriment of safety. Taking risks or shortcuts may lead to serious injury, or damage to plant.

Regular scheduled maintenance ensures equipment is kept in optimal condition at the lowest overall operating cost. Improvements in energy efficiency can be made in electronic design by using energy efficient components, such as microcontrollers or up-to-date power management systems.

Energy costs can be addressed through advanced power management programmes that predict, assess and audit usage and then create plans and procedures to help optimise energy usage. They can also be addressed using **Variable Speed Drive** (VSD) technology or intelligent motor control systems.

Tip: Electric motors consume an estimated 65% of the electrical energy used in industry. Variable Speed Drives may be configured to both power a motor and during braking, re-circulate stored energy back into the mains supply, reducing energy consumption and significantly reducing energy bills. High energy costs coupled with reducing costs of VSDs are leading to faster **Return on Investment** (ROI). By being able to vary the speed and torque of an electric motor, and in turn the driven load, the following benefits will be realised:

Substantial energy savings – especially when considering either new installations or equipment packages, replacing oversized and underloaded motors. Rather than having an electric motor running continuously at full speed, a VSD allows the user to slow it down or speed it up depending on the demand.

Reduced need for maintenance – Being able to vary speed and torque of an electric motor means there is less wear and tear on the motor and the driven machine. For example, the ability to bring a 'process' up to speed slowly prevents the sudden shock loading that can damage a motor and the driven machine over time.

The VSD represents technology that boosts the performance of an electric motor and saves energy. VSDs enable more cost-effective production, reduce the greenhouse effect and play a part in meeting emissions targets. Despite this, less than one in ten electric motors in the world is fitted with a **VSD** - financially it would be justified to install a VSD on at least one in three electric motors.

One of the biggest benefits of controlling the speed of an electric motor according to demand is the energy saving opportunity over other control methods that are used in combination with motors running at fixed speed **e.g.** Direct On Line (DOL) starting, or Star-Delta starting.

How does a VSD save energy?

A VSD regulates the speed of the motor, and in turn the speed of the pump, by controlling the energy that goes into the motor, rather than restricting the flow of a process running constantly at full speed. In the example below, from the stopped position, once the run switch to the VSD is activated and the pump starts running at its minimum speed, the Flow Transmitter (FT) measures the water flow in the pipe and signals the reading to the PID controller which in turn will increase the speed of the motor via the VSD, the pump will turn faster to maintain the water flow at the PID controller set point required: 100 L/min. If the water flow rate overshoots the 100 L/min set point, the PID controller will react to this error and reduce the speed of the motor again to correct the error.

VSD Example:

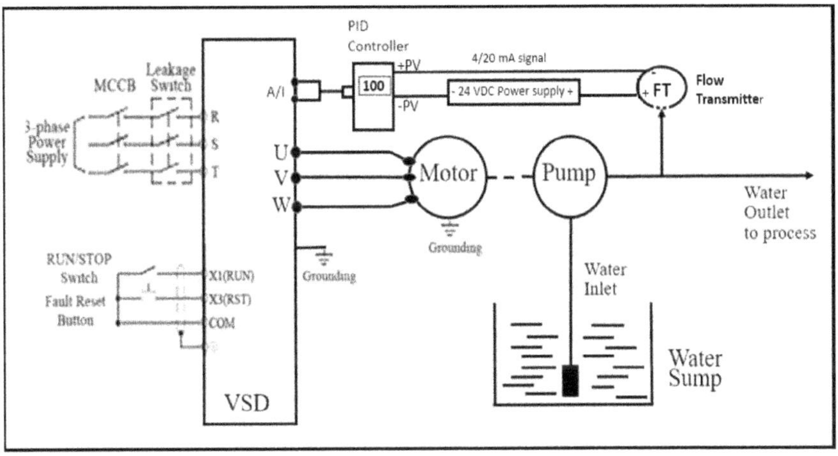

Running a motor at full speed while throttling the output is like driving a car with one foot on the accelerator and the other on the brake; a part of the produced output immediately goes to waste. Where fluid flow is controlled by dampers or valves, much of this potential for energy saving is lost. The application of a VSD, on the other hand, enables the motor to respond to the changing flow requirement, and therefore directly translates this potential into electricity savings.

A VSD can save over 60% of the energy as it controls the energy at source, only using as much as is necessary to run the motor with the required speed and torque - much in the same way as the accelerator in the car controls the engine revs.

VSD's can dramatically reduce energy consumption in pump and fan applications. The power required to run a centrifugal pump, or a fan is proportional to the **cube of the speed**. This means that if 100% flow requires full power, 75% requires 0.753= 42% of full power, and 50% flow requires 0.53= 12.5% of the power.

E.g. If Power is directly proportional to the speed cube, formula would be:

Power = (Fluid speed)3

Power max (100%) = (1)3 = 1

So, Power consumption at 75% flow:

Power? % = (0.75)3 = 42.18% of max. power

A small reduction of the speed can make a big difference on the energy consumption. Since many fan and pump systems run at less than full capacity much of the time, a VSD can make huge savings compared to a motor driving a load under mechanical control.

A VSD can also make it possible to stop a motor completely when it is not required as re-starting with a VSD causes far less stress than starting direct-on-line. Soft starting is an inherent feature of the VSD. Regulating the motor speed has the added benefit of easily accommodating capacity rises without extra investment, as speed increases of 5-20% is no problem with a VSD as long as there is enough spare capacity in the system.

Other advantages of Variable Speed Drive Control …

- Lower starting current
- Reduced mechanical stress
- Flexibility of operation
- Reduced noise levels
- Reduced hardware requirements – starters, power factor correction, metering/monitoring no longer required.

Tip: To avoid failure, VSD's kept in storage for more than two years, or the duration is unknown, **must** have their capacitors charged at a controlled rate as the VSD's contain electrolytic capacitors which lose their ability to withstand rated voltage if not periodically charged to rated voltage (consult the VSD manufacturers' guidelines).

Lighting:

Tip: Lighting in commercial and domestic buildings uses a lot of electricity. A typical office building using T8 conventional fluorescent lamps and ballasts can save up to **60%** by:

- Using dimming controllers and dimmable ballast in areas with natural daylight, can save up to 30-60%.
- Adding occupancy detectors in meeting rooms to switch off lights can save up to 40%.
- Switching the T8 tubes to T5 lamps can save 10-15% with typical return on investment in 3 months.
- Replace conventional fluorescent lamps with LED lighting.
- Replace high bay lighting with Induction lighting.

LED (light-emitting diode) Lighting

LEDs are commonly used in a whole host of applications such as monitors and flat screen televisions, smart phones and tablets, car headlights and indicator lamps, street lights, traffic lights and Christmas tree lights, signage, and entertainment applications. As energy-efficient LED technology develops, so do our uses for LEDs.

Interesting LED Facts

- Heat is what causes traditional light bulbs to shine (incandescence), whereas electrical voltage itself makes LEDs shine (electroluminescence).
- LEDs low UV content makes them far less attractive to insects than traditional lights.
- Good quality LEDs can easily exceed 50,000 hours' lifetime – that's over 5 ½ years of continuous use!

- Blue LEDs can help keep food fresher. They have a strong antibacterial effect on foodborne pathogens leading to their increased usage in fridges.
- You might think 'white is white' but there is a vast range of white hues and tones to choose from since white LEDs are not actually white. You can select the right kind of white for purpose by using colour temperature. Warm whites have a yellow-red tone and are cosy and calming. These are often used in homes, hotels and restaurants and imitate a candlelight tone. Cool whites have an icy blue tone. These are harsher and tend to be used in industrial environments and street lighting. Neutral whites sit between the two and are the preferred choice for classrooms, retail, and office environments.

LED drivers and power supplies

One of the first factors to consider (having chosen your desired LED) is what type of driver you need. Whether you need a constant current (CC) or constant voltage (CV) depends on the module or LED you need to power, and this information will be supplied with the product.

LED connectors and heat sinks:

To ensure your LEDs work with maximum effectiveness and efficiency, you'll need the right infrastructure to go around them. In addition to the actual LEDs, you must consider what other components you will require. These may include power supply, connectors, cables, mounts, or holders. It is a common misconception that LEDs do not produce heat. In fact, most of the electricity used in an LED (around 70%) is transformed into heat rather than light. The heat must be removed at source because if your LED runs too hot it will fail, using correct heat sinks will ensure maximum operating life and efficiency.

This is what makes the heat sink so important. In general, heat sinks can remove heat in three ways:

1. **Conduction:** transfer of heat from one solid to another
2. **Convection:** transfer of heat from a solid to a moving **fluid** (in LED terms the fluid is most likely to be air)
3. **Radiation:** transfer of heat through thermal radiation

Standard office lighting Vs LED lighting systems example:

Standard office lighting normally consists of recessed ceiling panel lights with x4 bulbs in each light. If a space has x10 of these panels with standards bulbs of 18W each, that's a total consumption of 720W per hour @ €0.2168 p/kW, **equates to €6.24 p/week.**

By replacing the 4*18 Watt bulbs with 295 x 295 ceiling tile **LED**, each at a total of 13W, the consumption per hour would only be 130W @ €0.2168 p/kW, **equates to €1.13 per week.** **That's a massive 89% cost saving.**

Tip: You can now use a wireless infrared portable remote control device to configure and modify certain parameters of LED luminaires without the need to access the light, and all at your fingertips. The signal range is approximately 20 metres. The simple menu screen allows you to make setting changes to the luminaires e.g. you can modify the intensity parameters of the luminaire and the time duration of the supply in conjunction with a motion sensor.

Incandescent bulb versus LED bulb:

Incandescent Bulb	LED Bulb	Lumens
25W	3 - 5W	250
40W	6 - 9W	450
60W	8 - 15W	800
75W	12 - 17W	1,100
100W	15 - 19W	1,600
125W	20 - 25W	1,850
150W	Up to 30W	2,600

Tip: 'Out of sight' plant rooms in factories, offices and other such places should have push button timer lighting control with key switch override for long maintenance periods. For safety, strategically placed low energy LED's should be illuminated - 24/7/365. Replace halogen lamps with LED retrofit lamps. Typical life for LEDs is 50,000 hours+ over that of a halogen lamp which is only 2,000 hours.

Induction Lighting:

Induction lamps are expected to last over 100,000 hours in continuous service. They are instant start/re-strike, vibration resistant, maintenance free, will save an average of over 50% in real energy costs, emit better visual effective lumens than most other lamps, are 100% recyclable at end of life, use a solid amalgam mercury pellet instead of harmful liquid mercury found in almost every other type of fluorescent lighting. This can be removed safely from the product at the end of its life eliminating the need for costly waste disposal.

Induction lighting is used successfully in warehousing, manufacturing, floodlighting, street lighting, architectural lighting and many more.

Examples of other Energy Saving Technologies:

Heat pump technology offers an opportunity to implement a renewable, low maintenance and consistent heating system. Heat pumps work by running a low-temperature, lower pressure refrigerant fluid in heat exchanger coils through a heat source, such as the ground, outside air, or circulating water.

Ground Source heat pump: Heat pumps are used for space heating and cooling, as well as water heating. They operate on the fact that the earth beneath the surface remains at a constant temperature throughout the year, and that the ground acts as a heat source in winter and a heat sink in summer. Heat pumps take advantage of this by transferring the heat stored in the earth to buildings in winter and the opposite in summer for cooling. Through compression, heat pumps can 'pump up' heat at low temperature and release it at a higher temperature so that it may be used again. A heat pump looks similar and can perform the same functions as a conventional gas or oil boiler for space heating and sanitary hot water production. For every unit of electricity used to operate the heat pump, up to four units of heat are generated. Therefore, for every unit of electricity used to pump the heat, 3-4 units of heat are produced. Ground source heat pumps use a 'closed loop' system of water/anti-freeze to collect the soil heat.

Air Source heat pump: delivers maximum performance and comfort, year-round. The energy consumption is put to a minimum by continually optimising the three key performance parameters of air flow (variable-speed EC fan), heating circuit flow (electronic expansion valve) and heat distribution flow (Optimum technology). An Air source heat pump absorbs heat from the outside air, which is then used for heating of hot water and hydronic heating systems, delivering efficient energy savings at temperatures as low as -20°C. This means you can reduce your energy consumption for heating by up to 75%.

Water Source heat pump: operates much like a traditional air source heat pump except that it extracts and dissipates heat by way of a body of circulating water instead of outside air.

Clean Energy: Costs for wind and solar technologies, electric vehicles (EVs), and light-emitting diodes (LEDs) have dropped by as much as 94% since 2008.

Key Elements of an Energy Efficiency Strategy

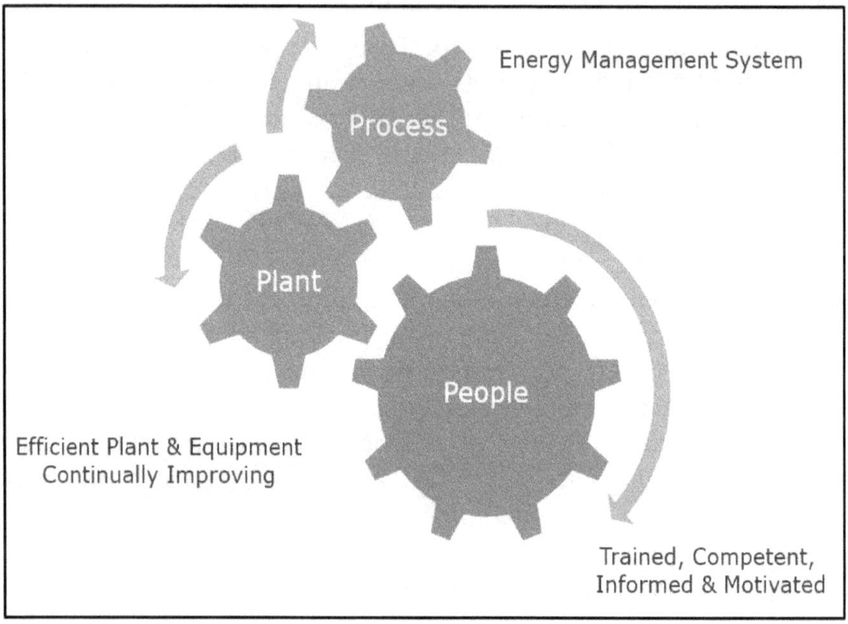

15. HELPFUL TIPS

Domestic Hot and Cold-Water Schematics Example:

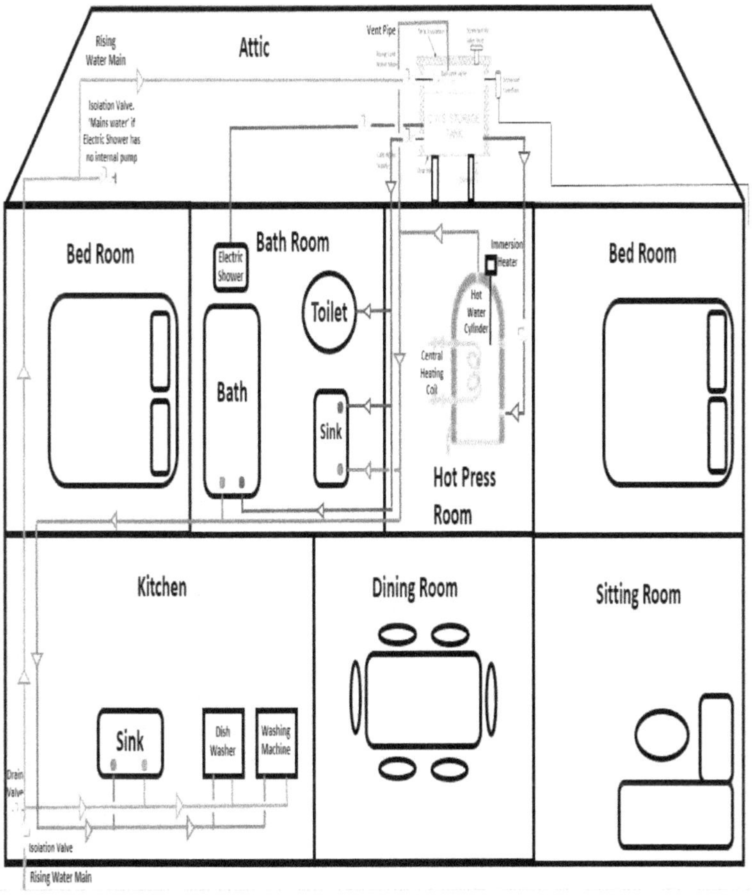

When was the last time the **mains water** valve was checked for operation into your property?

- Do you know where it is?
- Is it a good quality hand valve and capable of handling severe frost without cracking and leaking?
- Does it open and close, and properly isolate the mains water from your property when the valve is closed?

A simple check on this valve should be done at least twice a year and remember when the valve is turned fully open, give one half turn back the opposite way (if it's a wheel valve), to prevent seizing.

Questions you must ask yourself regarding water isolation valves and water tanks on your property:

- Do you know where all water valves are situated?
- Do you know where all water tanks (if any) are located?
- Are they labelled?
- Do you know what their function is?

If the answer is 'No' to any of these questions, locate and identify them. **Check that they open and close easily and are not seized,** try to ensure you do not have to go looking for them if you have a water leak and it is dripping onto your new carpet.

Ensure all individual water valves are clearly labelled, if not, label them or get someone to do it for you.

A Domestic 'Open' (Open to atmospheric pressure) **Central Heating Radiator Schematic example:**

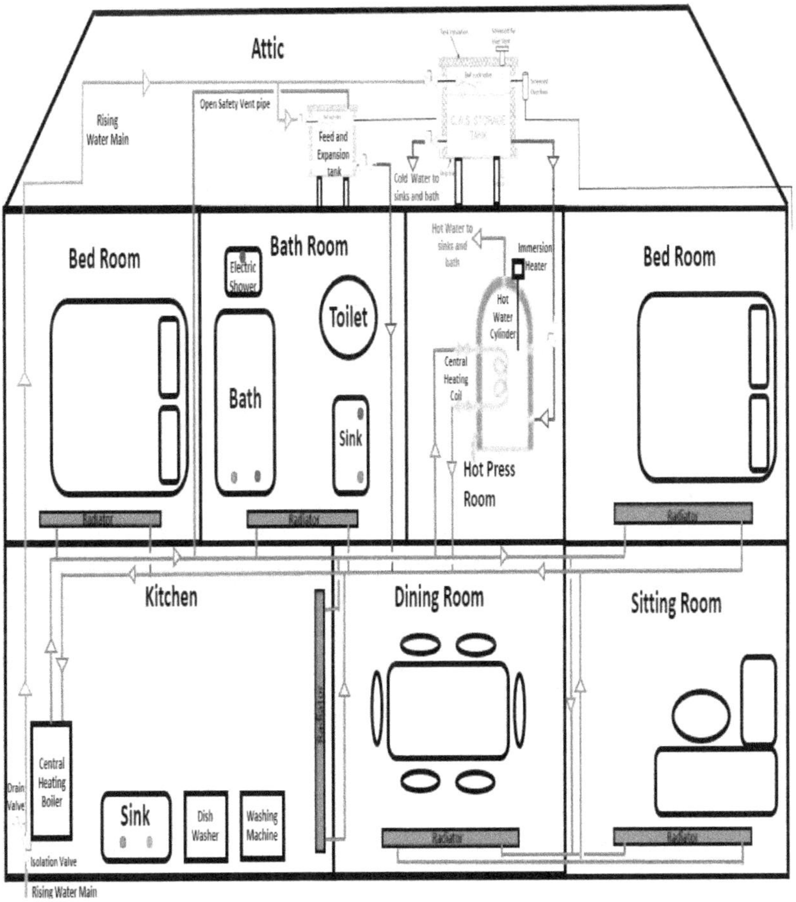

Operation: A Natural Gas fired central heating boiler maintains a hot water temperature set point of 70°C (158°F) which is recirculated throughout the entire central heating system via its own internal circulating pump.

Problem: Radiator in one bedroom is only slightly warm, which is normally indicative of a circulatory flow problem and possibly caused by air locking, rust, black sludge (magnetite), or scale deposits which are all enemies of any heating system. A logical approach should now be implemented in identifying the actual problem itself. Check other radiators, are they at the same temperature or hotter. If all other radiators are hotter, the problem is localised to this one radiator. Check are the flow (Thermostatic if fitted) radiator valve and the return radiator valve both open.

If both are open, use a radiator key and bleed the radiator of any gases (these gases are normally a by-product of rusting) that might be trapped via the bleed screw (situated at the top of the radiator itself). Be careful, this venting gas or water can be very hot and can cause skin burns.

If water is present and the radiator is still only lukewarm, it could be a partially blocked or malfunctioning Thermostatic Radiator Valve (TRV) which would have to be removed and cleaned. It may be that the pumps' differential pressure is not capable of delivering the hot water required to heat the radiator. The hot water may also be taking the least line of resistance, thus, bypassing the luke warm radiator, in this case the radiator system will need to be balanced. Start throttling down the radiator valves to other hotter radiators to force more hot water flow to the lukewarm radiator. If this does not rectify the problem, the pumps' impellor itself may need to be replaced or a larger pump capable of delivering the differential pressure required to heat all the radiators may need to be installed (this might happen on a new installation or if new radiators were added to the original system where the maximum differential pressure the pump is capable of delivering was not considered).

If you find a radiator that is hot on the bottom and cold at the top, bleed it using a properly fitting radiator key and check if water comes out (be very careful when doing this as this water is normally very dirty and can stain carpets …. have an old cloth readily available to prevent dirty water being bled spilling onto the carpet). Check another radiator on the same floor and see is the problem universal to all radiators or localised to one. If you bleed another radiator and no water comes out, turn off the heating system immediately. Investigate why there is no water present. The first thing to do is check the heating systems' **Feed & Expansion Tank** (in the attic or loft) if it is not a pressurised system. This keeps the water topped up in the entire central heating system. If this tank is empty, either the ball cock is faulty, or the water is turned off, check both regularly. Once the fault has been rectified, water is re-established and back in the system, bleed the radiators again until all air is removed, and water is present, switch the power back on to the heating system and check all radiators are heating.

If a radiator is found to be cold at the bottom and warm at the top, it is normally a sign that there is black sludge in the radiator. Close both flow and return valves to radiator, completely isolating it from the system. The water will need to be completely drained from the radiator by loosening the screwed fitting between radiator itself and the radiator hand valves. Have an adequately sized container available when draining it and don't allow any spillages (to remove the water quickly, open the bleed screw slightly). The radiator must now be taken outside the building, flushed with a water hose (common garden hose will do). Before doing so, ensure the open ends of the radiator are stuffed with old rags to prevent any residual dirty water draining out while taking it outside the building (this dirty water will destroy carpets if it spills on them). Stand the radiator up, remove the old rags and give it a good flush until all silt is removed. Let it drain completely and reinstall, ensuring all compression fittings are water tight. Open hand valves to the radiator and allow to fill, when water comes out of the bleed valve, close immediately, it should now be fully heating. Check the radiator over the next 24 hours for any water leaks or droplets and tighten fitting to maintain water seal.

N.B. Do not over tighten the compression fitting, you could possibly shear the threads, use a proper sealant such as PTFE (Plumbers) thread seal tape to ensure a water tight seal if need be.

Note: Above is a typical example of a domestic heating system, now, imagine being in an industrial setting where there is a heating system one hundred or maybe one thousand times larger with all the vessels, expansion tanks, buffer tanks, heat exchangers, associated pipe work, circulating pumps, control valves, instruments etc. being controlled and operated by a distributed control system (DCS). Regardless of the size or scale of such an installation the fundamental fuel burning, heat exchanging and flow and return pumping principles are basically the same, just a bigger version.

Tip: If a central heating system is only heating certain areas of a house, check to see are there any motorised valves fitted which switch heating on throughout the house at various times of the year via a controller/time clock: immersion/cylinder tank only (Summer months); radiators downstairs only. Check to see is there any indicator light on the motorised valve (this will prove that there is power being provided). If the indicator light is on and it is still not working, its internal motor could be burned out or the valve itself could be mechanically seized. Repair or replace accordingly.

N.B. Heat flows naturally from a warmer to a cooler place and is lost from a building through ceilings, walls, and floors, it is important to insulate properly. Heating accounts for up to 70% of energy bills in buildings and without adequate insulation, much of the energy that is being paid for is wasted.

TIP: Hot water cylinders/calorifiers should be fitted with a thermostat to ensure that the water is not heated more than necessary. The hot water temperature should be checked to ensure that it is maintained at 60°C. When replacing an old hot water cylinder, install a cylinder with factory applied insulation, which will significantly increase heat retention of the cylinder.

Tip: Installing a modern high efficiency **A Rated** boiler in a building which operates at efficiencies of > 90% can give an annual saving of up to 25% on fuel costs.

Boilers should be serviced at least annually to ensure efficient and safe operation. A poorly maintained boiler can often use 10% more energy than necessary and may also be less reliable.

E.g. An 'oil fired' heating boiler temperature set point is set at 70°C and is struggling to reach this temperature although it's consuming the same amount of fuel it normally would to maintain this required temperature in the previous 12 months of operation. Have the boiler checked and serviced as soon as possible by a registered service engineer, the boilers' internal heat exchanger is more than likely coated with ash deposits and needs to be cleaned. These ash deposits are preventing efficient heat exchanging taking place which leads to poor heat transfer between the burning fuel and the water being circulated through the boiler and then around the heating system. This results in more fuel being expended by the boiler, incurring higher financial costs on the property owner whilst trying to deliver the same heat output.

A boiler efficiency test including the adjustment of the air / fuel ratio should form part of each service visit.

N.B. Listen out for unusual **boiler noises**, this may be indicative of sludge or lime scale build up. Remember that the presence of foreign matter in a heating system can adversely affect the operation of the boiler (e.g. overheating and noisy operation of its heat exchanger).

TIP: There are companies who specialise in removing undesirable scale or sludge deposits by power flushing, once the scale or sludge deposits are loosened, the unwanted debris is purged from the heating system with clean water. The system's clean water will also be chemically treated (Boiler, radiators, and associated pipe work) with corrosion inhibitor to prevent further problems. This type of treatment will restore circulation and efficiency to a system.

Install a **Magnetic Central Heating Filter.** It is a multi-element magnetic filter designed to increase the life of central heating systems and deliver maximum efficiency, it includes a dirt trap for non-magnetic particles. Easy to install, simple to maintain and suitable for all central heating systems. It also acts as a water treatment dosing point for liquid chemicals. Recommended in the Domestic Building Services Compliance Guide to help maintain the efficiency and reliability of central heating systems.

Just like a car is serviced at regular intervals, so should a building. Your property should always be maintained in a **weather proof** state. Circumstances outside your control can happen e.g. severe frost, flooding, lightning strike …. but at least try to have mitigating measures in place that are within your control to protect the property.

When moving into a new property insist from the builder or architect (before handing over any money) on having a set of schematics of all the utility runs both internal and external: Electricity cables; Natural Gas/ L.P.G. pipes; oil lines; water pipes; drain pipes; foul sewer pipes; air ducts; vents. Keep these schematics in a safe place and use them as references if any further renovation is to take place or simply hanging a picture, this may save you a lot of time and money in the future.

Tip: When it comes to pipe or cable laying, whether it be an industrial or domestic setting. Take as many digital photos as possible of the exact layout of the various pipes or cables before they are **covered in** using a point of reference for each photo e.g. a street light pole, a building, stair case (it will be time well spent). Put the photos in an archive folder for future reference.

N.B. When water (which is virtually incompressible) freezes, **it expands its volume by up to 9%.** As pipes are already full of water, the pipes (e.g. copper and galvanised pipes) burst due to the large pressures being exerted on the walls of the pipes and joints due to the water expansion itself. The water leaking problem may not occur when the water pipes are frozen, **it will occur mainly when the frozen water inside the pipes starts to defrost** and begins to leak where the pipes have split, or pipe joints have cracked. Properly fitted insulation on pipes is essential in preventing such occurrences. **Heat tracing** should be considered, also known as electric heat tracing or surface heating, is a system used to maintain or raise the temperature of pipes and vessels. Trace heating takes the form of an electrical heating element run in physical contact along the length of a pipe. The pipe must then be covered with thermal insulation to retain heat losses from the pipe. Heat generated by the element then maintains the temperature of the pipe.

Provision must be made to protect equipment which is not self draining against frost damage in environments where it may be exposed to temperatures below freezing point.

Water **expands its volume by approximately 4%** when its temperature rises. An expansion tank or expansion vessel is a small tank used to protect **'Closed'** (not open to atmospheric pressure) water heating systems from excessive pressure. The tank is partially filled with air, whose compressibility cushions shock caused by water hammer and absorbs excess water pressure caused by thermal expansion.

Expansion Vessel Example:

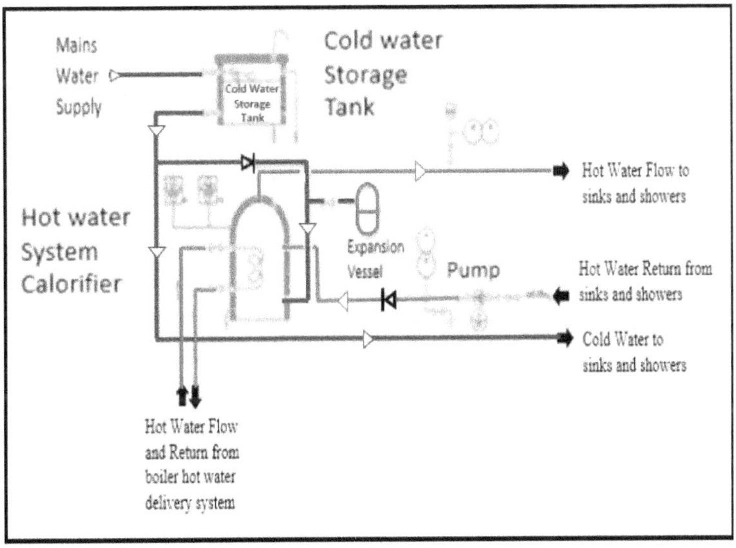

Tip: Ensure attics/lofts and basement areas are adequately insulated. Properly insulate any associated water tanks and pipe work in the areas as well. Check the condition of any metal/galvanised/copper water tanks on a property for rusting, decolourisation, pitting, or pinholes. Consider replacing with a new one or with a plastic tank. Check the condition of the associated copper pipe work also, replace as necessary or consider installing an established **suitable for purpose** plastic pipe product. The plastic pipe manufacturer will give expert advice needed for a proposed installation. Keep **plastic pipe joints** to a minimum where possible, especially in exposed areas. Plastic pipe will normally accommodate the increase in volume that happens when water changes to ice and then return to its original shape during a thaw, ready to accommodate further freezing. Plastic pipe and its associated fittings are suitable for most domestic and commercial plumbing applications.

Tip: Modern boilers incorporate a built-in frost thermostat which automatically turns on the boiler and pump if the water in the boiler falls below 5°C, providing the electrical supply is on. The boiler will operate until the water temperature in the system reaches approximately 10°C measured on the flow sensor in the boiler. Any other pipework outside of the boiler should be protected from the risk of freezing and protected with insulation. Additional protection from external frost thermostat and pipe thermostat should also be considered.

Tip: Where mobile homes, caravans, summer houses are concerned that are not occupied or heated during the winter months; **ensure** to turn off/disconnect the mains domestic water supply to the property. Drain all water tanks and pipe work where possible, open dedicated water drain points as recommended by the appliance vendors (e.g. gas water boilers, shower mixers) and if necessary, loosen actual water pipe joints to remove any residual water left in the system and then retighten. Refer to the mobile home vendors' instructions regarding **central heating systems** as it may have an anti freeze/water mixture which would **not** require a drain down.

N.B. Consider applying this tip when going on a winter holiday if you are not intending to leave the heating 'on' in your house (at a minimum, **turn off** the mains water isolation valve into your house and drain the water header tank in the attic by opening the cold and hot taps, and then closing them once the water is drained). You may think that this is too much trouble and a waste of water, but consider it only takes 10 minutes to carry out this tip and may save thousands of gallons of water being emptied into your living room. Coming home to find a house flooded with water from a burst pipe due to freezing conditions will quickly dissipate any happy holiday memories you may have. P.S. Once you return, don't forget to **turn on** the mains water isolation valve first before turning anything else back on e.g. heating system.

If there is **Natural Gas** supplying the cooking and heating needs, turn off the main gas valve into the property as well.

If you have Natural or Bottled Gas coming into your property, would you know?

- Where the main gas shut off valve is situated if you sensed an unexplained strong smell of gas and had to turn it off in an emergency?
- Do you know or have you or members of your family easy access to the **emergency phone number** for the gas company? Other emergency **call-out** phone numbers such as electrical power supply or water supply companies should be up on a wall (in the utility room or kitchen) for all to access in the event of a breakdown of any one of the essential services to your home.

Tip: If there is no running water to a house over a sustained period due to frozen pipes and the toilet can't be flushed (or the cistern flushing mechanism is not working). If possible, capture any water from the drain pipes running off a roof and redirect into a 10 litre bucket, it's a good way of retaining water for flushing toilets. Pour this water directly into the toilet bowl itself quickly, and in one go. The same flushing vacuum principle will apply and remove the contents of the toilet bowl.

Pipe Freezing (Civil/Commercial/Industrial works)

Pipe freezing is a non-intrusive, controlled method to isolate a section of pipeline by freezing the contents at either end through the application of Liquid Nitrogen (Liquid Nitrogen has a temperature of -196°C. With it you can freeze oil, water, and most other liquids). The pipework freezing process begins by drawing out the heat from the product within the pipe until it reaches a temperature below its freeze point until eventually the product freezes to form an internal ice plug.

The major advantage of liquid nitrogen is that a 'break-in' into a pipe can be performed without incurring the costs and down-time of a complete system drain down.

The Benefits of Pipe freezing:

Pipe freezing is widely acknowledged as a safe, reliable, and highly cost-effective method of temporary pipeline isolation and offers several major benefits some of which are:

- No need to drain down systems, thus avoiding costly and time-consuming venting procedures
- No need to arrange for the bulk transfer/storage of fluids
- Cuts losses of expensive liquids, such as treated water and system inhibitors.
- Overcomes the problems of handling hazardous materials and enhances safety when working on lines containing contaminated or volatile liquids.

Tip: How to free out a 'mechanically seized' water valve:

Apply penetrating oil spray to any part that looks as if it may move. Wipe off any penetrating oil spray remaining on the surface, it will achieve nothing unless it can get to the threads. Normally, penetrating oil spray containers come with a very fine tube attachment, which enables you to get into the tightest of places. Once you have done this, work your way through the following steps:

- Try tapping (shocking) the body of the valve with the handle of a wrench or adjustable spanner while you turn the valve will sometimes free it up.
- Warm up the body of the valve with a hair dryer, or (carefully) with a blowtorch which might expand the body of the valve enough to allow the gate (or other moving parts) to move more freely.
- As a last resort try to attach a spanner or wrench to the handle of the valve and extend the handle of the wrench with something slipped over it to give you a little more leverage. While you are using this extra leverage, you must hold the valve body tightly in position to stop any of the joints to and from it, twisting.

Tip: How to free out a 'mechanically seized' nut and bolt:

If you have tried all less destructive methods (using lubricants and penetrating oils or heating the bolt gently to try and soften it a bit so you can twist and free it) and nothing works, try a '**Nut Splitter**'. It is slipped over the seized nut, then a screw on the side of the splitter is tightened which forced a wedge into the side of the nut. Keep cranking the nut splitter screw until the nut makes an audible pop and you get through the nut. Back off the nut splitter screw and clean the bolt of the split nut. When dealing with a nut that's pretty much rusted through or just won't budge, this should do the job. Unfortunately, this method destroys the nut and occasionally, the bolt too, depending on its condition.

Underground Water Leak Detection - Ultrasonic Water leak detectors detect the ultrasonic sound of a leak. You are probably familiar with the hissing sound a large leak makes. Smaller leaks also emit sound; however, the frequency is too high for our ears to detect it. Ultrasonic Leak Detectors translates the ultrasonic hissing sound to a lower frequency where it can be heard through headphones, and leads you to the source.

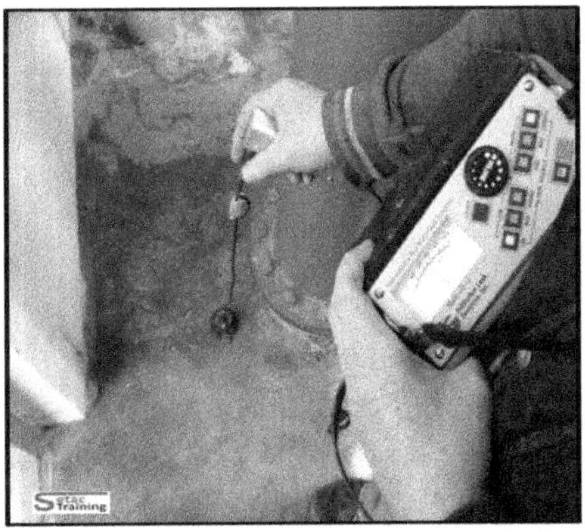

For a more 'difficult to detect' issue in a large underground piping network, a Water Leak Detection Specialist using **tracer gas** might need to be employed. If the noise from the leak cannot easily be heard then tracer gas can be a very useful technique to locate it. This process is up to 100 times more sensitive than common leak detection methods.

Example: Using an as-built P&ID, they will identify pipe routes and then individually isolate. A series of deep 10 mm holes will then be drilled in the road every metre along the identified water pipe route. The isolated section of pipe will be pressurised with Tracer Gas (comprises of 5% Hydrogen and 95% Nitrogen) via the Water Main hydrant (H18). The pressurised tracer gas contains hydrogen, which is the lightest and smallest molecule in the atmosphere, will exit at the leak point and make its way to the surface and come up through the holes drilled in the concrete where it will be detected by the highly sensitive gas detector. Once the leak has been pinpointed along the pipe route, repair works can begin.

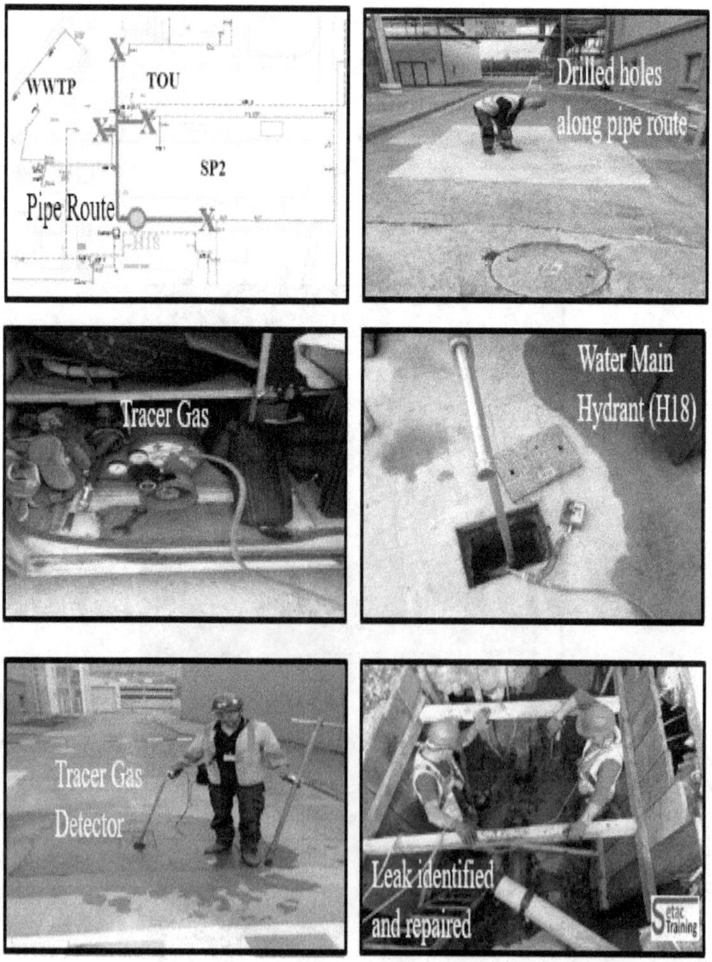

Problem: All electrical sockets are not working in a Commercial/Domestic dwelling.

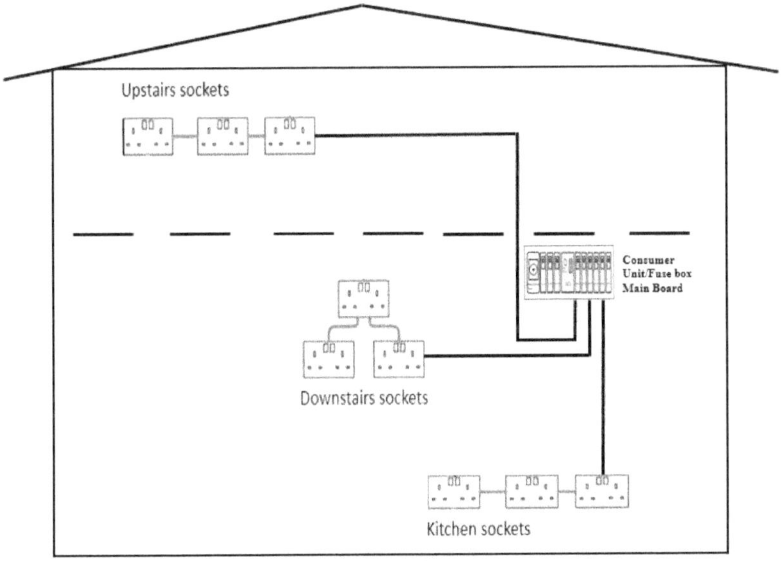

Do you know where the main Consumer Unit/Fuse box Main MCB Board is located? Try to get familiar with the different MCB's (fuses in some cases, especially older buildings) and not to have to go looking for it if mains power is lost or if the main circuit breaker trips which feeds the entire building and are left in the dark. Ensure all individual MCB's are clearly labelled, if not, label them or get someone to do it for you.

If in the event of a main MCB board RCD (Residual Current Device) trip and all RCD protected circuits (e.g. sockets) are not working, normal practice in older commercial/domestic installations in some countries is to use a single RCD for all RCD protected circuits. Sockets are on the RCD, lights usually aren't. Electric showers will have their own dedicated RCD.

Investigate thoroughly for actual fault before attempting to reset the RCD. Observation is critical, be satisfied after doing a sweep of the building that everything is ok, watch out for anything unusual.

If the RCD trips for an unknown reason and will not reset, a logical approach should now be implemented:

Expanded View:

Consumer Unit/Fuse box Main MCB Board:

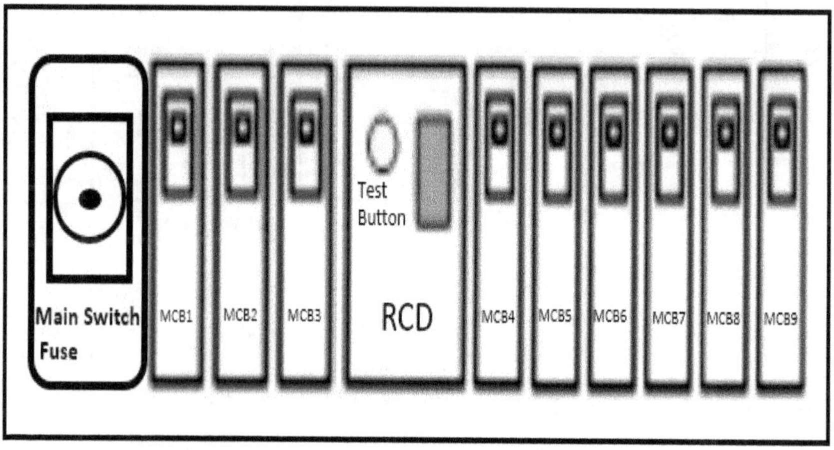

Tip: Instead of going around and unplugging all electrical appliances plugged into the socket outlets (some may be inaccessible due to kitchen cabinets, washing machines, dish washers), resetting the RCD and then **plugging in** each electrical appliance one by one until the faulty appliance is found. Switch off all individual circuit breakers at the main board that are affected by the RCD (Expanded View: MCB's 4 - 9), reset the RCD, and turn each individual circuit breaker back on, one at a time (allow up to 60 seconds between switching, as some equipment may not start up immediately, it may a **run up time**). The socket circuit that causes the RCD to trip can be found much more quickly and the rest of the socket circuits can be brought back into service as soon as possible (especially fridge freezers).

Leave the circuit breaker that protects the faulty socket circuit in the **Off position**, reset the RCD, reset all other individual circuit breakers, all other socket circuits will now be back online, investigate what appliances are plugged into the faulty circuit and unplug them.

Reset the circuit breaker, the RCD should not trip, if it does, there could possibly be a cable problem that feeds the sockets or possibly a rodent problem (mice will eat anything). Assuming the RCD has reset, before you start plugging back in each appliance and turning each one on individually, check for any sign of damage to the cable feeding the appliance or the appliance itself. When the RCD trips again remove the faulty appliance and do not plug back into the socket unless appliance is repaired or replaced. Another example why an RCD may trip could be water ingress into the socket outlet itself or electrical appliance.

N.B. All circuit breakers perform the following functions:

- SENSE when an overcurrent occurs.
- MEASURE the amount of overcurrent.
- ACT by tripping in a timely manner to prevent damage to the circuit breaker and the conductors it protects.

Loose Connections such as an untightened screw in a terminal, heat up and expand when energised.

A loose conductor termination causes arcing at the connection and can result in a potential fire hazard especially when under a large electrical load e.g. an electric shower.

Tip: If using a coiled extension cord to power an electric heater, kettle, transformer etc, ensure the cord is suitable for purpose and fully unwound from the coil/drum retaining it before using. The cord itself can over heat (burn marks or staining may become evident); this in turn will damage the cord and can also cause a fire. Some temperature rise may occur on the cord while a large power load is operating because of the heavy conduction current. An uncoiled and correctly rated extension cord does not normally overheat.

N.B. Ensure the cord **is not** plugged into the mains power socket when coiling or uncoiling the cord. Observe that the cord is not cut or damaged, if so replace/repair immediately, e.g. a **powered up** damaged or cut power cord lying in water is extremely dangerous and can cause instant death.

Tip: When screwing a screw into a piece of wood or into a wall plug in concrete, place a little soap on the tip of the screw before inserting, it will be much easier to screw in by hand or if a power screw driver is being used, it will extend the battery charge as less power will be needed.

Tip: To prevent a piece of wood from splitting when hammering a nail into it, turn the nail upside down, place the head of the nail in the spot where it is to be driven. Gently hit the pointed end of the nail with the hammer into the wood. Turn the nail the right way around and then hammer the nail home. The nail will cut its way through the wood grain, rather than wedging a pathway. The wood should not split.

Tip:

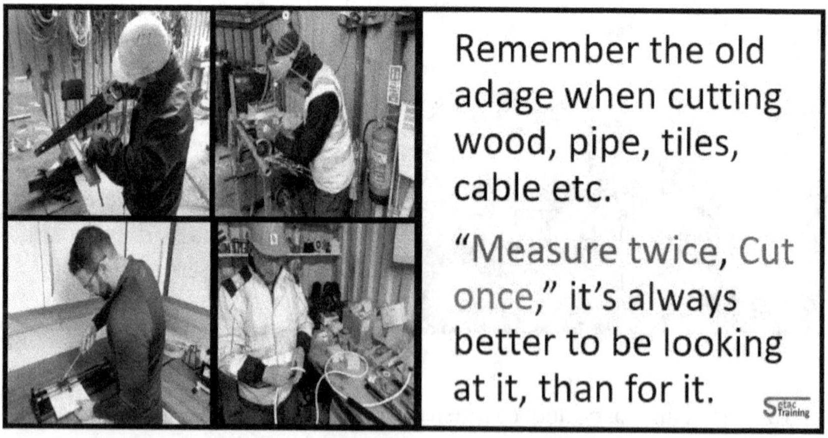

Remember the old adage when cutting wood, pipe, tiles, cable etc.
"Measure twice, Cut once," it's always better to be looking at it, than for it.

Tip: If cutting or grinding metal with a grinder over a finished floor: wood; linoleum; carpet .... ensure the immediate surrounding area of the floor itself is protected with a non-flammable cover. The flying sparks and miniature pieces of metal being generated by the grinder can leave permanent minute burn stains on the floor (ceramic tiles included) which can prove impossible to remove.

Tip: If living in an old property, try to find or map out where hidden electric cables, gas and water pipes are located **before** deciding to screw, nail or drill into walls and floors. This can prevent personal injury or a very costly repair job. This may include breaking open a section of a living room wall to get at the punctured pipe, repair it and then afterwards having to have it patched and plastered, then painted …. a lot of unnecessary expense and hardship because of one nail.

Tip: If you do have the misfortune of driving a nail or screw into a water pipe, to do a quick repair, wrap some PTFE (plumbers) tape on a self tapping screw and screw it into the hole to seal it, this should provide a temporary solution while getting a proper fitting to carry out a permanent repair.

How to charge a car battery using another vehicle

If the **car battery** in your car has gone flat/dead, follow the instructions below to recharge it using another car:

Park a second car that is running, close enough to your car so a **red and black** set of jumper cables can be attached. Use heavy duty jumper cables at all times, invest in a good set, it will be money well spent. Turn off all unnecessary power consumption devices in your car: lights; heater fan; radio/CD systems; heated rear windows …. so as much electricity as possible will be directed into your car battery when being charged. Proceed then to turn off the ignition of the running car ensure to **remove the keys from the ignition in both cars** and keep them in your pocket (especially your car) as the doors can automatically lock when power is restored to the battery and you can be left with a different problem, possibly having to break the car door window to gain access.

Be careful that the two cars are not touching. Identify clearly the positive (+) and negative (-) terminals on both sets of batteries in the respective cars, clear around the general battery area if need be (N.B. always have a working portable battery flashlight, a screw driver and a few old rags in your car).

N.B. Attaching the jumper cables the wrong way around could damage both cars, ensure to connect the red cable to the positive terminal and the black cable to the negative terminal, **under no circumstances should the cables be reversed**. Sometimes battery connections can be badly corroded, you may have to clean or scrape them down to get a good electrical connection between the battery terminals and the portable lead clamps.

Take your time, ensure it is clear in your head what you are doing, make sure you are in a well-ventilated area as a charging battery gives off hydrogen gas and is explosive.

N.B. Never allow the exposed metal ends of the positive and negative jump cables to touch/short circuit when connected to a battery (always keep them fully insulated from one another) large sparks will be generated and possibly ignite the surrounding area causing severe burns to you or damage to the car itself. This **short circuit** can cause severe damage to the battery also.

1. Attach the (+) red clamp of the jumper lead to the positive (+) battery terminal of your car. Always be cognisant of where the other positive (+) red clamp end of the jumper lead that you are attaching to the battery of the other car is situated, preferably in your other hand **i.e. not left in a puddle of water (water conducts electricity) and certainly not left placed on the engine of the working car.**
2. Attach the opposite end of the same jumper lead (+) red clamp to the positive (+) battery terminal of the working car.
3. Attach the (-) black clamp of the jumper lead to the negative (–) battery terminal of the working car.
4. Attach the opposite end of the same jumper lead (-) black clamp to an unpainted metal surface on your car to provide a solid ground connection (refer to your cars' O&M manual to source where a good ground connection can be found). **Do not attach it to the negative (-) battery terminal of your car.** Ensure that the jumper leads will not be in the way of or get entangled with any moving parts.

5. Start the working car and wait several minutes for the battery of your car to charge, ***don't be in a hurry,*** only after this charging period of several minutes should you try and start your car. If the car doesn't start, check the connections between the battery terminals and the portable leads again to ensure there is a good electrical connection between them, wait a few minutes and try again. (**Tip:** The pitch or sound of the engine in the working car will normally drop when it is transferring an electrical charge to your car).
6. Once your car starts**, remove the jumper cables in the opposite order from when you put them on,** again, being aware of where the jumper leads are situated at all times.
7. Let your car run for a while or better still, drive around for 30 minutes or so before turning the engine off, this should be sufficient time to fully charge the battery. **The battery needs time to charge from the alternator if you want to be able to start the car again later.**

If you find that it is getting harder and harder to start your car and you find the charge in your battery is not holding or is insufficient to start the car. Have an **Auto Electrician** check it, the car electrics could have an internal fault with a component e.g. car alarm, which is drawing excess power from the battery. It may also be just time to have a new **like for like** battery fitted. Don't wait until it completely fails before fitting a new one.

Tip: Car tyres

Properly inflated, well threaded tyres on a car not alone make for a safer, comfortable, fuel efficient drive but also the tyres themselves act as very efficient water pumps when driving on a wet surface. A small water wave forms in front of car tyres when driving in wet weather, if tyres cannot dispel these miniature water waves, it will try to go over them (water is virtually incompressible) and therefore the car loses traction on the road due to poorly threaded tyres.

A midsize car weighs approximately 1,500 kg's which is constantly pressing down on its tyres; the better the threaded tyre, the safer and more efficient it is. A properly threaded tyre running at 100 km's per hour will dispel up to 6 litres of water a second away from the tyre giving the car better road holding. The less water between the tyre and the road, the better the grip.

Inspect tyres regularly; remove small stones and debris from the threads. The tyre threads are also water channels, if damaged, partially blocked or worn, they lose their effectiveness. It is very important the entire tyre threads are kept at the manufacturers recommended depth for safe driving and are kept clear.

Tip: Tyres that are running on the road at 0.5 Barg below recommended pressure increase fuel consumption by 2-3%.

Tip: If thinking about selling a car, to add a little more value, ensure to retain all the paperwork associated with the car: N.C.T/M.O.T certificates; any works done on the car including details and dates of any service carried out. Insist on the paperwork and the associated costs being supplied by the mechanic. Having a full service record and associated original car instruction manuals to hand over will make the sale of a car a lot easier and will be much appreciated by the buyer.

16. AUDIT INSPECTION READINESS GUIDELINES

A company's overall aim is to ensure the sufficiency and robustness of its

- **Manufacturing**
- **Engineering**
- **Environmental**

systems and to ensure compliance with the associated regulations and standards.

Companies operating in today's process industries are challenged with maximising profitability in the face of continuously increasing regulatory pressures. Compliance with these regulations requires complete and accurate engineering information to be controlled throughout the asset lifecycle.

Main regulatory requirements for managing engineering information in process industries are:

1. **ISO 55000** - is an international standard covering management of physical assets. **Asset Management (AM) involves:**

- The generation and implementation of AM policy, strategy, objectives and plans as well as optimising performance, cost, and risk.
- Asset-related risk identification, analysis, control and management processes and contingency planning.

- Understanding asset lifecycle management and the role of appropriate tools to optimise the performance, risk, and cost of assets.
- Identifying the information and performance measures needed for effective asset management and evaluating the effectiveness of asset information systems.
- Understanding the financial implications of asset and asset management decisions for the organisation and translate technical issues into business implications.

2. **ISO 15926** - is a standard for data integration, sharing, exchange, and hand-over between computer systems.

3. **ISO 9000** - is a series, or family, of standards. ISO 9001 is a standard within the family. The ISO 9000 family of standards also contains an individual standard named ISO 9000. This standard lays out the fundamentals and vocabulary of Quality Management Systems (QMS).

4. **FDA 21 CFR Part 11** - is the part of Title 21 of the Code of Federal Regulations that establishes the United States Food and Drug Administration (FDA) regulations on electronic records and electronic signatures (ERES). Part 11, as it is commonly called, defines the criteria under which electronic records and electronic signatures are considered trustworthy, reliable, and equivalent to paper records (Title 21 CFR Part 11 Section 11.1 (a)).

An **Audit** is an evaluation of an organisation, its associated systems, and processes. Audits are performed to ascertain the validity and reliability of information and to provide an assessment of a **system's** internal control to ensure it is compliant, transparent, effective, and efficient. The goal of an audit is to express an opinion on the organisation or system in question under evaluation, based on work done on a test basis.

A **Good Manufacturing Practice Audit** evaluates if a company has its facility, manufacturing processes and associated analytical methods under control in meeting specified requirements, thus demonstrating that its products are consistently produced and controlled to the quality standards appropriate for their intended use and conform to the regulatory requirements stipulated by health authorities.

Quality Management System (QMS) - Are the Quality Systems compliant and can they stand up to scrutiny? E.g. **People** are trained, qualified and competent to carry out job responsibilities. Integrity of all **records** both (Manual and electronic) are accurate and true (Data Integrity is critical). **Equipment** is qualified and maintained to meet its designed requirements.

Quality Risk Management (QRM): is now a regulatory expectation, and it makes good business sense. The goal of the risk assessment is to increase process understanding using QRM tools (e.g. FMEA, FTA, PRA, HAZOP, HACCP) to identity the risks and develop a strategy to minimise or control them and deliver safe and effective products.

Elements of a Quality Management System – Vendor Audit Qualification, & Bespoke QMS

Maintain a Tender content library to include:

- Company organisation chart
- Financial Accounts/Statements
- Insurances
- Accreditation Documents
- References
- Key Staff Biographies
- Buying, Maintaining and Certification of test equipment.

Good Documentation Practices: e.g. the results of inspections and testing must be documented and signed using **ALCOA:**

A: Attributable – Who & When

L: Legible – Clear & understood

C: Contemporaneous & Complete – Timeliness/ Initialed & Dated

O: Original – Preserved in its original form

A: Accurate – Conform with protocol.

Example: Marking up of documents and drawings shall follow site GDP.

1. All entries shall be in ball point pen.
2. All entries shall be initialed and dated.
3. Any additions to the document/drawing shall be marked in blue.
4. All deletions shall be marked in red.
5. All comments shall be marked in green.

Good Manufacturing Practices are fundamental to any manufacturing industry and are often required to be implemented in plants and factories by national governments.

An **Engineering Audit** usually involves checking performance and evaluating deficiencies and inefficiencies against requirements. When a company carries out an internal Engineering audit, it will be taken for granted that site procedures are being adhered to and a good understanding of the hazards on plant and the range of equipment are known. A test is then performed to ensure a good understanding of requirements at all levels in the organisation exists and a confirmation that appropriate systems are in place.

An **Environmental audit** is an evaluation intended to quantify a company's environmental performance/compliance. These audits are intended to review the company's **legal compliance status.** Compliance audits generally begin with determining the applicable compliance requirements against which the systems and operations will be assessed.

The auditor is not looking for faults in a system; they are looking for compliance to the standard. If a non-conformance is found, it should be viewed as an opportunity to improve, not as a reason to reprimand. To benefit a company, auditing will not only report non-conformances and corrective actions but also highlight areas of good practice and provide evidence of conformance. In this way, other departments may share information and amend their working practices thus, enhancing continual improvement.

A company's internal audit should do more than just check that legislation and company procedures are complied with, the following should also be carried out:

- An objective assessment of performance relative to best practice
- Recommendations of where and how improvements may be made
- An improved understanding of legislation and best maintenance practice by engineering personnel
- An appraisal of equipment and associated systems with assurance that they are fit for purpose for a defined period
- A commitment to improve.

A company should always have its facility in a **'prepared and audit ready'** state, it's good practice to have both internal areas (e.g. freshly painted and polished) & external areas (e.g. gardens, pathways, roadways) well maintained and free of debris. **All** restrooms must be kept in a clean, fully stocked (E.g. liquid soap, paper towels, toilet paper), fully functioning (e.g. no blocked toilets or urinals, hot water available, hand dryers operational) and presentable state - 24/7/365.

Tip: First impressions last – no matter what line of business you're in; if an auditor/inspector, visitor or a potential customer encounters a messy or dirty restroom, their perception of your business, your products and services will be tarnished.

Regulators can perform unannounced audits at any time.

Having a detailed preparedness and a systematic approach for any impending regulatory inspection is a must. Remember during an audit, it is not only the company image that is being projected; it is also the professional culture as well as knowledge base that leave a lasting impact.

Personnel must have the necessary skill sets and tools to participate as a member of an audit preparation team. A company must ensure to have a thorough and effective preparation programme developed for a regulatory body audit. Choose personnel who will be interacting with an auditor and train them on how to conduct themselves in front of the auditor. Prior to the audit, ensure that the team has reviewed any documentation that an auditor is likely to request.

Tip: 'A key to success during any audit is preparation.'

Regulatory Body Site Audit:

A company designate (e.g. QA/Eng/EHS Manager) is responsible for meeting the auditor/inspector on their arrival at the plant and act as a spokesperson for the company.

The company designate is responsible for ensuring that the auditor follows the company's procedures for visitor's onsite (signing in at security, wearing identification badges). The auditor must be escorted during the inspection where questions are understood and addressed specifically. First impressions count and it's important to convey to the auditors that you have your facility (effective implemented engineering and environmental management systems) and manufacturing processes under control and you know what you are doing.

Helpful hints for a company and its staff to remember during an audit are:

- Ensure the personnel who are in front of the auditors have the required technical knowledge and expertise, confidence, and presentation skills

- Personnel should always be polite and helpful
- When a document is requested, provide the requested document and no more
- Do not volunteer information that has not been requested unless it is an advantage to do so
- Do not guess an answer
- Don't think on your feet – get the SOP!!! use documents to aid your answer, which helps you, and gives a good impression
- Do not hide information
- Do not argue or display anger towards the auditor
- Never cause a deliberate delay. If for some reason you cannot deliver a copy of a document quickly, explain the reason for delay
- Look confident and smile.

There are a couple of things that onsite personnel shouldn't do:

- Be obstructive or argumentative
- Say something when being given the **silent treatment** from the auditor
- Provide answers to a question not related to an area of their responsibility/expertise, especially when their knowledge may be limited.

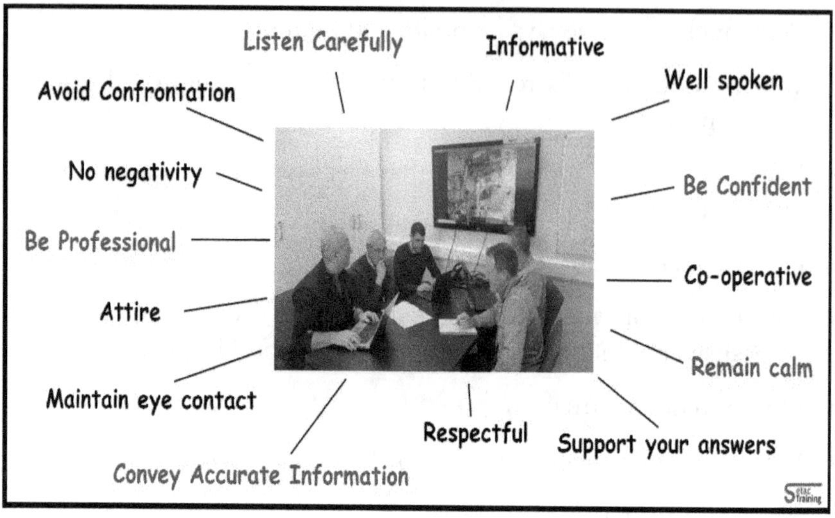

Once the auditor has completed the audit, they will hold a short debriefing meeting with the company designate(s) at which they will give the company a verbal summary of the main findings of the audit. The auditor will prepare a final report as soon as possible after the audit. This report generally will set out the purpose of the audit:

1. What was audited.
2. Who was present.
3. Summary of the main findings.
4. Recommendations.
5. Description of what was found and observed and recommendations.

Some of most common audit observations for both life sciences and food industry relating to calibration issues are:

- No formal procedure in place for calibration execution and deviation management.
- Not following procedures.
- Missing documentations and certificates.

The auditor normally sends the final report to the company. The company is then required to reply within the period specified in the final audit report to the recommendations setting out what it has done, or proposes to do, to satisfy those recommendations.

Assign one person to be the company's contact for receiving the audit report and answering any follow up questions that the auditors may have after leaving the site. A person should also be delegated the responsibility for coordinating any corrective actions and compiling the audit report. It is important to be co-operative and to commit to providing a written audit response to the auditing body as to the audit findings

17. GENERAL TEMPLATES

Trouble Shooting (Short Version)

Guidelines:

When problem solving, an organised approach must be undertaken. Sometimes it is possible that two relatively obvious problems combine to provide a set of symptoms that can mislead the faultfinder. Be careful of this possibility and avoid solving the wrong problem. Try to avoid being caught in the trap of blindly following instructions without having a full understanding of the task to be completed. The steps to be taken fall into the following categories.

- The problem and its limits must be defined
- All possible causes must be identified
- Make the necessary corrections
- Try to establish what was the **initiating factor** that caused the fault in the first place and make the necessary corrections.

The first step in effective problem solving is to define the limits of the problem. When a problem develops, compare all information with normal conditions. Knowledge and consistent records are the basis for avoiding the unusual.

Make a list of all deviations from normal operation. Do not rush in and make wild guesses. Use the systematic approach.

Delete any items not relating to the symptom and separately list those items that might. Use this list as a guide to further investigate the problem.

The second step in problem solving is to decide which items on the list are possible causes and which items are additional symptoms.

The third step is not to alter several things at once; it may never be known what caused the problem. You may then find you have the right answer to the wrong problem. Do not adjust settings and parameters. Do not disconnect or withdraw pieces of apparatus from a unit without ensuring they can be put back to their original state.

In solving a wrong problem, a new one can be created.

Identify the most likely cause and take action to correct the problem. If the symptoms are not relieved move onto the next item on the list and repeat the procedure until the cause of the problem is found. Once identified and confirmed:

- Make the necessary corrections
- Keep incident reports
- Learn from them and use for future reference.

N.B. "IF IN DOUBT, ASK."

CALL IN EXPERT HELP AND LEARN FROM THEM

Trouble Shooting & Fault Finding

Troubleshooting is the process of diagnosing, locating and correcting malfunctions. Problem solving requires understanding, brain storming, solution seeking and proper implementation. It's about working through an issue using a logical and systematic approach.

First - Define the problem. Check for any alarms (visual or audible), Operation & Maintenance manuals, decision trees, training presentations, Supervisory Control And Data Acquisition (SCADA) data trends etc. for more information. Figure out what you know and what you don't know.

Understand; then investigate. Compare all performance parameters with normal conditions, examine the facts first. Using your intuition; determine causality and judge accordingly. Do the basics very well, differentiate between what you know and what you don't.

Never rush to a conclusion, try not to make decisions or form opinions before you have all the pertinent facts. Make a list of the symptoms, consider their significance and then systematically delete them from your investigation. Don't be in a rush to find the answer, understand the question first.

Identify the likely cause of the malfunction using soft data and factual hard data; The more facts you have, the better! Make sure facts are facts… not assumptions; decisions based purely on assumption, are usually wrong.
Hard Data + Soft Data + Wisdom = Good Decision Making

Give direction to others so they can correct the problem. If the problem is corrected, **keep good records and archive,** learn from them. Knowledge and consistent records are the basis for avoiding the unusual.

During **fault finding,** investigate and risk assess first. Exercise caution and be systematic. Use logical and deductive reasoning. Do not alter several things at once. In solving a wrong problem, a new one can be created. Don't make difficult decisions in isolation. "If in doubt, ASK."

Collect the Hard Evidence. What is it telling you? Review the anecdotal evidence also i.e. analyse, understand. and consider its significance. Be patient, as impatience leads to mistakes being made. Apply good 'Root Cause Analysis' techniques.

Analyse the evidence. Before carrying out any invasive works – disassembling and reassembling equipment. Look at the symptoms first i.e. Stop, look, listen, touch, smell, and learn. Don't suffer from 'Change Blindness'. It's easy to miss something you are not looking for.

Locate the fault. Use your experiential knowledge, technical know-how, and analysis to guide you. Take advantage of the resources available to you e.g. Test equipment, system intelligence & the informed opinion of Subject Matter Experts (SME's).

Determine and remove the cause. Once you have located the fault, and before making the repair, you must ask why it has occurred, then find and remove the 'root cause'. If this is not done, the fault is likely to reappear.

Once the fault is rectified, check it. The system must be 'tested and proven' thoroughly to ensure that it is working properly and fit for purpose. N.B. Learn from Mistakes i.e. Review the What, Why, and How, & Contributing factors.

Trouble Shooting Guide (Short Version)

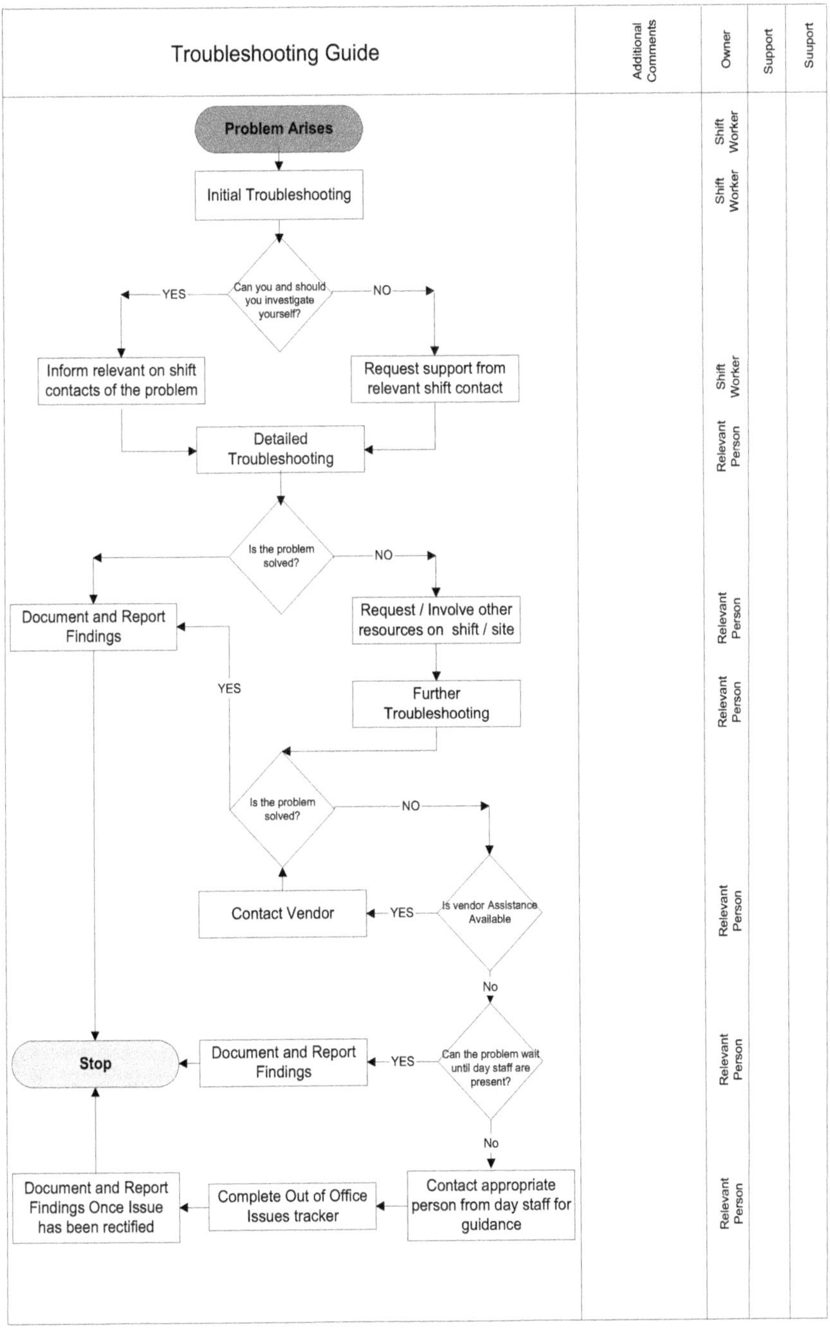

Trouble Shooting Report

Name: **Department:**
Time: **Date:**

Initial Troubleshooting Questions:

1. State problem:..
   a. What is the effect of the problem?..
   b. Is there more than one effect?...
2. When did the problem start?...
   a. Do trends / measurements identify a time when the problem started?...
   b. Do trends / measurements identify a behaviour change?..
3. Where is the problem?..
   a. Is it local to a piece of equipment?...
   b. Is it local to a step in a process / system?...
   c. Is it observed on more than one step of the process or more than one piece of equipment?...

Possible Causes	Facts that prove either it is the root cause or it is not

Detailed Troubleshooting

Please tick the appropriate boxes that were considered / referenced during the trouble shooting process. Please attach a copy of any relevant information found.

	Description/Comments
SOPs / WI's	
Trends via automation system	
Historical communications	
Automation documentation such as FDS	
Engineering documentation such as P&ID drawings	
Electrical drawings	
Mechanical drawings	
Technical reports/Operational issues	
PDA's/ Job cards	
Emails	
Spare parts not on site Personnel onsite availability Necessary skill not on site Other	

Findings

Write a short summary of findings

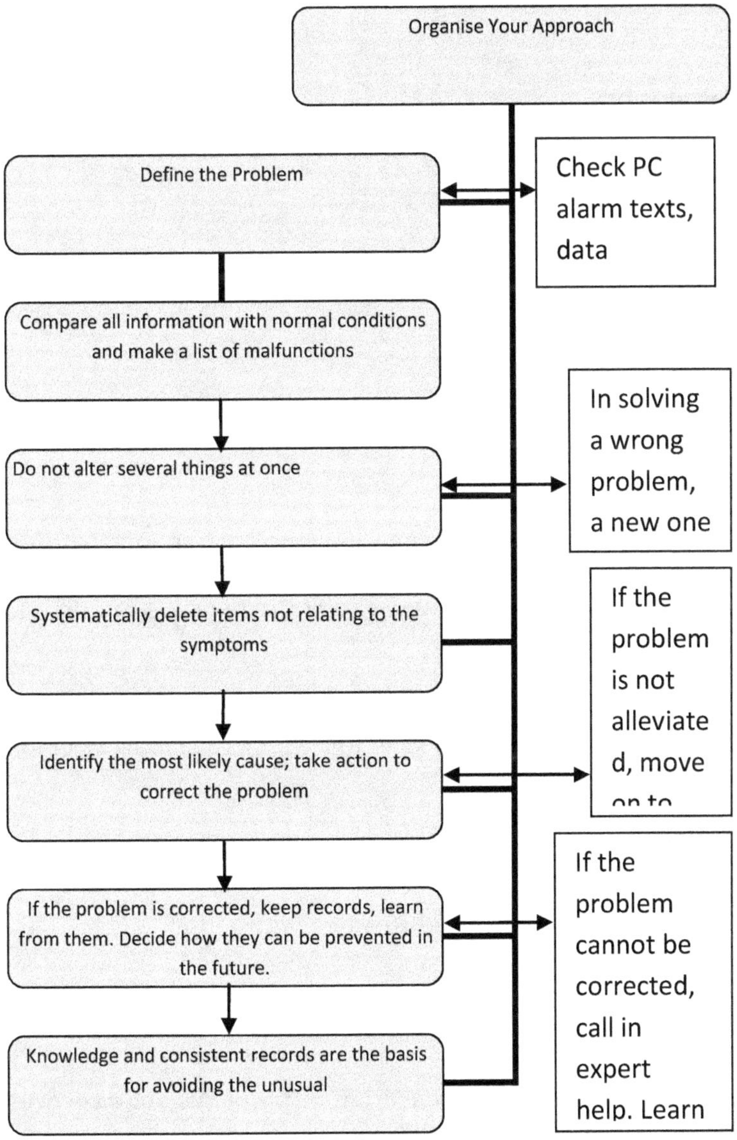
Organise Your Approach
Define the Problem
Check PC alarm texts, data
Compare all information with normal conditions and make a list of malfunctions
Do not alter several things at once
In solving a wrong problem, a new one
Systematically delete items not relating to the symptoms
If the problem is not alleviated, move
Identify the most likely cause; take action to correct the problem
If the problem is corrected, keep records, learn from them. Decide how they can be prevented in the future.
If the problem cannot be corrected, call in expert help. Learn
Knowledge and consistent records are the basis for avoiding the unusual

Think Safety & Environment

If in doubt,

Stop,

Think,

Ask,

Seek advice,

Never take a chance.

Engineering Goal

The Engineering Department is central to an organisation and what it does, as a group, by providing services, documents and information is crucial to people outside the department who are also performing crucial functions.

1. All tasks must be accomplished in a lean, efficient manner and at a high level of quality.
2. Be proactive in failure prevention.
3. New **value-added** opportunities must be seized as they develop.
4. Uncover problems and difficulties early, before they become major crises.
5. Share knowledge and expertise and feel committed to carrying out decisions.
6. Strive for excellence and ensure that problems are dealt with and objectives met.

Example of a LO/TO flow chart Page 1 of 2:

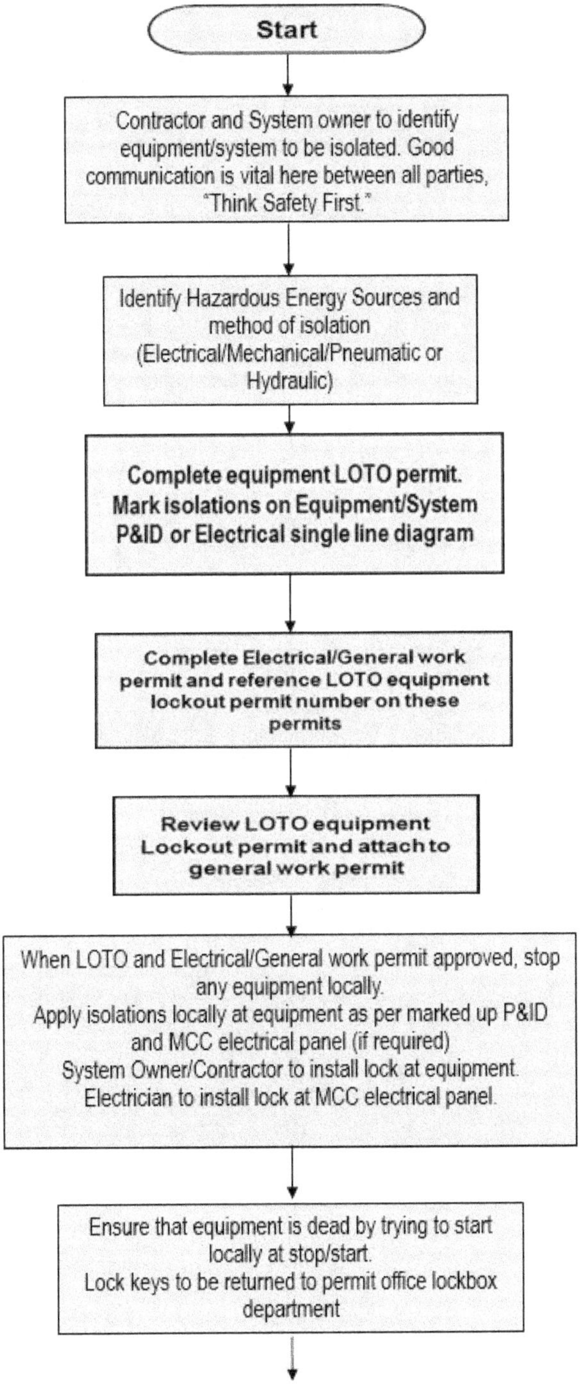

Example of a LO/TO flow chart Page 2 of 2:

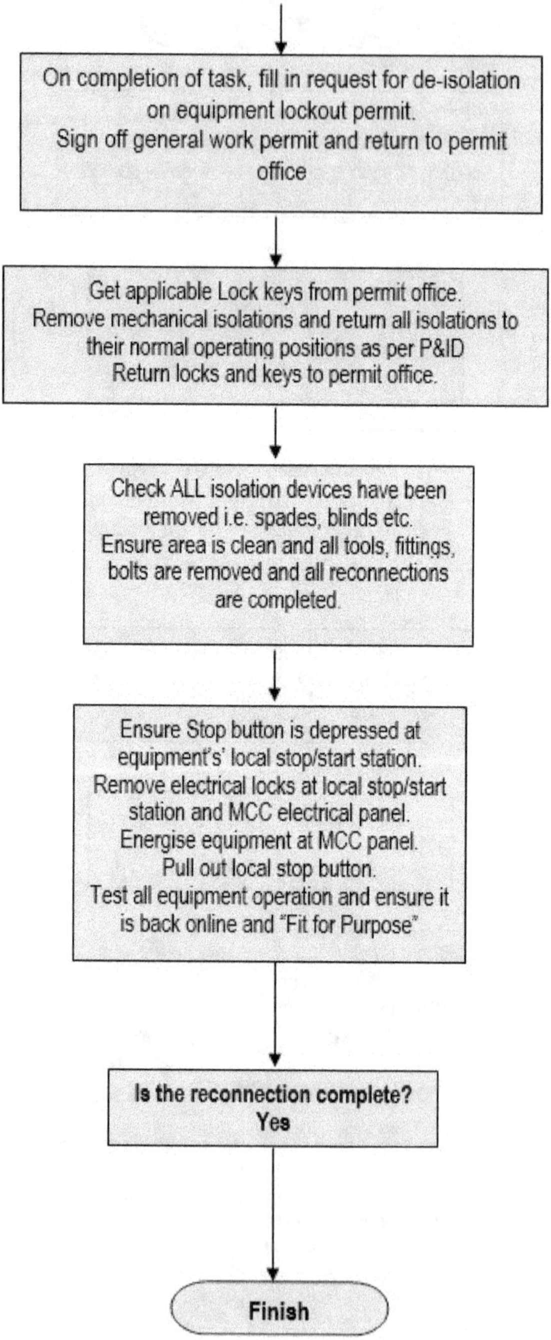

On completion of task, fill in request for de-isolation on equipment lockout permit.
Sign off general work permit and return to permit office
Get applicable Lock keys from permit office.
Remove mechanical isolations and return all isolations to their normal operating positions as per P&ID
Return locks and keys to permit office.
Check ALL isolation devices have been removed i.e. spades, blinds etc.
Ensure area is clean and all tools, fittings, bolts are removed and all reconnections are completed.
Ensure Stop button is depressed at equipment's' local stop/start station.
Remove electrical locks at local stop/start station and MCC electrical panel.
Energise equipment at MCC panel.
Pull out local stop button.
Test all equipment operation and ensure it is back online and "Fit for Purpose"
Is the reconnection complete?
Yes
Finish

Maintenance Management Vs Facilities Management

Setac Training

Maintenance Management	Facilities Management
A **Maintenance Manager** oversees the safety, installation, repair and preventative maintenance regimes of machines, electrical systems, mechanical systems, drainage systems, buildings and other structures.	A **Facilities Manager** is focused on providing the optimal safe working environment to the End User's workplace needs and demands whilst ensuring maximum 'uptime, reliability, and efficiency' of the key utility services.
Challenges: Increase productivity; minimise overhead costs; reliability; efficiency; reduce cost of repairs and servicing; planning; scheduling; maintain maximum 'uptime' of machinery; solve problems; calibrations; regulation & compliance.	A strong **Organisational Culture** must be maintained e.g. High performance staffing; workforce learning; accountability; continuous improvement; sustainability; reliability; knowledge management.
A **skills analysis program** must be in place to establish strengths and weaknesses within the engineering team including initial training, cross-skilling, up-skilling and knowledge sharing.	Proven **Business Processes** are key when contracting for products and servicing; vetting; budgeting and procurement; compliance monitoring; energy monitoring; bench marking.
A **Computerised Maintenance Management System** helps in managing maintenance activities, reducing downtime, controlling costs, minimising the investment in engineering spare parts.	Proven **Technical Processes** must be in place e.g. Scheduling; systems operation and maintenance; building management systems; data management; project management.
A **Calibration Management Software** system meets all requirements regarding calibration management and implementation. It involves scheduling, documenting and controlling calibration activities.	Good **Facilities Asset Management** is vital when it comes to automation hardware and software; communications and network equipment; vehicles; equipment; tools.
Commonly used **Precision Maintenance** programmes: Preventive Maintenance (PM); Predictive Maintenance (PdM); Condition-based Maintenance (CbM); Condition Monitoring (CM)	**Emergency Planning:** An emergency, such as a Power Outage, a Natural Gas outage, Flooding etc. needs to be dealt with immediately. Have an adequate action plan ready and be prepared.

Energy Management – PDCA Cycle

PLAN: An Energy policy must be put in place. Senior Management's commitment is vital when it comes to establishing a new Energy Management System (EnMS). An EnMS champion and his/her team must direct activities throughout the whole organisation.

Do: implement the Energy Mgt. action plans by conducting an energy review, and establish the baseline energy performance indicators (EnPI's), objectives, targets and action plans necessary to deliver results in accordance with opportunities to improve energy performance and the organisation's energy policy.

Check: monitor and measure processes and the key characteristics of its operations that determine energy performance against the energy policy and objectives; review the results. Action only makes sense if it leads to the desired result.

Act: The constant measurements are broken down in reports. Reporting of results and management review is crucial. Take actions to continually improve performance and the Energy Management System.

The EnMS cross-discipline team must have visibility i.e. **Smart metering** of all utilities in place e.g. Electricity, Water, Natural Gas; Oil, Compressed Air. "If you don't measure it, you can't manage it."

ISO 50001 standard: aims to help companies continually reduce their energy use, and their energy costs and their greenhouse gas emissions.

Building Energy Performance must be a top priority. The performance gap between the predicted and actual energy performance of buildings will not reduce unless all 'company owners' and 'end users' see the issue as a priority. All occupants are Energy Managers, all consume energy in some form.

If you want a quick win, it will be by revisiting existing complex buildings, retuning them using system intelligence via a **digital platform** (BMS), and training and educating people who use them.

Identify and prioritise continuous improvement **energy efficiency projects** that can generate capital. This improves your performance and productivity thus improves your company's bottom line.

Energy Budget Development - Anticipate your spend with greater accuracy by closely managing your energy budget. e.g. daily monitoring of energy usage to maintain budgets.

The 3 A's of Energy Management:

Audit - Energy audit; monitoring energy input.

Analysis - Measure, analyse, and manage.

Action - Evaluation and decision making.

"If you can see where you're using it, you can see where you're wasting it."

Monitoring & Targeting (M&T): predicting levels of energy performance against which future energy performance can be measured, thus allowing the impact of energy-saving actions to be evaluated.

Audits & Auditors

An **Audit** is an evaluation of an organisation, its associated systems, and processes. Audits are performed to ascertain the validity and reliability of information and to provide an assessment of a system's internal control to ensure it is compliant, transparent, effective, and efficient.

A **Good Manufacturing Practice Audit** evaluates if a company has its facility, quality management system (QMS) and manufacturing processes under control in meeting specified requirements.

A **Health and Safety Audit** is an objective assessment of current arrangements and performance standards in your organisation. It is a useful tool for establishing legal compliance and health and safety performance levels.

An **Environmental Audit** is an evaluation intended to quantify a company's environmental performance/compliance. They are intended to review the company's legal compliance status.

An **Engineering Audit** usually involves checking performance and evaluating deficiencies and inefficiencies against requirements and to prove that the systems are compliant, transparent, effective and efficient.

Remember during an audit, it is not only the company image that is being projected; it is also the **professional culture** as well as knowledge base that leave a lasting impact.

Auditors are not looking for faults in a system; they are looking for compliance to the standard. If a non-conformance is found, it should be viewed as an opportunity to improve, not as a reason to reprimand.

Ensure the 'key' company personnel who are in front of the auditor have the required technical knowledge and expertise, confidence, and presentation skills. Personnel should always be polite and helpful, look confident and smile.

When the auditor requests a document, provide the requested document and no more. Do not volunteer information that hasn't been requested unless it's an advantage to do so.

Personnel must not guess an answer; use SOP's/documents to aid your answer, which helps you, and gives a good impression. Do hide information, argue or display anger towards the auditor.

Never cause a deliberate delay. If for some reason you can't deliver a copy of a document quickly, explain the reason for the delay. A key to success during any audit, is preparation.

Personnel must not provide answers to a question to the auditor not related to an area of their responsibility/expertise, especially when their knowledge may be limited.

18. PEOPLE SKILLS

Having technical skills and knowledge and being dedicated to continual learning and development is a prerequisite for any engineering related activity. Equally as important though, is that you acquire the relevant **soft people skills** needed to help you interact professionally, efficiently, and effectively with people within your own department and with other departments within the company or within the global company network. The same applies when it comes to outside vendors and clients in the pursuit of maintaining the company's goals and objectives. There are 4 areas, which, if addressed properly, will help any business to be more successful, they are:

1. **Communication.**
2. **Open mindedness.**
3. **Team work.**
4. **Change Management**

1. Communication is the **creation of understanding** by the transfer of knowledge using verbal, past experiences, stories, script, illustrations, and examples. It will play a key role in any engineering discipline. Social awareness is a skill in which many of us are lacking. When a relevant message is being delivered verbally either in a 1-1 situation or in a team brief; **know when to talk and know when to listen**, you cannot multi-task speaking and listening. When you are thinking intently about what you want to say, you're not listening to what is being said.

Don't wait for information to be supplied, go after it, you may need to obtain information from several sources to address your present needs. How specific knowledge is accumulated, identifying the people who have this knowledge and how to go about attaining it from them will determine how easy or hard your task is going to be. The first challenge to be faced is how to use this specialist knowledge to complete a required task and secondly, maintaining effective communication with the person who assigned the task to you in the first place.

2. **Open-Mindedness:** A company should develop a culture of open-mindedness to ensure personnel operate in a good collaborative environment. It is important to realise that few people are the way we would like them to be.

'Soft skills' are essentially people skills -- the non-technical, intangible, personality-specific skills that determine your strengths as a leader, listener, negotiator, and conflict mediator. It's how you empower, motivate, and inspire others to achieve a common goal.

'Hard skills', on the other hand, are more likely to appear on your resume -- your education, experience, and level of expertise.

Are you an agreeable person? Conscientious? Do you communicate effectively? Solve problems efficiently? These are the types of questions aimed at uncovering the strength of your soft skills.

Conflict naturally occurs in every workplace, but it can have a toxic effect if it is not effectively handled. People with different personalities or competing agendas can lead to the festering of personal animosity. Having good conflict management skills will lead to less mental stress and a healthier working environment. Showing diplomacy, respect, and being sensitive to other people's points of view can make solving a problem must easier to resolve. Never use aggressive or confrontational methods to get information. All the required information may not be on site, knowing where, who to contact and how to source it as quickly as possible is key.

When in need, ask for advice, support, or guidance. Don't assume others will notice when this is the case. Work colleagues will be focussed on their own tasks and maybe haven't much spare time to answer your questions. The success of your 'day to day work activities' may depend on the assistance of others, remember that up to 80% of the success in any job you do is based on your ability to deal with people. Sometimes, although rarely, you may have to deal with an aggressive, short tempered person who may dismiss you when you ask for information. Stay calm but at the same time responsive. Have the right mindset to resolve the issue at hand. When faced with a difficult person, it is easy to want only to be right. Acknowledge there may be a need to compromise to progress forward. Timing is everything, **pick the right moment,** become aware of how other people react when approached. Try to anticipate and calibrate their mood, they might be having a bad day and are irritable. A little humour in the workplace can improve your work culture and productivity but remember that offensive behaviour among colleagues carries a high price, because 'civility' at work means more than just good manners.

Tip: "One of the only things in life that will get you in trouble 'daily', is what comes out of your mouth. Learn to be mindful, less controversial, considerate, and to control what you say; your life will be a lot easier." Words are powerful, choose them well. Be more reflective; if you say only what you think, don't expect to hear only what you like. You can't talk your way out of problems you talk (or behave) yourself into. Never miss an opportunity to be silent, speak only when you to have something brilliant to say.

Positive working relationships are important in any work place. Respectful treatment creates the foundation for cooperation between individuals and collaboration among departments. When people are rude and disrespectful to one another, they are much less likely to share information, contribute helpful suggestions, or offer assistance. Be aware of how defensive you become to negative feedback. Never reject a piece of constructive criticism completely without acknowledging that at least part of it is helpful. Never assume that every critic is a hater, some people are telling you the truth.

Do's and Don'ts of Criticism

- **Do** listen objectively, seek first to understand
- **Do** ask for specifics and watch your tone of voice

- **Do** get a second opinion and do your own research
- **Do** apologise, if warranted, take ownership and responsibility
- **Do** show that you are taking feedback into consideration
- **Do** take corrective action.

- **Don't** ignore the criticism, especially if its constructive
- **Don't** get defensive, angry, or rude; watch your body language
- **Don't** waste time making excuses, if you're wrong, your wrong
- **Don't** blame others, accept responsibility
- **Don't** dwell on the error, learn from it, and move on
- **Don't** engage in a cover-up
- **Don't** react in haste before considering the best plan of action

Take it on the chin

Criticism can result from perceived inadequacies, actual wrongdoings, or unmet expectations. It can also stem from pettiness, jealousy and 'having a bad day' syndrome. Criticism can only impact you negatively if you continue to berate yourself. So, focus on your successes. Develop a thick skin and be prepared to deal with criticism, whether it is justified or not. Be objective, take action and let it go.

Differences in perspectives and priorities can lead to differing views on the contributions a person makes at work. If you have been given a list of work to be completed within a certain period and difficulties are been experienced completing the tasks due to factors outside your control, keep your manager informed of the situation. Ask them for direction; don't assume that they already know problems are being encountered.

3. Team Work: The benefits of team work cannot be under estimated. If one person on the team comes up with a great idea or figures out how to solve a major problem, everyone benefits, so it important to support each other. You cannot do everything yourself. Accepting help or delegating responsibility is not a sign of weakness, but a growing process that will help you both personally and professionally. You may only be just starting out on your career or are already at the upper-management level. It doesn't matter how smart or capable you are, **if you work well with others,** you will find that the overall results come much faster.

Whether in a 1-1 meeting or a group session with other team members where there must be some 'straight talking' done, you must have the ability to communicate clearly, specifically and unambiguously no matter how frustrated you may become at other people's apathy or lack of knowledge. **Never** be disrespectful, inconsiderate or **talk down** to others. "Praise loudly, blame softly." Never criticise another person publicly.

Tensions may be high, and arguments may arise during a crisis, it's tempting to make things personal and make statements to assert your claims or defend yourself. Focus and be patient; disentangle the person from the problem. "Shrewd, strategically asked, open ended questions" work better to calm any situation.

4. Change Management: "Change is inevitable."

Managing change successfully requires effectively balancing soft and hard skills, often in emotionally trying and technically challenging circumstances. Change management is a process which, when applied properly, is self-reinforcing and can be the start of ongoing improvement and learning for an organisation. When applied incorrectly, it can vary from a waste of time to utter disaster.

As a 'Manager' of change, you must adopt a **rational approach** for implementing change within your organisation. Understand the change process on why we/others resist being changed, but also how to convert our natural response to change into support and enthusiasm for the change. Develop a **way of thinking** about organisational change that encourages the formation of teams, respects intelligent criticism, and invites participation.

Successful Change Management is more likely to occur if the following are included:

- Effective Communications that informs various stakeholders of the reasons for the change (why?), the benefits of successful implementation (what is in it for us, and you) as well as the details of the change (when? where? who is involved? how much will it cost?).
- Devise an effective education, training and/or skills upgrading scheme for the organisation.
- Counter resistance from the employees of companies and align them to overall strategic direction of the organisation.
- Provide personal counselling (if required) to alleviate any change-related fears.
- Monitoring of the implementation and fine-tuning as required.

A **Subject Matter Expert (SME)** is a person who is an authority in a particular area or topic. They exhibit the highest level of expertise and experience in performing a specialised job, task, or skill within the organisation.

Good communication, decision-making, collaboration and planning are required when resolving a difficult issue.

- Are all the Subject Matter Experts in situ?
- What are the established facts and what are the assumptions?
- Prioritise, are we in control?
- What is the plan of action and who will do it?
- What is the next step and how are we going to go about it?
- Who will monitor and record progress?

Planning is the process of thinking about and organising the activities required to achieve a desired goal. It involves the creation and maintenance of a plan. Planning is one of the most important project management and time management techniques. Planning is preparing a sequence of action steps to achieve some specific goal. If you do it effectively, you can reduce much of the unnecessary time and effort of achieving the goal. A plan is like a map. When following a plan, you can always see how much you have progressed towards your project goal and how far you are from your destination. Knowing where you are is essential for making good decisions on where to go or what to do next. One more reason you need planning is, the 80/20 Rule. It is well established that for unstructured activities, 80 percent of the effort gives less than 20 percent of the valuable outcome. You either spend much time on deciding what to do next, or you are taking too many unnecessary, unfocused, and inefficient steps.

Planning is also crucial for meeting your needs during each action step with your time, money, or other resources. With careful planning you often can see if at some point you are likely to face a problem. It is much easier to adjust your plan to avoid or smoothen a coming crisis, rather than to deal with the crisis when it comes unexpected.

How should we communicate? – Today many negotiations are conducted by phone, email, or text and the interaction can be quite different than meeting face-to-face. The short length and abbreviations used in texting greatly increase the likelihood of unintended misunderstandings and misinterpretation of what a person means.

The lack of audio and visual cues in texts and email make it much harder to hear or interpret the emotional undertones of communication, which can feed our tendency to hear the worst. Furthermore, not having the other person in front of us reduces or eliminates the impact of "mirror neurons" in our brain, which normally increase our empathy and sense of human connection with our counterpart.

Tip: Use your non-dominant hand when texting, it makes your brain more focussed on what it's doing.

Try to communicate in person, emails are an excellent means of communication and transferring files, information …. but, it's always better to interact with people "face to face" where possible.

Reserve email for those not in the general workplace. Pick up the phone instead or leave a place of work to deliver the message. Try get along with others, sometimes personalities clash, but stay professional. Never be afraid to have an **honest conversation** with your colleague(s), it leads to a better working environment when problem solving and making better judgement calls. Sometimes, "If you want to get along, you have to go along."

Tip: Have you ever said something or sent a message in anger and regretted it afterwards? Before making a possible bold contentious statement in a meeting or in a 1-1 discussion with members of your team, peers, or manager, consider the possible repercussions (if any) and can you handle them? If your natural instincts tell you not to say it, trust them. Discuss it with a trusted colleague first and ask them for their advice and try to establish do 'they see it, the way you see it'. The same applies when sending a controversial email; get a trusted colleague to proof read it first before sending. If you are left with no alternative and must say or send something that may cause offence or get a negative reaction, try to adopt the **sandwich approach technique**. 'Give a person a compliment, make your comment and then give another compliment'. The message is delivered between two pieces of niceness.

Think positive – Having the right **can do/will do** attitude helps you to think, prioritise, solve problems, make decisions and be creative; it can help you in making an objective appraisal of any situation and less likely to make mistakes. Avoid making the same mistake twice, can you avoid or go around the problem and try to turn it to your advantage. ***Crisis Mindset*** - the ability to make effective decisions about what to change and how to change it during a crisis separates a person who has done their research, armed with theory, knowledge, and experience from the rest. Have confidence in yourself and you'll inspire others to have confidence in you.

Learn to **manage** a given situation. Plan for the unexpected, this will keep surprises to a minimum. If you've thought of the things that could go wrong with any proposed work planned, confident decisions on corrective actions can then be made when necessary.

Tip: Whatever job title and role you currently have, ensure you know the responsibilities of the role and what is expected of you. The same applies if you are promoted, ensure you are also aware of all the extra responsibilities that come with your new role **e.g.** managing and being responsible for company personnel, site contractors and specialist vendors under your remit.

E.g. You are a newly promoted team leader of a crew of 10 personnel, all involved in resolving a major issue. Regardless of how large or small a problem maybe, it's not the problem that matters; it's how you respond to it. Know your **'Call to Action'**: Stay focused, show leadership, confidence, control, patience, organisation, strategic planning, and be able to shift gears to quickly find solutions to problems regarding personnel and equipment. Good tactical advice requires knowledge of specific circumstances. Be confident, yet humble. Ask questions and really listen, not half listen. Lead and manage by expectations and deliverables. Work hard, treat people well, and look out for one another. In turn, you won't have to worry about how to command respect, you'll have earned it.

Give clear direction and work through each issue until it is resolved. "Be aware of any potential dangers, maintain a **'Heightened Situational Awareness'** and don't suffer from '**Situational Blindness'**." During periods of high activity – your brain sometimes ignores things but doesn't tell you that. You only get parts of the information and your brain fills in the gaps of what you expect.

Whatever environment you or your team are in, be ever mindful of changing conditions: breathable air conditions; deteriorating weather conditions; tidal conditions or heavy machinery moving into the area. Never put the lives of your team or your own at risk regardless of the time constraints or rising costs involved in trying to get the problem resolved.

Don't blame others for your mistakes, own up to any errors you make. Offer explanations, not excuses. An error doesn't become a mistake until you refuse to correct it. Try to fix problems and don't let pride keep you from speaking up and seeking advice from others if you can't resolve an issue. Take responsibility for your decisions and actions, you'll be respected for it, and maybe even set an example. With responsibility comes accountability and a person must be responsible for their mistakes.

When people work together, honest mistakes and disappointments happen, and it's easy to blame someone who causes these. However, when everyone starts pointing fingers, an unpleasant atmosphere can quickly develop. This lowers morale, undermines trust, and is ultimately unproductive.

Instead, encourage everyone in your team to think about the mistake in a constructive way and don't be afraid to 'message up' bad news. Talk openly about the mistake no matter how uncomfortable it may be and what you can all do to fix what happened and move forward together. Also discuss what actions need to be taken and implemented to make sure that this mistake doesn't happen again.

The team must have confidence in you and your abilities to lead them in both normal day to day activities, and through a crisis. If they think you are reckless and take unnecessary risks regarding their welfare, safety, or the work to be carried out, they may not follow you.

Remember, actions speak louder than words, always ***lead from the front and protect those who are behind you.*** Accept that it is inevitable that one of your crew may make a mistake, if they are to be reprimanded, do it in a 1-1 scenario, never in front of a group of people, you will lose their respect and that of those around you. Tact is the art of making a point without making an enemy. Avoid criticising others; keep your personal opinions to yourself. If it's professional criticism, keep it constructive and always think twice before offering it.

Tip: If you must speak about a person to others, imagine that person is beside you as you're speaking; you may be more considerate and selective with your words.

Don't be intimidating towards your team, you'll find you will never get a straight answer and end up defeating your own purpose. Nothing lowers the tone of a conversation more than a raised voice; a good argument has no need of a loud voice. In saying that, don't ever be afraid to challenge either. Let them know the standards that must be maintained always regarding:

- Being flexible and cooperative
- Being responsible and accountable for their acti ons
- Health & Safety practices
- Housekeeping practices
- Personal Attitude
- Professional behaviour
- Professional capabilities of their actual 'day to day' work activities
- Interaction and communication with other team members
- Willingness to share both tacit and explicit **work related** information
- Being helpful to others.

N.B. "Don't take risks, if in doubt 'ASK'. Stop !!! seek advice, and don't change anything. Don't be afraid to ask dumb questions, they are easier to solve than dumb mistakes."

People are better decision makers when they have a breadth of knowledge and experience to draw upon. Effective decision makers realise this and balance several elements. These include the need for hard, accurate, and reliable data which is rigorously analysed. Without this, any **decision** is nothing more than a guess or gamble! Although hard data is important, decision making will always require some element of intuition, sometimes all the facts may not be available when making difficult decisions. The 'Intuitive decisions' you have to make will represent the most efficient use of your accumulated experience.

You may be faced with a situation that must be resolved quickly, and there may be only anecdotal information available. Once you have good knowledge of control systems and equipment in general, your wisdom and proven trouble shooting and fault finding methods should take over and enable you to try new approaches and seek creative new solutions to a problem. You must learn to adapt to whatever situation arises. Keep an open, ever learning mind where new thinking may be required and not become entrenched in the way you always do things.

People don't come to work to fail but you must **take action when performance slips.** Good team leaders never rely on wishful thinking or the innate good in humankind to solve a problem.

Being a team leader needs all the above attributes, but, just as important is having knowledge on disciplinary and grievance procedures.

Try to ensure you have the right mix of people on the team, who maybe are stronger in certain areas than you are, the ability to let go and to delegate to others is a quality that develops over time. Trust your people and they will start to trust you, delegating will make you a much better leader because it shows your trust in others and allows them to shine as well. Let them get on with the job and get out of their way, only act in an advisory or overseeing capacity.

Learn from mistakes and get acquainted with as many talented people as possible. The key is to learn from others' experiences. Proper people skills techniques must be honed, practiced, and implemented. Personnel across the company should be encouraged to read as many articles as possible on the subject, especially **Emotional Intelligence,** it will be time well spent.

Looking After Yourself!

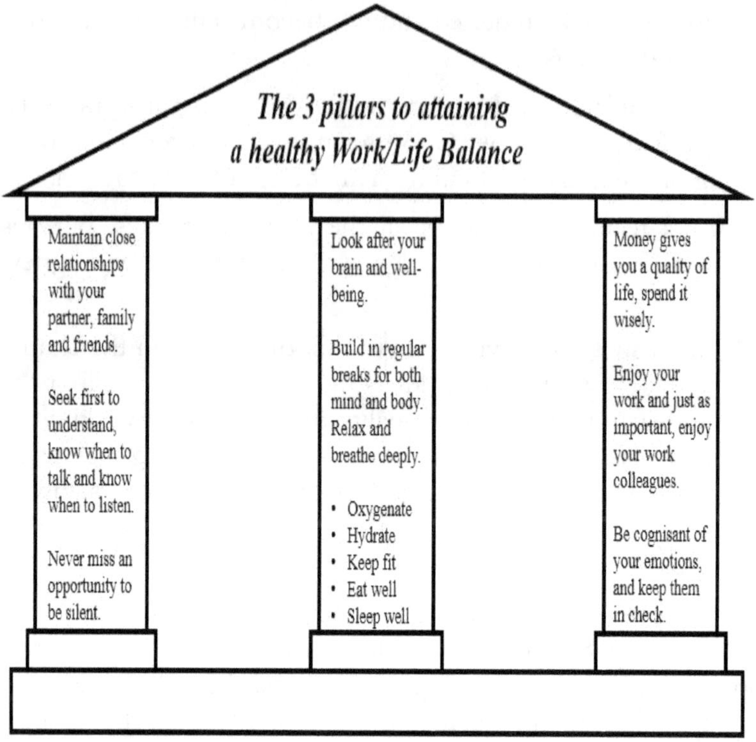

The 3 pillars to attaining
a healthy Work/Life Balance
Maintain close relationships with your partner, family and friends.
Seek first to understand, know when to talk and know when to listen.
Never miss an opportunity to be silent.
Look after your brain and well-being.
Build in regular breaks for both mind and body. Relax and breathe deeply.
• Oxygenate
• Hydrate
• Keep fit
• Eat well
• Sleep well
Money gives you a quality of life, spend it wisely.
Enjoy your work and just as important, enjoy your work colleagues.
Be cognisant of your emotions, and keep them in check.

Taking care of yourself is just as important as taking care of your business. Having a good **Work/Life Balance** is vital and recognising the importance of looking after your **Brain and Well-Being.** We spend a third of our lives sleeping, an activity as crucial to our health and well-being as eating. 'Our thinking, decision making, and problem solving abilities are directly affected by what we eat and drink'.

Brain Facts - Your brain has over 86 billion brain cells (neurons), capable of many trillions of connections which controls both our conscious and unconscious thoughts and decisions. Put into context, your brain has the capacity to store up to 10 million books, each with a thousand pages crammed full of text. It consumes more energy than any other organ in the body and it continues to form new connections between neurons throughout your life, your brain never stops changing.

Hydration - Your brain is made up of about 75% water. Water makes up two thirds of your weight and is involved in so many of the critical functions of your body such as:

- Body Temperature
- Brain Function
- Waste Removal

So constant hydration and replenishing what is lost on a day to day basis should be a top priority of anyone who wants to stay ahead of the game. For clear thinking it is essential that you drink sufficient quality drinking water to keep your brain hydrated. An Adequate Intake (AI) for men is roughly about 13 cups (3 litres) of total beverages a day. The AI for women is about 9 cups (2.2 litres) of total beverages a day.

Oxygenation – Your brain uses 20% of the total oxygen and blood in your body ~ keep fit! A human inhales and exhales some 20 kilograms of air daily.

Diet – The human brain is the fattest organ in the body and may consist of at least 60% fat. The brain consumes up to 40% of our carbohydrate intake ~ keep it well nourished. It requires about 25% of the body's glucose.

Sleep Well, Think Well - The importance of adequate sleep in decision making and error reduction cannot be over emphasised. Build in breaks for both mind and body. The brain can process information and prepare for actions during sleep, effectively making decisions while unconscious.

Fatigue can kill: Be aware of the symptoms and look for them in yourself and others, we are not good judges of our own impairment.

Fatigue can affect performance in several ways:

1. It can slow **reaction times** - reduced hand-eye coordination or slow reflexes.
2. It can degrade **decision making** - short term memory problems and an inability to concentrate.
3. Noticeably reduced capacity to **engage** in effective interpersonal communication.

Reduce your 'work related stress' levels. Build in regular breaks for both mind and body. We are not machines – we do need recovery time! Sometimes you just need to remove yourself completely from a stressful environment and recharge **e.g.** take some 'time out' to listen to some 'peaceful and relaxing' music, take a hot bath, read a book, call or meet up with a friend, practice yoga or meditation, keep fit (it is recommended to get at least 150 minutes of moderate aerobic activity or 75 minutes of vigorous aerobic activity a week, or a combination of moderate and vigorous activity). If your stress levels remain high, you may need to seek professional help and advice so you can learn new coping mechanisms in how to deal with what's causing them to escalate.

N.B. A company's focus must be to continually provide optimal working conditions for all its people. Heating, noise, and smells can affect workplace performance. Scientific studies show that improving employees' physical and emotional response to their working environment – from the look and feel of a workplace to non-visual sensory inputs – such as smell, noise, and temperature – can significantly improve productivity and cognitive performance.

19. TERMS AND DEFINITIONS

AAV (Automatic Air Venting): When liquid is pumped through a system at start up, the initial air inside the piping is pushed into the air vent by the pressure of the flow. After the initial air venting, liquid flows into the air vent. The float rises with the rising liquid and closes the valve.

Accidentology: the study and analysis of the causes and effects of accidents.

ACOP: (Approved Codes of Practice) are guidance with specific legal standing. They deal with a wide range of hazardous materials and working practices. Employers who are prosecuted for a breach of health and safety law, who have not followed an ACOP, are likely to be found at fault by the courts.

Accuracy: nearness of a sensor or indicator measurement reading to the real value of the quantity being measured – usually expressed as a percentage error.

Adage: a proverb or short statement expressing a general truth.

After Action Review (AAR): is a simple process used by a team to capture the lessons learned from past successes and failures, with the goal of improving future performance. It is an opportunity for a team to reflect on a project, activity, event, or task so that they can do better the next time.

AHU (Air Handling Unit): is used to regulate and circulate air as part of a Heating, Ventilating, and Air-Conditioning (HVAC) system. An AHU is usually a large unit containing a fan, heating/cooling elements, filter racks or chambers, sound attenuators, and dampers. AHU's connect to a ductwork ventilation system that distributes the conditioned air through the building.

ALARP (as low as reasonably practicable) is a term often used in the regulation and management of safety-critical and safety-involved systems. The ALARP principle is that the residual risk shall be reduced as far as reasonably practicable.

Ambient Temperature: The average or mean temperature of the surrounding environment in contact with sensor or equipment concerned.

A/I (Analogue Input): is a continuous variable signal. Typical analogue signals vary from 0 to 20 milliamps, 4 to 20 milliamps, or 0 to 10 volts. It can have any state between 0 and 100% e.g. a 4/20 mA analogue input signal to a PLC from a flow transmitter in a process line indicates that a flow is 40% of maximum possible flow.

A.I. -Artificial Intelligence - (sometimes called Machine Intelligence): is the capability of a machine to imitate intelligent human behaviour.

Analogue Instrument: if a flow rate must remain constant in a process stream at 40% of its maximum flow, opening the valve to 40% will not guarantee this as the upstream pressure may vary and therefore increase or decrease the flow through the valve. The way to guarantee 40% flow rate is to constantly measure the flow in the process stream using a flow meter and vary the valve opening.

Anecdotal: (of an account) not necessarily true or reliable, because based on personal accounts rather than facts or research.

Air Change: is how many times the air enters and exits a room from the HVAC system in one hour. Or, how many times a room would fill up with the air from the supply registers in sixty minutes.

Air Conditioning: the process of treating air to meet the requirements of a conditioned space by controlling its temperature, humidity, cleanliness, and distribution. **Note:** TM44 Regulations - Inspection of Air Conditioning Systems. The Energy Performance of Buildings Directive (EPBD) requires Member States across Europe to put in place 'measures to establish a regular inspection of air conditioning systems of an effective rated output of more than 12 kW's'. If you have commercial size cooling or air conditioning plant in your buildings, you are required by law to carry out an energy survey according to TM44.

Analogue loop: typically consisting of an analogue input and an analogue output e.g. where a flow meter (analogue input) and a control valve (analogue output) form a loop. The control valve is opening and closing proportionately to vary the flow rate in a process line where the flow meter measures the flow rate to achieve the desired amount of liquid as directed by a PLC control system.

A/O (Analogue Output)**:** is a continuous variable signal. Typical analogue signals vary from 0 to 20 milliamps, 4 to 20 milliamps, or 0 to 10 volts. It can have any state between 0 and 100% e.g. a 4/20 mA analogue output signal from a PLC is sent to a VSD which in turn varies the speed of a motor to a set speed.

ASI (Actuator Sensor Interface): is an industrial networking solution (physical layer, data access method and protocol) used in PLC, DCS and PC-based automation systems. It is designed for connecting simple field I/O devices (e.g. binary ON/OFF devices such as actuators, sensors, rotary encoders, analogue inputs and outputs, push buttons, and valve position sensors) in discrete manufacturing and process applications using a single two-conductor cable.

Asset Management: is inherently a cross-functional activity involving aspects such as concept, new project design, construct, commissioning, operation & maintenance, upgrading, decommissions and demolition. Getting the right dialogue taking place between these groups is necessary so that joined up thinking takes place and the life cycle costs of assets are minimised. It is the practice of managing assets to achieve the greatest return and the process of monitoring and maintaining facilities systems, with the objective of providing the best possible service to its end users. It is vitally important for all capital intensive industries which rely on cost effective and trouble free operation of their plant and equipment in order to deliver products and services to their customers.

Asset Obsolescence: can be defined as structures, systems or components ending their usefulness as a result of changes in knowledge, standards, technology or needs. It can sometimes be identified by the absence of necessary spare parts or technical support.

Backup and Disaster Recovery – for full protection of critical data and infrastructure. If your systems are ever struck by a malicious attack, it is vital you can restore data to get you up and running quickly.

BAS (Building Automation System): Building automation is the automatic centralised control of a building's heating, ventilation and air conditioning, lighting, and other systems through a building management system (BMS) or building automation system (BAS).

BAT (Best Available Technology): is a term applied with regulations on limiting pollutant discharges with regard to the abatement technology.

Bench Marking: is the process of comparing one's business processes and performance metrics to industry bests and/or best practices from other industries. Dimensions typically measured are quality, time, and cost. Improvements from learning mean doing things better, faster, and cheaper.

BLEVE (Boiling Liquid Expanding Vapour Explosion): is an explosion caused by the rupture of a vessel containing a pressurised liquid above its boiling point.

Building Automation: is the automatic centralised control of a building's heating, ventilation and air conditioning, lighting, and other systems through a **Building Management System** (BMS) or **Building Automation System** (BAS).

BIM (Building Information Modelling): is an intelligent 3D model-based process that equips architecture, engineering, and construction professionals with the insight and tools to more efficiently plan, design, construct, and manage buildings and infrastructure.

BSC (Balanced Score Card): is a strategic planning and management system that is used extensively in business and industry to align business activities to the vision and strategy of the organisation, improve internal and external communications, and monitor organisation performance against strategic goals.

Building Manager: supervises the hard and soft services of a property(s) and whose main purpose is overseeing the safety and maintenance of those designated properties and its bespoke equipment and ensuring that they are in compliance with all applicable regulations. Recommending and coordinating improvements to the property as needed to ensure a safe, functional, and appealing space.

Building Management: is a discipline that comes under the umbrella of facility management. It is a particular economic activity, a set of property maintenance, operation, repair and maintenance. This is a legal and technical set of operations required for building maintenance and preservation of usable condition, as well as functionally required for the maintenance of the land to ensure that property is used in accordance with the purpose.

CAPA (Corrective Action and Preventative Action)

Calibration: is the process of comparing the accuracy of an instrument reading to known standards. It is a demonstration that an instrument or device produces results within specified limits by comparison with those produced by a reference or traceable standard over an appropriate range of measurement.

Calibration Range: The calibration range is defined as: the region between the limits within which a quantity is measured, received, or transmitted, expressed by stating the lower and upper range values. The limits are defined by the zero and span values. The zero value is the lower end of the range. Span is defined as the algebraic difference between the upper and lower range values.

Calorifier: A Calorifier is a heat exchanger which heats water indirectly by circulating it over a heating coil or multiple coils. The source of heat can be water or steam, heated by an external heat source, contained within a pipe immersed in the water.

CAPEX (Capital Expenditure): money spent by a business or organisation on acquiring or maintaining fixes assets, such as land, buildings, and equipment.

CAT: Critical Assessment Team

Causality: the relation between a cause and its effect or between regularly correlated events or phenomena.

Cause and Effect Diagram: A graphic tool that helps identify, sort, and display possible causes of a problem or quality characteristic. A Cause and Effect Diagram helps determine root causes; encourages group participation; uses an orderly, easy-to-read format; indicates possible causes of variation; increases process knowledge and identifies areas for collecting data.

Cavitation: as liquid pressure falls below its vapour pressure, bubbles form and implode on impellers and interior surfaces, damaging pump internals, disrupting flow and leading to seal failure.

CbM (Condition based Maintenance)**:** main goal is to increase reliability and availability of machinery thus minimising system downtime, spare parts, and labour costs. CbM can be performed by scheduling downtime, labour and materials based on machinery health.

CBT (Computer Based Tool): Virtually any programme or utility that helps programmers or users develop applications or maintain their computers can be called a tool.

CE Marking: The letters 'CE' are the abbreviation of French phrase 'Conformité Européene' which literally means 'European Conformity'.

CE Marking of Machinery: All machines supplied in the European Economic Area (EEA) from January 1st, 1995, must comply with the European Machinery Directive and be safe.

CHP (Combined Heat and Power): is on-site electricity generation that captures the heat that would otherwise be wasted to provide useful thermal energy—such as steam or hot water—that can be used for space heating, cooling, domestic hot water and industrial processes. CHP technology allows a company to generate its own Electricity and Heat and by doing so reduces its energy costs considerably. Gas turbine development in recent years has made the combined heat and power generation more attractive to commercial and industrial installations. By changing from separate Electricity and Heat Generation sources to a single combined Heat and Power unit based on its site, a company will enjoy considerable energy cost savings along with a secure source of energy provided by a proven technology.

CI (Continuous Improvement): is an ongoing effort to improve products, services, or processes. These efforts can seek 'incremental' improvement over time or 'breakthrough' improvement all at once.

CMMS (Computerised Maintenance Management System): A maintenance software system allows the protection of a company's investment in assets and equipment. The system will help to manage maintenance, reduce downtime, control maintenance costs and minimise the investment in engineering spare parts. This type of system often becomes the showcase for maintenance departments during client and regulatory audits, it allows full visibility and traceability on all engineering and maintenance activities.

Coaching: is a form of development in which a person called a coach supports a learner or client in achieving a specific personal or professional goal by providing training, advice and guidance. The learner is sometimes called a coachee. Coaching is linked to leadership, communication, role model, leadership motivation and employee development. Coaching is a required competence of every leader. Every coaching interaction is different, and every coach is different. To be an effective coach you must recognise and understand who you are and how you engage most effectively with your team. Coaching is releasing a person's potential to maximise their own performance. It is about helping you to learn rather than teaching you.

CoP (Code of Practice): can be a document that complements occupational health and safety laws and regulations to provide detailed practical guidance on how to comply with legal obligations and should be followed unless another solution with the same or better health and safety standard is in place.

COP (Coefficient of Performance): of a heat pump is a ratio of heating or cooling provided to work required. Higher COPs equate to lower operating costs.

C&Q (Commissioning & Qualification): **Commissioning** is a methodical, documented process to ensure that facilities, systems, and equipment meet established design requirements and stakeholder expectations. The commissioning process verifies the following: what was specified was installed; that it functions properly; and that it was successfully turned over to the user. **Qualification** is a process that extends beyond commissioning because it is primarily concerned with verifying facility and system aspects that can affect product quality.

Facilities and System Qualification consists of three parts:

1. ***Installation Qualification (IQ):*** Adheres to, and is installed, as per approved specifications.
2. ***Operational Qualification (OQ):*** Operates as intended throughout all anticipated ranges.
3. ***Performance Qualification (PQ):*** Performs as intended, meeting predetermined acceptance criteria over time.

The basic difference between commissioning and qualification is that the former is concerned with good engineering practice, whereas the latter primarily verifies facility and systems aspects that can affect product quality.

Competency: Competency is the ability to satisfactorily perform the required functions of a job, usually defined in terms of tasks or levels of skill for a specific job. Certified competence is a credential or certification that an individual has the basic skills, knowledge, and experience to carry out tasks on a piece of equipment.

Compliance: means conforming to a rule, such as a specification, policy, standard or law. Regulatory **compliance** describes the goal that organisations aspire to achieve in their efforts to ensure that they are aware of and take steps to comply with relevant laws, polices, and regulations. **'Compliance costs Money'** whether in a GMP or EH&S environment. A compliance cost is expenditure of time or money in conforming with government requirements such as legislation or regulation (**e.g.** an environmental air monitoring system can have a €60,000+ price tag) companies must be fully aware of this fact.

Condensation: is caused by warm moist air coming into contact with a surface that is colder than the air's dew point. **E.g.** So what happens in an electrical enclosure? In humid conditions, the warmer the air is, the more water vapour it holds. When warm air inside an enclosure comes into contact with the enclosure's colder casing, it cools. And it may chill to a point where it can no longer hold water vapour. The vapour, suspended as a gas, turns into water. The air has reached its 'dew' point and condensation occurs.

CM (Condition Monitoring): is the process of monitoring a parameter of condition in machinery, such that a significant change is indicative of a developing failure e.g. overheating and vibration is monitored for early signs of impending failure. Equipment can be monitored using sophisticated instrumentation such as vibration analysis and ultrasonic monitoring equipment or the human senses. Where instrumentation is used, actual limits can be imposed to trigger maintenance activity. It is a major component of Predictive Maintenance.

Confined Space: means any place, including any vessel, tank, container, vat, silo, hopper, pit, bund, trench, pipe, sewer, flue, well, chamber, compartment, cellar, or other similar space which by virtue of its enclosed nature creates conditions which give rise to a likelihood of accident, harm or injury.

Control loop: is a process management system designed to maintain a process variable at a desired set point. Each step in the loop works in conjunction with the others to manage the system.

Corrective Maintenance: (a.k.a. Reactive Maintenance) – action is initiated when the unscheduled event of an equipment failure occurs and can be defined as a maintenance task performed to identify, isolate, and rectify a fault so that the failed equipment, machine, or system can be restored to an operational condition within the tolerances or limits established for in-service operations. Corrective Maintenance is carried out on all items where the consequences of failure or 'wearing out' are not significant and the cost of this maintenance is not greater than Preventive Maintenance.

CPU (Central Processing Unit): is a microprocessor system that contains the system memory and is the PLC's decision-making unit. The CPU monitors inputs, outputs, and other variables and makes decisions based on instructions held in its programme memory.

Critical Instrument: An instrument where failure or inaccuracy may lead to an EHS or GMP non-compliance.

CRU (Condensate Recovery Unit): is a vessel which receives condensate from all practical points in a process plant or heating system where condensate is produced. The vessel stores the condensate and mixes it with fresh make up water prior to it being used as boiler feedwater.

CTP: Commissioning Test Pack: e.g. All new construction projects require a large commitment to paperwork, particularly at Testing/ Commissioning stage.

CUB (Central Utility Building): a remote structure housing all chillers, steam boilers, pumps, air compressors, vacuum systems, and the central facility control centre.

Current - Symbol (I) - measured in Amperes - Electrical formulae:

(**V/R**) or (**P/V**) or √(**P/R**).

CV: Control Valve: is a **valve** used to **control** fluid flow by varying the size of the flow passage as directed by a signal from a controller. This enables the direct **control** of flow rate and the consequential **control** of process quantities such as pressure, temperature, and liquid level.

Cyber-attack: is any type of offensive manoeuvre employed by nation-states, individuals, groups, or organisations that **targets computer information systems**, infrastructures, computer networks, and/or personal computer devices by various means of malicious acts usually originating from an anonymous source that either steals, alters, or destroys a specified target by hacking into a susceptible system. The perpetrators can look for ransoms from users looking for access to regain access to their computers. Detecting a cyber-attack early is key.

Data Analysis: is a process of compiling, extracting, analysing, cleansing, transforming, and modelling data with the goal of discovering useful information, informing conclusions, and supporting good decision-making.

Data Analytics: is a broader term of which data analysis forms a subcomponent. It is the process of examining data sets using proven tools and techniques in order to draw conclusions about the information they contain, increasingly with the aid of specialised systems and software.

Data Integrity: is the maintenance of, and the assurance of the accuracy and consistency of, data over its entire life-cycle, and is a critical aspect to the design, implementation and usage of any system which stores, processes, or retrieves data.

Data Mining: is the analysis step of the 'knowledge discovery in databases' process, or KDD. It is the extraction of patterns and knowledge from large amounts of data.

DCS (Distributed Control System): is a control system for a process or plant, wherein control elements are distributed throughout the system. In a DCS, a hierarchy of controllers are connected by communications networks for command and monitoring. A typical DCS consists of functionally and/or geographically distributed digital controllers capable of executing from regulatory control loops in one control box. The input/output devices (I/O) can be integral with the controller or located remotely via a field network. These controllers have extensive computational capabilities and, in addition to Proportional, Integral, and Derivative (PID) control, can generally perform logic and sequential control.

Defibrillator: an apparatus used to control heart fibrillation by application of an electric current to the chest wall or heart. When someone suffers a sudden cardiac arrest, it delivers a shock to the heart to allow it to resume its normal rhythm.

DI (Digital Input): can have only 2 states **i.e.** on or off. **E.g.** an on/off actuated valve with feedback micro switches fitted to the on/off actuator on the valve signals to a PLC of its position – opened or closed.

Digital instrument: an on/off valve can only be opened or closed; a switch can only be on or off.

Digitalisation: is the use of digital technologies to change a business model and provide new revenue and value-producing opportunities; it is the process of moving to a digital business.

Digitisation: is the process of changing from analogue to digital form.

D.I.Y.: Do It Yourself

Digital loop: is a closed circular arrangement consisting of an input and an output e.g. an on/off valve with feedback (micro switches or proximity sensors fitted to the on/off actuator on the valve to tell control system of its position) forms a digital loop i.e. a valve is energised by a control system output and the valve sends a feedback signal to the control system input to confirm its open or closed status. When the control system directs the valve to 'open', it expects to get a feedback to say the valve has opened. This is a loop.

DO (Digital Output): A digital output can have only 2 states **i.e.** on or off. **E.g.** an on/off solenoid actuated valve opens and closes via a digital output signal from the control system i.e. the actuated valve opens or closes as the digital output signal energises/de-energises the solenoid.

DP (Differential Pressure): is the pressure increase provided by the pump between the pump inlet and outlet. It is measured as the difference between the pressure at the pump's discharge flange and the suction flange.

D/Q (Design Qualification): the purpose of DQ is to compare, using a structured approach, the proposed design with the user requirements to provide assurance the proposed system 'design critical' items will satisfy the needs of the system owner while conforming to all pertinent regulations.

EED (Energy Efficient Design): is a methodology that assists organisations to design, construct and manage projects to achieve minimum energy consumption.

E.g.: means 'for example' and comes from the Latin expression exemplum/exempli gratia.

EH&S (Environmental, Health and Safety): is an umbrella term for the laws, rules, guidance, and processes designed to help protect employees, the public and the environment from harm.

e-learning: learning conducted via electronic media, typically on the Internet.

EL (Electroluminescence): is an optical phenomenon and electrical phenomenon in which a material emits light in response to the passage of an electric current or to a strong electric field. Electroluminescence is the non-thermal conversion of electrical energy into light energy.

Electronic Record: closed systems used to create, modify, maintain, archive, retrieve, or transmit electronic records must have procedures and controls to assure the integrity, authenticity, and confidentiality of electronic records.

ERP (Enterprise Resource Planning): is business process management software that allows an organisation to use a system of integrated applications to manage the business and automate many back-office functions related to technology, services, and human resources. ERP software integrates all facets of an operation — including product planning, development, manufacturing, sales, and marketing — in a single database, application, and user interface.

Embedded systems: An embedded system is a computer system with a dedicated function within a larger mechanical or electrical system, often with real-time computing constraints. It is embedded as part of a complete device often including hardware and mechanical parts. Embedded systems control many devices in common use today.

Environment: is a very broad term, it includes all things that are above, below, and around us; be it air, water, or plant and animal life.

EPA (Environmental Protection Agency): is responsible for environmental research development, monitoring, licensing and regulation as well as the enforcement of environmental law.

etc. – et cetera - is a Latin expression that means: 'and the rest'; and others; and so forth: used at the end of a list to indicate that other items of the same class or type should be considered or included.

Explosive Atmosphere: means a mixture, under atmospheric conditions, of air and one or more dangerous substances in the form of gases, vapours, mists or dusts in which, after ignition has occurred, combustion spreads to the entire unburned mixture.

Facts: are supported by clear, undeniable evidence and may be confirmed or disproved.

FAT (Factory Acceptance Test): is usually preformed at the vendor's premises prior to shipping to a client. The vendor tests the system in accordance with the clients approved test plans and specifications to show that system is at a point to be installed and tested on site. It's an essential aspect of the whole system lifecycle and should be performed by experienced personnel. Time spent doing a proper FAT will lead to fewer problems when the equipment is installed on your site.

Fault tolerance: is a system's ability to tolerate faults and continue operating properly.

FDS (Functional Design Specification): is a document used by a company in a pre-development phase to translate all notes, concepts, and scope into a complete requirements document.

FGas: Fluorinated gases (F-gases) are man-made gases. There are four types: **hydrofluorocarbons (HFCs),** perfluorocarbons (PFCs), sulphur hexafluoride (SF_6) and nitrogen trifluoride (NF_3). **Hydrofluorocarbons (HFCs**) are used in various sectors and applications, such as refrigerants in refrigeration, air-conditioning and heat pump equipment; as blowing agents for foams; as solvents; and in fire extinguishers and aerosols.

First Aid: is the assistance given to any person suffering a serious illness or injury, with care provided to preserve life, prevent the condition from worsening, or to promote recovery.

Flow Chart: A diagram that uses graphic symbols to depict the nature and flow of the steps in a process. A flow chart also promotes process understanding and identifies problem areas and improvement opportunities.

Flywheel backup: The integrated flywheel energy storage is inherently reliable, delivering predictable, consistent backup power. The normal state of Uninterruptible Power Supply is with the flywheel spinning constantly, storing kinetic energy. When called upon during a power outage, the flywheel is ready to assume the load. By contrast, battery failures are the leading cause of UPS load loss and system downtime, because failures are inherently difficult to predict.

FMECA (Failure Modes Effects and Criticality Analysis): is a technique which allows the most appropriate maintenance tasks to be selected for equipment. It is a rigorous technique originally developed for the Nuclear Industry in the 1960s but since that time it has also been applied successfully to a wide range of industries. It is particularly useful for heavily-regulated environments such as pharmaceuticals, oil and gas, and airlines. It is also very important for organisations who wish to move toward a leaner approach to manufacturing. FMECA should be carried out by the people who know their plant and equipment the best – the operators and technicians.

FMEDA (Failure Modes Effects and Diagnostics Analysis): This is a detailed analysis of the different failure modes and diagnostic capability for a piece of equipment. This is an effective method for determining failure modes and failure rates, a requirement for certification against IEC 61508 in most certification agencies.

FT (Flow Transmitter): Turbine flow meters measure the rate of flow in a pipe or process line via a rotor that spins as the media passes through its blades. A **mass flow meter**, also known as an inertial **flow meter** is a device that measures **mass** flow rate of a fluid traveling through a tube. The mass flow rate is the mass of the fluid traveling past a fixed point per unit time. A **magnetic flow meter** (mag flowmeter) is a volumetric flow meter which does not have any moving parts and is ideal for wastewater applications or any dirty liquid which is conductive, or water based.

FTA (Fault Tree Analysis): is a top-down, deductive analysis which visually depicts a failure path or failure chain.

GAMP (Good Automation Manufacturing Practice)**:** is largely about automated system validation. It is a formal process of thorough documentation, testing, and logical process steps that validate clients' required specifications. The process begins with a user requirements specification for the machine from which a functional requirement and a design specification are created. These documents then form the basis for the traceability matrix and for the formal testing of internal acceptance, factory acceptance, and site acceptance. Categorising software is used to support the approach to validation based on the difficulty and individuality of the computerised system.

GDP (Good Documentation Practices): is a term in the pharmaceutical industry to describe standards by which documents are created and maintained.

GEP (Good Engineering Practice): is engineering and technical activities that ensure that a company manufactures products of the required quality as expected (e.g. by the relevant regulatory authorities). Good engineering practices are to ensure that the development and/or manufacturing effort consistently generates deliverables that support the requirements for qualification or validation.

GMP (Good Manufacturing Practice): are guidelines that outline the aspects of production and testing that can impact the quality of a product.

GPR (Ground Penetrating Radar): is a geophysical technique that produces continuous high-resolution profiles of the subsurface by measuring the travel time of an electromagnetic pulse between a transmitter, a reflective boundary, and a receiver. The velocity of an electromagnetic pulse varies with the dielectric property of the penetrated material. E.g. Non-destructive testing survey in the detection of adequate reinforcement **rebar** in concrete.

HACCP (Hazard Analysis and Critical Control Points): is a systematic preventive approach to food safety from biological, chemical, and physical hazards in production processes that can cause the finished product to be unsafe and designs measures to reduce these risks to a safe level.

Hard Data: information such as numbers or facts that can be proved.

Hazard:

- **Physical:** are conditions or situations that can cause the body physical harm or intense stress. Physical hazards can be both natural and human made elements.
- **Chemical**: means the physico-chemical or chemical property of a dangerous substance which has the potential to give rise to fire, explosion, or other events which can result in harmful physical effects of a kind similar to those which can be caused by fire or explosion, affecting the safety of a person.
- **Biological:** are biological agents that can cause harm to the human body. Some biological agents can be viruses, parasites, bacteria, food, fungi, and foreign toxins.

Hazop (Hazard and operability study): is a structured and systematic examination of a complex planned or existing process or operation in order to identify and evaluate problems that may represent risks to personnel or equipment. HAZOP is a common hazard analysis method for complex systems. It can be used to identify problems even during the early stages of project development, as well as identifying potential hazards in existing systems. The technique is based on breaking the overall complex design of the process into a number of simpler sections called 'nodes' which are then individually reviewed. It is carried out by a suitably experienced multi-disciplinary team during a series of meetings. The HAZOP technique is qualitative and aims to stimulate the imagination of participants to identify potential hazards and operability problems.

HP (Horse Power): is a unit of measurement of power (the rate at which work is done). The mechanical horsepower, also known as imperial horsepower, of exactly 550 foot-pounds per second is approximately equivalent to 745.7 Watts.

HR (Human Resources): are the people who work for the organisation; human resource management is really employee management with an emphasis on those employees as assets of the business.

HR (Human Relations): refers to the researchers of organisational development who study the behaviour of people in groups, in particular workplace groups and other related concepts in fields such as industrial and organisational psychology.

HMI (Human Machine Interface): the user interface is (a place) where interaction between humans and machines occurs. The goal of interaction between a human and a machine at the user interface is effective operation and control of a machine and feedback from the machine itself which aids personnel in making operational decisions.

Human Performance Analysis: assists organisations in performance improvement through the understanding and leveraging of the individual, process, and organisational behaviours necessary to facilitate safe operating performance. The interaction of people and technology in safety-sensitive organisations leads to inherent risks.

HVAC (Heating, Ventilating & Air Conditioning): is the technology of indoor and vehicular environmental comfort. Its goal is to provide thermal comfort and acceptable indoor air quality. HVAC is an important part of residential structures such as single family homes, apartment buildings, senior living facilities, hotels, medium to large industrial plants, office buildings such as skyscrapers and hospitals, on-board vessels, and in marine environments, where safe and healthy building conditions are regulated with respect to temperature and humidity, using fresh air from outdoors.

Ventilating or Ventilation (the V in HVAC) is the process of exchanging or replacing air in any space to provide high indoor air quality which involves temperature control, oxygen replenishment, and removal of moisture, odours, smoke, heat, dust, airborne bacteria, carbon dioxide, and other gases. Ventilation removes unpleasant smells and excessive moisture, introduces outside air, keeps interior building air circulating, and prevents stagnation of the interior air.

Ventilation includes both the exchange of air to the outside as well as circulation of air within the building. It is one of the most important factors for maintaining acceptable indoor air quality in buildings.

Hydronics: is the use of a liquid heat-transfer medium in heating and cooling systems. The working fluid is typically water, glycol, or mineral oil.

IBC totes: An intermediate bulk container (IBC), IBC tote, or pallet tank, is a reusable industrial container designed for the transport and storage of bulk liquid and granulated substances, such as chemicals, food ingredients, solvents, and pharmaceuticals.

ICT (Information Communication Technology). Everyday usage of digital technology includes when you use a computer, tablet, or mobile phone, send email, browse the internet, make a video call - these are all examples of using basic ICT skills and technology to communicate.

i.e.: is a Latin expression meaning **'that is'** which written out fully in Latin is 'id est'.

Immersive Learning: places individuals in an interactive learning environment, either physically or virtually, to replicate possible scenarios or to teach particular skills or techniques. Simulations, role play, and virtual learning environments can be considered immersive learning.

Incandescence: is the emission of light by a solid that has been heated until it glows or radiates light. When an iron bar is heated to a very high temperature, it initially glows red, and then as its temperature rises, it glows white.

In situ: is a Latin expression meaning **'in the place'** e.g. In the aerospace industry, equipment on board an aircraft must be tested *in situ*, or in place, to confirm everything functions properly as a system.

Interlock: safety/mechanical interlocks are put in place to prevent circumstances, which pose a risk to personnel, the environment and equipment. Electrical interlocks provide electrical isolation by means of auxiliary contacts in relays and magnetic motor starters. Process interlocks are used to prevent things happening that would affect the quality of the product.

IIoT (Industrial Internet of Things): is a term for all of the various sets of hardware pieces that work together through the internet of things connectivity to help enhance industrial processes. It encompasses applications, including robotics, medical devices, and software-defined manufacturing and production processes. IIoT offers innovative ways in doing things better and utilising assets that already exist, it's focus is on unlocking the hidden potential of connected devices.

IoT (Internet of Things): is the internetworking of physical devices, vehicles, buildings, and other items—embedded within electronics, software, sensors, actuators, and network connectivity that enable these objects to collect and exchange data.

Inspection: An inspection is, most generally, an organised examination or formal evaluation exercise, the grade of inspection may be visual, close, or detailed. In engineering activities, inspection involves the measurements, tests, and gauges applied to certain characteristics in regard to an object or activity. Inspections are usually non-destructive.

Instrument: A device for recording, measuring or controlling a physical characteristic of a system, e.g. temperature, pressure, flow rate, etc.

IQ (Installation Qualification): is the documented verification that a system has been installed as per the approved design, manufacturer's instructions, and the systems owner's requirements.

Equipment components are identified and checked against the manufacturers' component listing. The working environment conditions are documented and checked to ensure that the components are suitable for the operation of the equipment itself.

Isolating Device: an isolating device is a mechanical unit that physically blocks or interrupts the flow or release of hazardous energy e.g. Electrical – a circuit breaker that interrupts all phases, Mechanical – locking bolt or brake, Hydraulic – line valve or spade blank.

Isolock: an isolock is a device that enables several people to lock out the isolation device with their own lock.

Isometric (ISO) drawing: is way of presenting designs/drawings in three dimensions.

IT (Information Technology): the study or use of systems (especially computers and telecommunications) for storing, retrieving, and sending information.

Kinaesthetic: or tactile learning is a learning style in which learning takes place by students carrying out physical activities, rather than listening to a lecture or watching demonstrations.

Knowledge: facts or experiences known by a person, a state of knowing information on a subject.

KPI (Key Performance Indicator): is a measurable value that demonstrates how effectively a company is achieving key business objectives. Organisations use KPIs to evaluate their success at reaching targets.

KVA: Kilo - Volt - Ampere is simply 1,000 volt amps. A volt is electrical pressure. An amp is electrical current. A term called apparent power (the absolute value of complex power, S) is equal to the product of the volts and amps.

KWh: Kilo - Watt – hour is a derived unit of energy equal to 3.6 megajoules (3,600,000 joules). If the energy is being transmitted or used at a constant rate (power) over a period of time, the total energy in kilowatt-hours is the power in kilowatts multiplied by the time in hours.

Lean Maintenance: Maintenance engagement with a lean manufacturing programme accelerates the development of an organisation's lean capability. Maintenance plays a vital role in supporting Lean Manufacturing and this requires a change in emphasis from a traditional approach to maintenance. Lean manufacturing requires a greater level of reliability and dependability from equipment.

Level Sensor: is a device for determining the level or amount of fluids, liquids or other substances that flow in an open or closed system. There are two types of level measurements, namely, continuous and point level measurements.

LIDAR (LIght **D**etection **A**nd **R**anging): is a surveying technology that measures distance by illuminating a target with a laser light.

LOPA (Level of Protection Analysis): is a methodology for hazard evaluation and risk assessment.

LT: Level Transmitters are sensors with an electrical transmission output for remote indication of liquid, powder, or bulk level. Capacitive, float, mechanical, submersible, and ultrasonic models are the different types available of level transmitters.

LV: Low Voltage

Maintenance: is the in-depth inspection of a plant's machinery and associated components and assemblies, it involves the routine actions taken to preserve the fully serviceable condition of the installed equipment.

Maintenance – 5S: is the name given to the Lean Manufacturing method for the clearing out of all unnecessary things to allow room for the acquisition of tools and parts in the fastest and easiest manner. A comparison of 5S methodology with an evaluation and optimisation of a Preventive Maintenance programme at a plant quickly shows how similar these processes are. **5S** is the name of a workplace organisation method that uses a list of five Japanese words: Seiri, Seiton, Seiso, Seiketsu, and Shitsuke. Translated into Roman script (Sort, Straighten, Sweep, Standardise, Sustain), they all start with the letter 'S'. The list describes how to organise a work space for efficiency and effectiveness by identifying and storing the items used, maintaining the area and items, and sustaining the new order. The decision-making process usually comes from a dialogue about standardisation, which builds understanding among employees of how they should do the work.

Maintenance Planning: Planning decides what, how and time estimate for a job. Scheduling decides when and who will do the job. Planning of a job should be done before Scheduling a job. The benefits of good maintenance planning and scheduling are numerous and include:

- Increased productivity of tradespeople
- Reduced equipment downtime
- Lower spare parts holdings
- Less maintenance rework

MAGNEHELIC Pressure Gauge is the industry standard to measure fan and blower pressures, filter resistance, air velocity, furnace draft, pressure drop across orifice plates, liquid levels with bubbler systems and pressures in fluid amplifier or fluidic systems.

Manual Call Point: manual fire alarm call points for fire alarm systems are devices that enable people to raise a fire alarm in the event of a fire incident by pressing or breaking an element to activate the fire alarm system.

Manometer: is a device in which columns of a suitable liquid is used to measure the difference in pressure between two points or between a certain point and the atmosphere. Likewise the widely used **Bourdon gauge** is a mechanical device, which both measures and indicates the pressure of gases or liquids and is probably the best known type of pressure gauge.

MCB (Miniature Circuit Breaker): A circuit breaker is an automatically operated electrical switch designed to protect an electrical circuit from damage caused by excess current, typically resulting from an overload or short circuit. Its basic function is to interrupt current flow after a fault is detected. Unlike a fuse, which operates once and then must be replaced, a circuit breaker must be manually reset to resume normal operation.

MEWP (Mobile Elevated Working Platform): are widely used in the workplace and are replacing ladders and tower scaffolds as the preferred method of access.

Mistake Proof Engineering: mistake-proofing should be considered during the development of a new product to maximise opportunities to mistake-proof through design of the product and the process (elimination, replacement, prevention, and facilitation). Once the product is designed and the process is selected, mistake proofing opportunities are more limited (prevention, facilitation, detection, and mitigation).

MCC (Motor Control Centre): is an assembly of one or more enclosed sections having a common power bus and principally containing motor control units. Motor control centres normally consist of an assembly of several motor starters which can include variable speed drives, programmable controllers, and metering.

Modification: change to the design of the equipment which affects material, fit, form or function.

Medium Voltage (MV): Voltages 600 V and below are referred to as 'low voltage' voltages. From 600 V- 69 kV are referred to as '**medium voltage**'. Voltages from 69 kV-230 kV are referred to as 'high voltage' and voltages 230 kV-1,100 kV are referred to as 'extra high voltage,' with 1,100 kV also referred to as 'ultra-high voltage'.

MRO (Maintenance, Repair & Overhaul): involves fixing any sort of mechanical, plumbing or electrical device should it become out of order or broken (known as repair, unscheduled or casualty maintenance). It also includes performing routine actions which keep the device in working order (known as scheduled maintenance) or prevents trouble from arising (Preventive Maintenance).

MSDS (Material Safety Data Sheet): contains basic information intended to help personnel work safely with a material.

MTBF (Mean Time Between Failures): is the predicted elapsed time between inherent failures of a system during operation. MTBF can be calculated as the arithmetic mean (average) time between failures of a system. The term is used in both plant and equipment maintenance contexts.

MTTR (Mean Time to Repair): is a basic measure of the maintainability of repairable items. It represents the average **time** required to **repair** a failed component or device.

N.B.: is a Latin expression that means **'Note Well'** which written out fully in Latin is 'Nota Bene'.

Non-Critical Instrument: is an instrument where failure or inaccuracy will not result in an EH&S or GMP non-compliance. Instruments are typically operational instruments used for commissioning or engineering purposes.

Odometry: is the use of data from motion sensors to estimate change in position over time.

ODS (Ozone Depleting Substance)

OEB (Occupational Exposure Band): A mechanism to accurately assign chemicals into 'category's or 'bands' based on their health outcomes and potency considerations.

OEE (Overall Equipment Effectiveness): is a critical methodology to drive improved efficiency, equipment availability, performance, higher quality, and reduced costs for companies who are looking to maximise productivity while minimising operational costs. It is a powerful KPI in providing a metric that can be used by operations, engineering, maintenance, quality, and continuous improvement teams. Companies are now seeing the benefits of OEE from the energy consumption viewpoint as well. OEE highlights the impact of issues such as unplanned stoppages, slow running equipment, high reject rates etc. The first step in the process is to establish the baseline OEE metric, this invariable reveals significant scope for improvement initiatives. Utilising a proven problem solving methodology, improvement initiatives are implemented to improve the OEE metric and consequently the energy efficiency of the plant.

OEM (Original Equipment Manufacturer)

Off Spec: failure to meet the prescribed specifications or standards.

One-on-One meeting: is a forum for communication from supervisor to employee, as well as from employee to supervisor. Unlike 'status reports or tactical meetings', the One-on-One meeting is a place for coaching, mentorship, giving context, or even venting. Regular, effective meetings with direct reports also ensures accountability.

Operational Efficiency: represents the life-cycle cost-effective mix of Preventive, Predictive, Condition based and Reliability-centred maintenance technologies, coupled with equipment calibration, tracking, and computerised maintenance management capabilities all targeting reliability, safety, occupant comfort, and system efficiency.

Operational Maintenance: is the care and minor maintenance of equipment using procedures that do not require detailed technical knowledge of the equipment's or system's function and design. This category of operational maintenance normally consists of inspecting, cleaning, servicing, preserving, lubricating, and adjusting, as required.

OPEX (Operational Expenditure): an operating expense, operating expenditure, operational expense, operational expenditure or Opex is an ongoing cost for running a product, business, or system.

OQ (Operational Qualification): is the documented verification that the installed equipment or systems will operate throughout the design range. Equipment's functions are checked to ensure that they conform to the manufacturer's specifications. This includes the use of certified, traceable simulators and standards to verify that the equipment and its process instruments are processing input signals correctly. Confidence must be established that process equipment and sub-systems are capable of consistently operating within stated limits and tolerances.

Opinions: are preferences, beliefs, and points of view.

OSHA: Occupational Safety and Health Administration

OSH: Occupational Safety & Health

OSV: open safety vent

OT (Operations Technology): is hardware and software that detects or causes a change through the direct monitoring and/or control of physical devices, processes, and events in the enterprise.

Overhaul: action to examine and restore to a fully serviceable condition 'equipment' which has been in use or in storage for a period of time but which is not faulty.

PAT (Portable Appliance Testing): is to ensure that portable and transportable equipment are maintained in a safe condition so as to avoid any hazard to person's or property. It is carried out once a year on every portable appliance i.e. any appliance that has a plug fitted to it into a mains socket e.g. Laboratory equipment, power tools, office equipment. The results must be held for at least 5 years and must be made available for inspection by inspectors of the Health and Safety Authority.

PDA: Personal Digital Assistant i.e. digital note taking

PdM (Predictive Maintenance): is where engineering personnel can accurately predict the health and status of assets which enables informed decisions to be made regarding maintenance cycles and interventions. Predicting the status of equipment health, leads to direct cost saving benefits, and improves your bottom line. It is vital to gather the 'hidden' data, manage the field data effectively and how this data can be brought together as 'Information' to make informed decisions regarding asset status.

Permit Issuer: Person responsible for the issuing and control of the permit.

Permit Holder: Person completing the work under the conditions of the permit.

Person in Charge: An occupier of an installation or premises or a manager, supervisor or operator of an activity or a suitably qualified and experienced deputy who is present on the installation during its operation.

PFD (Process Flow Diagram): is a diagram commonly used in chemical and process engineering to indicate the general flow of plant processes and equipment. The PFD displays the relationship between major equipment of a plant facility and does not show minor details such as piping details and designations.

PID – Proportional – Integral - Derivative: A PID controller is an instrument used in industrial control applications to regulate temperature, flow, pressure, speed and other process variables. PID controllers use a control loop feedback mechanism to control process variables and are the most accurate and stable controller. A PID controller continuously calculates an error value *e(t)* as the difference between a desired set point and a measured process variable and applies a correction based on Proportional, Integral, and Derivative terms. The PID terms of a PID controller are just numbers associated with a particular loop and they are adjusted by engineers during commissioning to give optimal loop performance which in turn gives optimal process performance.

PLC (Programmable Logic Controller): is a digital computer used for automation of typically industrial electromechanical processes, such as control of machinery on factory assembly lines. PLCs monitor inputs and other variable values, make decisions based on a stored program, and control outputs to automate a process or machine.

P&ID (Process/Piping and Instrument Drawings): is a diagram in the process industry which shows the piping of the process flow together with the installed equipment and instrumentation.

PM (Preventive Maintenance): is maintenance that is regularly performed on a piece of equipment to lessen the likelihood of it failing. Each machine is set up on a maintenance plan, based on the manufacturer's suggested maintenance schedule. Instead of waiting for the machine to malfunction or stop working completely, maintenance and inspections are scheduled at regular intervals. This enables you to discover things that may become issues before they become an issue. Another benefit to this approach includes setting up orders for replacement parts well in advance of when you will need them. A plan must be in place to do Preventive Maintenance, it can have a significant impact in many ways. Besides increased asset life it can reduce the need for redundant equipment due to reductions in unplanned downtime. Reducing redundant equipment and the need for earlier replacement of equipment has a financial impact that goes all the way to company financial statements and operating efficiency.

PM (Project Management): is the discipline of initiating, planning, executing, controlling, and closing the work of a team to achieve specific goals and meet specific success criteria.

Power - Symbol (P) - measured in Watts - Electrical formulae: ($\mathbf{V*I}$) or ($\mathbf{R*I^2}$) or ($\mathbf{V^2/R}$)

PPE (Personal Protective Equipment): refers to protective clothing, helmets, goggles, or other garments or equipment designed to protect the wearer's body from injury or infection. The hazards addressed by protective equipment include physical, electrical, heat, chemicals, biohazards, and airborne particulate matter.

PQ (Performance Qualification): provides documented evidence that the equipment and ancillary systems when linked together can perform effectively and reproducibly to satisfy the requirements of the intended engineering/manufacturing process.

PRA (Preliminary Risk Analysis): is an assessment of the level of the qualitative and quantitative risk involved in a clearly defined situation that involves a potential hazard. The risk analysis is carried out prior to the activity in order to specify all potential problem areas and put in place a plan to meet any challenges that may arise.

PRV (Pressure Regulating Valve): A pressure regulator is a control valve that reduces the input pressure of a fluid to a desired value at its output. Regulators are used for gases and liquids and can be an integral device with an output pressure setting, a restrictor, and a sensor all in the one body, or consist of a separate pressure sensor, controller, and flow valve.

PRV (Pressure Relief Valve): is a type of valve used to control or limit the pressure in a system. The primary purpose of a safety valve is the protection of life, property, and environment. A safety valve is designed to open and relieve excess pressure from vessels or equipment and to reclose and prevent the further release of fluid after normal conditions have been restored. A safety valve is a safety device and, in many cases, the last line of defence. It is important to ensure that the safety valve is capable to operate at all times and under all circumstances. A safety valve is not a process valve or pressure regulator and should not be misused as such. It should have to operate for one purpose only: 'overpressure protection'.

PSSR (Pressure Systems Safety Regulations) The duties imposed by PSSR relate to pressure systems for use at work and the risk to health and safety. The pipework with its protective devices to which a transportable pressure receptacle is or is intended to be connected to; a pipeline and its protective devices.

Problem solving: is the ability to identify and define problems as well as to generate effective solutions.

Proportionate Valve: (also known as a control valve), which is driven by an analogue output from a control system, could be made to open 20%, or by any other amount from 0 to 100% of fully open.

PT (Pressure Transducer**):** is a transducer that converts pressure into an analog electrical signal. A 4-20 mA Output Pressure Transducer is also known as a Pressure Transmitter. Since a 4-20mA signal is least affected by electrical noise and resistance in the signal wires, these transducers are best used when the signal must be transmitted over long distances.

PTC (Positive Temperature Coefficient): refers to materials that experience an increase in electrical resistance when their temperature is raised. Materials which have useful engineering applications usually show a relatively rapid increase with temperature, i.e. a higher coefficient.

PTW (Permit to Work): refers to management systems used to ensure that work is done safely and efficiently. These are used in hazardous industries and involve procedures to request, review, authorise and document tasks to be carried out by frontline workers.

PWO (Plant Wide Optimisation): is to get all parts of a manufacturing environment working cohesively to maximise asset utilisation, productivity and uptime - and seamlessly integrate these with all other aspects of the organisation.

QA (Quality Assurance): the maintenance of a desired level of quality in a service or product, especially by means of attention to every stage of the process of delivery or production.

RADAR: (RAdio **D**etection **A**nd **R**anging) a device for determining the presence and location of an object by measuring the time for the echo of a radio wave to return from it and the direction from which it returns.

RCA (Root Cause Analysis): is a structured process to study and learn from failures that do occur; it is a logical process for analysing the cause-and-effect relationships of an event or situation. Problems are best solved by attempting to correct or eliminate root causes, as opposed to merely addressing the immediately obvious symptoms. By directing corrective measures at root causes, it is hoped that the likelihood of problem reoccurrence will be minimised.

RCD (Residual Current Device): is a device designed to cause the opening of one or more contacts when the residual current flowing in the 'circuit' protected by the RCD reaches the rated residual operating current of the device **e.g.** 30 mA is the maximum value permissible for personal shock protection.

RCM (Reliability Centered Maintenance): is a process to ensure that assets continue to do what their users require in their present operating context. It is generally used to achieve improvements in fields such as the establishment of safe minimum levels of maintenance, changes to operating procedures and strategies and the establishment of capital maintenance regimes and plans. Successful implementation of RCM will lead to an increase in cost effectiveness, machine uptime and a greater understanding of the level of risk that the organisation is presently managing.

REACH: is the European Regulation for the Registration, Evaluation, Authorisation and Restriction of Chemicals (EC) No 1907/2006. It is the main EU law on chemicals, covering substances on their own or in mixtures or in articles for industrial, professional or consumer use.

Redundancy: means duplication or triplication of equipment that's needed to operate without disruption if primary equipment fails during operation.

Refractory: Refractories are heat-resistant materials that constitute the linings for high-temperature furnaces and reactors and other processing units. In addition to being resistant to thermal stress and other physical phenomena induced by heat, refractories must also withstand physical wear and corrosion by chemical agents. Refractories are more heat resistant than metals and are required for heating applications above 1000°F (538°C). In a boiler, the refractory protects the metal surfaces at critical points such as the rear door and in the furnace. This refractory should be inspected periodically to insure protection. Here is a list of what to look for and possible maintenance solutions:

- Visually inspect refractory. Look for large cracks or broken pieces. Small hairline cracks are to be expected.
- Wash coat the refractory with a high temperature bonding, air dry mortar.
- Face all cracks and joints with hi-temperature bonding cement.
- If any bricks have fallen out or show signs of excessive wear, replace them.
- Remember, once the repair is complete, it is important to follow the manufacturer's recommendation for curing the refractory.

Remote Off-site Backup Processes: a Computer Solutions vendor can implement nightly remote backup routines for all, or a selection of customer data where it will be fully managed and executed by the Computer Solutions vendors' engineers. Remote Off-site Back-up provides the protection a business needs.

Repair: action to restore faulty equipment to its fully serviceable condition complying with the relevant standard. Note: The relevant standard means the standard to which the equipment was originally designed.

RH (Relative Humidity): is a measure of degree of saturation with water vapour, it is a ratio that compares the amount of water vapour in the air with the amount of water vapour that would be present in the air at saturation at a particular temperature. It requires less water vapour to attain high relative humidity at low temperatures; more water vapour is required to attain high relative humidity in warm or hot air. It is usually expressed as a percentage with the symbol '%rh' e.g. 'the humidity is 51%rh'

Relevance of relative humidity:

- if relative humidity and temperature are high, the air feels damp
- a condition of 100%rh means the air is totally saturated with water vapour and will feel much hotter than the actual temperature
- relative humidity is strongly governed by temperature
- interaction of water vapour with materials is often in proportion to relative humidity

- lowering relative humidity increases evaporation and drying.

Imagine a parcel of air at known temperature and relative humidity, at 20°C and 50%rh. If we vary only temperature, without adding or removing water (or anything else), the relative humidity changes. The degree of saturation is increased or decreased simply by changing temperature. Relative humidity falls when temperature rises (and rises when temperature falls).

Resistance symbol **(Ω)** - The restriction to electrical current flow through a material, measured in Ohms. For a conductor wire, resistance is a function of diameter, length, and resistivity (resistance per unit length – a physical material property) - Electrical formulae: $(R=V/I)$ or (V^2/P) or (P/I^2).

Resistance decade box: or resistor substitution box is a unit containing resistors of many values, with one or more mechanical switches which allow any one of various discrete resistances offered by the box to be dialed in to an instrument.

Resistance thermometers: also called resistance temperature detectors or resistive thermal devices (RTD's) are temperature sensors that exploit the predictable change in electrical resistance of some materials with changing temperature. A resistance thermometer is an instrument or system incorporating a length of wire or film having predictable resistance vs temperature characteristics, forming a temperature sensor. Measurement of the resistance of the device yields its temperature.

RFID (Radio Frequency IDentification) reader is a device used to gather information from an RFID tag, which is used to track individual objects. Radio waves are used to transfer data from the tag to a reader.

Risk: means the likelihood of a person's safety being affected by harmful physical effects being caused to him from fire, explosion or other events arising from the hazardous properties of a dangerous substance in connection with work and also the extent of that harm.

Risk Assessment: a systematic process of evaluating the potential risks that may be involved in a projected activity or undertaking. A risk assessment must be conducted by internal trained and competent people, or external competent third parties.

Robot(s): can be defined as technologies, such as machine learning algorithms running on purpose-built computer platforms, that have been trained to perform tasks that currently require humans to perform.

RONA (Return On Net Assets): is a measure of financial performance of a company which takes the use of assets into account. Higher RONA means that the company is using its assets and working capital efficiently and effectively.

RSJ: Rolled Steel Joist

Run Chart: is a line graph of data plotted over time. By collecting and charting data over time, you can find trends or patterns in a process.

Safety Oversight: forms part of the safety regulatory process dedicated to ensuring that applicable safety regulatory requirements are met, and to the monitoring of the safe provision of services.

Self-contained emergency light: Self-contained emergency lighting luminaire means a luminaire providing maintained or non-maintained emergency lighting in which all the elements, such as battery, the lamp, the control unit and the test and monitoring facilities, where provided, are contained within the luminaire or adjacent to it (i.e. within 1 metre)

SIL (Safety Integrity Level): is defined as a relative level of risk-reduction provided by a safety function, or to specify a target level of risk reduction.

Safety Instrumented System (SIS): performs specified functions to achieve or maintain a safe state of the process when unacceptable or dangerous process conditions are detected. Safety instrumented systems are separate and independent from regular control systems but are composed of similar elements, including sensors, logic solvers, actuators, and support systems.

SAT (Site Acceptance Test): in engineering and its various sub-disciplines, acceptance testing is a test conducted to determine if the requirements of a specification or contract are met. It is the stage where the customer conducts testing for the components supplied under the project scope and tests the conformance of the delivered solution.

The Site Acceptance Testing stage includes 3 parts:

1. Integration Testing
2. Performance Testing
3. User Acceptance Testing

SCADA (Supervisory Control And Data Acquisition): is an industrial control system where a computer system monitors and controls a process giving transparent operating data. It reduces obligation to supervise.

SDS (Software Design Specification): is a description of a software system to be developed. It contains specific information about the expected input, output, classes, and functions. It lays out functional and non-functional requirements that describe user interactions that the software must provide.

Security and protection – to prevent Ransomware, DDoS, and other malicious viruses from infiltrating a business' systems, software, and infrastructure.

Sensor: is a device that converts a physical condition into an electrical signal for use by a controller, such as a PLC.

Sensorisation: defines the extent or the trend of embedding as many sensors as possible within a device or appliance. It defines how consumer technologies such as smart phones, tablet computers and smart TVs are integrated with multiple sensors and/or sensing technologies **e.g**. the iPhone, which is embedded with many sensors. Sensorization is primarily concerned with how a single device is embedded with sensors, and newer sensors are continuously added over time.

These sensors can include anything, some common ones being:

• Touch sensors

• Motion sensors

• Accelerometers

• Gyroscopes

• Compasses

Service factor (SF) - is a measure of periodically overload capacity at which a motor can operate without overload or damage **e.g.** a SF of 1.15 means the motor can take 15 percent more load than its rated capacity without breakdown.

Service Provider / Consultant: are usually a firm or individual that provides a service either on or off-site and are employed to provide a service, professional advice and/or training based on their professional experience and academic qualifications.

Shadow board: contains outlines of designated tools to show where they should be stored.

Shutdown: a shutdown is defined as an outage scheduled in advance, for maintenance or other purposes such as activities like modifications or new installations of equipment.

SME (Subject Matter Expert): is a person who is an authority in a particular area or topic.

Soft Data: is anecdotal, usually gathered in informal communications, and lacks the rigour that is implied in statistical **data**.

SOP (Standard Operating Procedure): is a set of step-by-step instructions compiled by an organisation to help workers carry out routine operations. SOPs aim to achieve efficiency, quality output and uniformity of performance, while reducing miscommunication and failure to comply with industry regulations.

Sparge Pipe: A horizontal pipe having fine holes drilled throughout its length so as to deliver a spray of water.

SRM: Standard Reference Method

Standard: a standard is an agreed way of doing something. It could be about making a product, managing a process, delivering a service, or supplying materials. Standards are written from the distilled wisdom of people with expertise in their subject matter, it is established by consensus and approved by a recognised body to provide requirements, specifications or guidelines that can be used consistently by manufacturers, sellers, buyers, customers, trade associations, users, or regulators.

State Based Control: is a method for designing plant automation based on the principle that all process facilities operate in a recognised, definable process state. These cover normal and abnormal conditions of the process. State Based Control provides an environment for 'knowledge capture' directly into the control design.

Statistical Analysis: Collect data, present the data, analyse and interpret the data and use it to help make decisions, solve problems, and design products and processes.

Statistics (Engineering): combines engineering and statistics using scientific methods for analysing data. Engineering statistics involves data concerning manufacturing processes such as: component dimensions, tolerances, type of material, and fabrication process control.

Statistical Methods: are used to help us describe and understand variability. It is important to understand the nature of variability in processes and systems over time.

Statistical Process Control (SPC): is an industry-standard methodology for measuring and controlling quality during the manufacturing process. Control charts are a very important application in this branch of statistics for monitoring, controlling, and improving a process.

Statistical Thinking: is one of the tools for process analysis. Statistical thinking relates processes and statistics and is based on the following principles: All work occurs in a system of interconnected processes. Variation exists in all processes. Understanding and reducing variation are keys to success.

Steam: Steam is efficient, economic to generate and provides excellent heat transfer. It is one of the most widely used media to convey heat over distances. It can hold five or six times as much potential energy as an equivalent mass of water. When water is heated in a boiler, it begins to absorb energy; depending on the pressure in the boiler, the water will evaporate at a certain temperature to form steam. Steam contains a large quantity of stored energy which will eventually be transferred to the process or the space to be heated. It can be generated at high pressures to give high steam temperatures i.e. the higher the pressure, the higher the temperature. More heat energy is contained within high temperature steam so its potential to do work is greater.

STP: Standard condition for temperature and pressure.

System Integration: is defined in engineering as the process of bringing together the component sub-systems into one system (an aggregation of subsystems cooperating so that the system is able to deliver the overarching functionality) and ensuring that the subsystems function together as a system.

System Owners: are the equipment owners responsible for the daily operations and use. They are also responsible for ensuring that there is appropriate participation during the CAT assessments and for approving the critical instrument list upon completion of the assessments.

Technological Literacy: the ability to use, manage, understand, and assess technology. Technology literacy is the ability to effectively use technology to access, evaluate, integrate, create and communicate information to enhance the learning process through problem-solving and critical thinking.

Template: A template is a form, mould, or pattern used as a guide to make something.

Thermocouple: an electrical circuit comprising two dissimilar materials. A voltage is generated that is dependent on the temperature at the junctions forming the limits of the dissimilar materials. The reference junction at one end of the conductors is usually maintained at 0°C to allow the measuring junction to be used as a temperature sensor.

TRV (Thermostatic Radiator Valve): is a valve that controls the flow of hot water to the radiators depending on the temperature in the room. A TRV contains a plug, typically made of wax, which expands or contracts depending on the room temperature. This plug is connected to a valve.

TI (Thermal Imaging): Infrared Thermal Imaging allows us to see and measure heat. An unaided human eye cannot see infra-Red heat.

Torque: is a measure of how much a force acting on an object causes that object to rotate. It is a turning force applied to a shaft, tending to cause rotation. Torque is normally measured in pound/feet and is equal to the force applied times the radius through which it acts.

TPM (Total Productive Maintenance): is a programme for planning and achieving minimal machine downtime. Equipment and tools are literally put on 'proactive' maintenance schedules to keep them running efficiently and with greatly reduced downtime. Machine operators take far greater responsibility for their machines upkeep. Maintenance technicians are liberated from mundane, routine maintenance, enabling them to focus on urgent repairs and proactive maintenance activities. A solid TPM programme allows you to plan your downtime and keep breakdowns to a minimum.

Transmitter: a device incorporating an electrical circuit that converts a signal from a transducer into a standard transmittable form – typically, a two wire DC current output ranging from 4 – 20 mA.

Transducer: a device that converts variations in a physical quantity, such as temperature, pressure, or brightness, into an electrical signal, or vice versa.

TT (Temperature Transmitter): A temperature transmitter is an electrical instrument that interfaces a temperature sensor (e.g. thermocouple, RTD, or thermistor) to a measurement or control device (e.g. PLC, DCS, PC, loop controller, data logger, display, recorder). Typically, temperature transmitters isolate, amplify, filter noise, linearise, and convert the input signal from the sensor then send (transmit) a standardised output signal to the control device. Common electrical output signals used in manufacturing plants are 4-20mA or 0-10V DC ranges. For example, 4mA could represent 0°C and 20mA means 100°C.

UPS (Uninterrupted Power Supply): it's vital to provide continuous power. A bad power supply can cause unexpected behaviour to running microprocessor- based equipment. Therefore, the control system is only as reliable as the power provided to it. The key is to attach the output power of the UPS to the primary controller, which filters surges and minimises system recovery when power is re-established.

URS (User Requirement Specification): is raised by the system owner at the very beginning of the project with assistance from relevant subject matter experts stating what the system owner requires the equipment to do, the URS should express requirements and not design solutions.

The URS provides the key reference point for testing throughout the qualification process, so it is essential that they are written in a form that would support this process. The following elements should be included:

- **Unique Reference Number:** This is required to provide traceability throughout the qualification activities
- Each requirement must be **SMART** (Specific, Measurable, Achievable, Realistic and Testable/verifiable)
- **Concise description**: Restrict description to a bullet point, focus on key items, be specific and ensure that the requirement is realistically achievable and where possible measurable
- **Reason / Driver indication**: Is the requirement compliance driven (mandatory), required to provide a fully functional system (Required) or for user convenience (Desired)

Vendor: is a part in the supply chain that makes goods and services available to companies or consumers. The term 'vendor' is typically used to describe the entity that is paid for goods that are provided, rather than the manufacturer of the goods itself. It is possible, however, for a vendor to sometimes operate as both a supplier (or seller) of goods and a manufacturer.

Voltage - symbol (V) - measured in Volts - Electrical formulae: (**I*R**) or (**P/I**) or √**(P*R)**

Voltage Drop: describes how the supplied energy of a voltage source is reduced as electric current moves through the passive elements (elements that do not supply voltage) of an electrical circuit. Voltage drops across internal resistances of the source, across conductors, across contacts, and across connectors are undesired as the supplied energy is lost (dissipated). Voltage drops across loads and across other active circuit elements are inevitable as the supplied energy performs useful work. **For example**, an electric space heater may have a resistance of 10 ohms, and the wires which supply it may have a resistance of 0.2 ohms, about 2% of the total circuit resistance. This means that approximately 2% of the supplied voltage is lost in the wire itself. Excessive voltage drop may result in unsatisfactory operation of, and damage to, electrical and electronic equipment.

VSD (Variable Speed Drive): is an electronic device that controls the characteristics of a motor's electrical supply allowing the speed and torque of the motor to be matched with the requirements of the machine it is driving. VSD's regulate the operation of electric motors and save energy by matching the output of motor-driven pumps, fans, conveyors, and similar equipment with the actual demand of the systems they support. In addition to saving money, VSD's can also improve process control, whilst reducing waste, maintenance costs, and carbon emissions.

W.I. (Work Instruction): is a tool provided to help someone to do a job correctly. This simple statement implies that the purpose of the **work instruction** is quality and that the target user is the worker.

Zeroth's Law states: If two thermodynamic systems are in thermal equilibrium with a third, they are also in thermal equilibrium with each other.

ZNE: Zero Net Energy - The ZNE consumption principle focuses on renewable energy harvesting as a means to cut greenhouse gas emissions. **E.g.** a building that contributes more than it consumes by installing energy generating technologies e.g. Wind and Solar (The amount of solar energy reaching the surface of the earth is 100,000 Terawatts. The amount of solar energy that civilisation is using for all its purposes today is 17–18 terawatts i.e. the maths add up. Up to now, Solar energy could not compete with fossil fuels because it was too expensive, but that is changing e.g. Silicon based solar panels).

20. SUMMARY

Preparation, Preparation, Preparation

You can't expect that nothing will ever go wrong. Have a plan and be prepared for the fact that despite your good planning and hard work, sometimes things go awry.

Contractor Management is the managing of outsourced work performed on your site. Managing contractors is a multi-step process, with each stage an important step in achieving success. Contractor management implements a system that manages contractors' health and safety information, insurance information, training programs and specific documents that pertain to the contractor and your company. It's important to build a good relationship and have mutual trust and respect with independent contractors and contractor companies.

When a contract company (consultants, engineers, tradesmen etc.) is employed to carry out a specialist role or project on your site and you are the **responsible person,** you must ensure you have it clear in your mind what you want done, and what you want them to do citing 'site standard' rules and procedures. Ensure you have the preparation planning done and be clear & concise in your instructions and leave no room for ambiguity.

Ensure all potential obstacles, both physical and red tape are dealt with and agreed by all the relevant onsite affected parties prior to any proposed work being carried out. Having contract personnel standing around while trying to get the **go ahead** to work on a piece of equipment because of lack of preparatory work on the company's behalf can be very expensive and possibly dangerous to the personnel carrying out the work due to the lack of an appropriate hazop or method statement of the intended work being considered and written. **N.B.** The cost of an accident is often many times greater than the cost of the analysis that could have stopped it.

Tip: Whether you have rectified a minor or major fault on a piece of equipment, replay your approach in your head on how you resolved the problem, write down the sequence, review the photos and learn from the experience. Make sure both the tacit and explicit knowledge gained, is digitally captured, and archived for future reference. Ask yourself, would you take a different approach the next time and maybe save valuable downtime and the use of costly spare parts.

N.B. With an incomplete understanding of a problem, it is very easy to jump to the wrong conclusions. **Dig for facts instead.**

Understanding, Understanding, Understanding, cannot be stressed enough and is vital in maintaining good trouble shooting and fault finding techniques. Understanding what makes a machine/system **work** i.e. ensuring it runs in a safe, environmentally friendly, efficient, and efficient manner to give maximum output with minimum downtime. It is important personnel understand how the equipment is designed to operate.

The better the access to knowledge, the better informed decisions can be made.

It is the difference between diagnosing and misdiagnosing a problem. Personnel have a much better chance of preventing any future potential problems or diagnosing an actual fault quicker, if they understand exactly how each component associated with the piece of equipment operates and how the individual components combine to make the machine do what it was designed to do. If possible, try to discover when a fault first started to manifest itself, use technology as much as possible especially if the pieces of equipments' parameters are monitored by a computerised data trending system. Great credit to all for resolving a fault and getting the plant back on line, more credit though for sourcing the cause of the problem, understanding why it happened, and implementing the follow up actions to prevent a reoccurrence.

Maintenance and Facilities Management key learning points:

- Think Safety
- Be Aware of any Environmental impact
- Maintain site Experiential Knowledge and Training
- Avoid Risk
- Focus on Facts first, then assumptions
- Don't make difficult Decisions in isolation
- If in Doubt, Ask
- Increase efficiency
- Reduce downtime
- Save money – Invest wisely
- Solve problems
- Improve productivity
- Keep good relationships

Quick Test No.1: Complete the Circuit – Answer:

Test: Cylinder 1A will not extend when push button 1S2 is pressed.

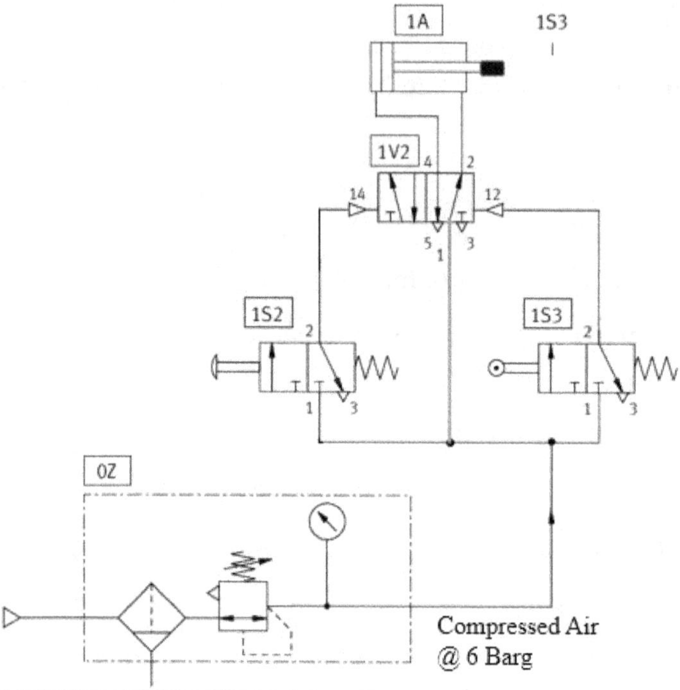

Quick Test No.2: Complete the Circuit - Answer:

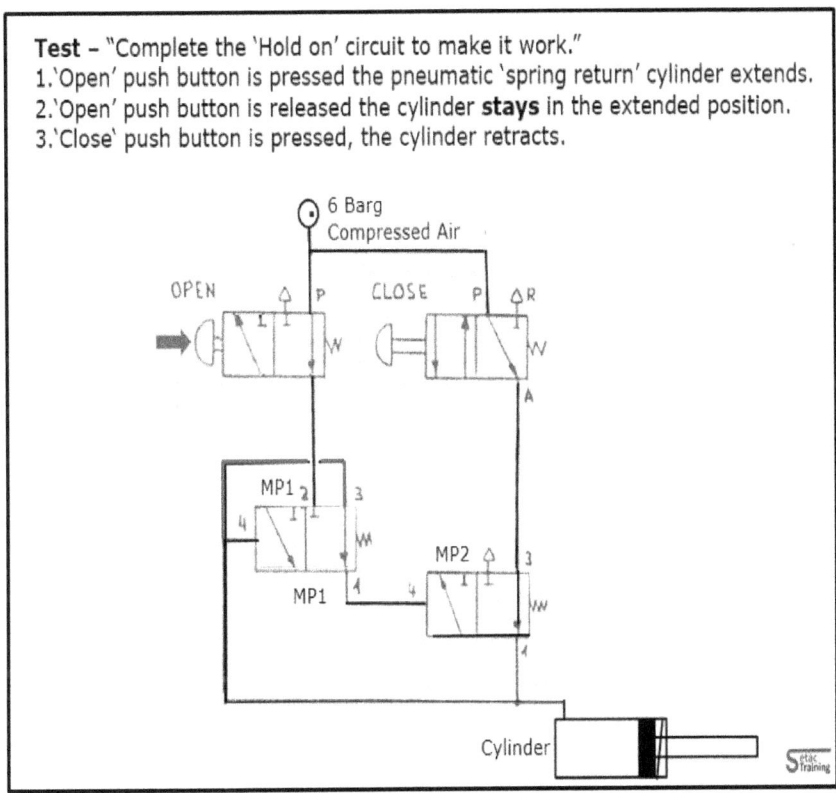

Test – "Complete the 'Hold on' circuit to make it work."

1.'Open' push button is pressed the pneumatic 'spring return' cylinder extends.
2.'Open' push button is released the cylinder **stays** in the extended position.
3.'Close' push button is pressed, the cylinder retracts.

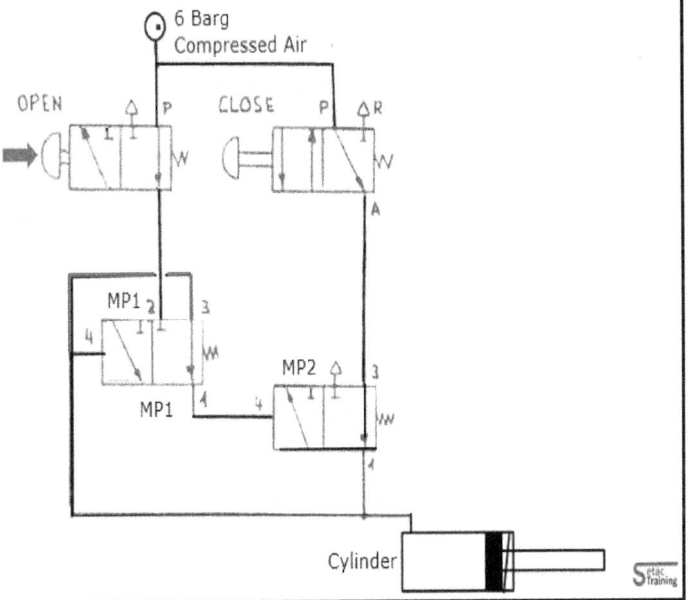

ABOUT THE AUTHOR

Brendan Shine has over 35 years 'hands-on' experiential knowledge (including electrical, ATEX, HVAC, electronic, pneumatic, hydraulic, and business qualifications) in the engineering industry. He is the Maintenance and Facilities Manager of a large and very successful pharmaceutical company who have been in operation for over 40 years. He is a member of the Association of Facilities Engineering. He is also an international lecturer in both Maintenance and Facilities Management where he works with senior engineering people focussing on improving both their technical and non-technical management skills.

Website: http://www.setac.ie/

www.ingramcontent.com/pod-product-compliance
Lightning Source LLC
Chambersburg PA
CBHW071352170526
45165CB00001B/15
* 9 7 8 1 5 4 8 6 0 4 4 3 1 *